Environmental Humanities in the New Himalayas

Environmental Humanities in the New Himalayas: Symbiotic Indigeneity, Commoning, Sustainability showcases how the eco-geological creativity of the earth is integrally woven into the landforms, cultures and cosmovisions of modern Himalayan communities.

Unique in scope, this book features case studies from Bhutan, Assam, Sikkim, Tibet, Nepal, Pakistan and Sino-Indian borderlands, many of which are documented by authors from indigenous Himalayan communities. It explores three environmental characteristics of modern Himalayas: the anthropogenic, the indigenous and the animist. Focusing on the sentient relations of human-, animal- and spirit worlds with the earth in different parts of the Himalayas, the authors present the complex meanings of indigeneity, commoning and sustainability in the Anthropocene. In doing so, they show the vital role that indigenous stories and perspectives play in building new regional and planetary environmental ethics for a sustainable future.

Drawing on a wide range of expert contributions from the natural sciences, social sciences and humanist disciplines, this book will be of great interest to students and scholars of environmental humanities, religion and ecology, indigenous knowledge and sustainable development more broadly.

Dan Smyer Yü is Kuige Professor of Ethnology in the School of Ethnology and Sociology, Yunnan University Kunming, China.

Erik de Maaker is Assistant Professor in the Department of Cultural Anthropology and Development Sociology, Leiden University, The Netherlands.

Routledge Environmental Humanities

The *Routledge Environmental Humanities* series is an original and inspiring venture recognising that today's world agricultural and water crises, ocean pollution and resource depletion, global warming from greenhouse gases, urban sprawl, overpopulation, food insecurity and environmental justice are all *crises of culture*.

The reality of understanding and finding adaptive solutions to our present and future environmental challenges has shifted the epicenter of environmental studies away from an exclusively scientific and technological framework to one that depends on the human-focused disciplines and ideas of the humanities and allied social sciences.

We thus welcome book proposals from all humanities and social sciences disciplines for an inclusive and interdisciplinary series. We favour manuscripts aimed at an international readership and written in a lively and accessible style. The readership comprises scholars and students from the humanities and social sciences and thoughtful readers concerned about the human dimensions of environmental change. For more information about this series, please visit: www.routledge.com/Routledge-Environmental-Humanities/book-series/REH

Environmental Humanities in the New Himalayas

Symbiotic Indigeneity, Commoning, Sustainability

Edited by
Dan Smyer Yü and Erik de Maaker

Routledge
Taylor & Francis Group
LONDON AND NEW YORK

First published 2021
by Routledge
2 Park Square, Milton Park, Abingdon, Oxon OX14 4RN

and by Routledge
52 Vanderbilt Avenue, New York, NY 10017

Routledge is an imprint of the Taylor & Francis Group, an informa business

British Library Cataloguing-in-Publication Data
A catalogue record for this book is available from the British Library

Library of Congress Cataloging-in-Publication Data
Names: Smyer Yü, Dan, editor. | Maaker, Erik de, 1963- editor.
Title: Environmental humanities in the new Himalayas : symbiotic indigeneity, commoning, sustainability / edited by Dan Smyer Yü and Erik de Maaker.
Description: Abingdon, Oxon ; New York, NY : Routledge, 2021. | Includes bibliographical references and index.
Identifiers: LCCN 2020056154 (print) | LCCN 2020056155 (ebook)
Subjects: LCSH: Human ecology--Himalaya Mountains Region. | Ethnoecology--Himalaya Mountains Region. | Traditional ecological knowledge--Himalaya Mountains Region. | Indigenous peoples--Himalaya Mountains Region--Social life and customs. | Himalaya Mountains Region--Environmental conditions. | Himalaya Mountains Region--Civilization.
Classification: LCC GF696.H55 .E59 2021 (print) | LCC GF696.H55 (ebook) | DDC 304.2095496--dc23
LC record available at https://lccn.loc.gov/2020056154
LC ebook record available at https://lccn.loc.gov/2020056155

ISBN: 978-0-367-69979-6 (hbk)
ISBN: 978-0-367-69981-9 (pbk)
ISBN: 978-1-003-14411-3 (ebk)

Typeset in Times New Roman
by SPi Global, India

Contents

Contributors

Bhargabi Das completed her Master's degree in sociology from Delhi School of Economics and is currently pursuing her Ph.D. in anthropology at National University of Ireland, Maynooth. Her research explores state politics and imagination among the residents of the river islands known as Chars in Assam, India. She documents and examines the intersections between the environment, the state, the politics of infrastructure, bureaucracy and citizenship in the context of South Asia. Her research is funded by the Irish Research Council. She has been honoured with the 2018 John and Pat Hume Award.

Alexander E. Davis is a lecturer in International Relations at The University of Western Australia. He received his Ph.D. in International Studies from the University of Adelaide. He is the author of the research monograph *India and the Anglosphere: Race, Identity and Hierarchy in International Relations* (Routledge 2018). His current research focuses on colonial legacies in India's borderlands, particularly the Indian Ocean and the Himalaya. He is head of the Australian Himalaya Research Network and a foundation member of the network New International Histories of South Asia.

Erik de Maaker is an assistant professor at the Institute of Cultural Anthropology and Development Sociology of Leiden University. His research focuses on perceptions of place and processes of place-making, as expressed in relation to land, ethnicity, indigeneity, heritage, environment and religious ontologies. His geographical specialisation is in the contiguous uplands of South and Southeast Asia (primarily Northeast India). He is a co-editor of *Media, Indigeneity and Nation in South Asia* (Routledge, 2019) and is an author prolifically published in journals such as *Asian Ethnology*, *South Asia: Journal of South Asian Studies*, *Visual Anthropology* and the *Journal of Borderland Studies*. He is also an award-winning visual anthropologist. He is one of the founding members of the Asian Borderlands Research Network (in 2008). He received his Ph.D. from Leiden University in 2006.

Rashila Deshar is an assistant professor at Central Department of Environmental Science, Tribhuvan University in Nepal. She received her doctoral degree in Environmental Science from the University of the Ryukyus in Okinawa, Japan with specialisation in forest ecology in 2006. Her recent publications are the co-authored chapters published by Springer, entitled "Indigenous Practice in Agro-Pastoralism and Carbon Management from a Gender Perspective: A Case from Nepal" (2020), "Climate Change and Gender Policy" (2020) and "Designing Water Resource Use for Poverty Reduction in the HKH Region" (2020).

Anwesha Dutta is a Senior Researcher with the Chr. Michelsen Institute in Bergen, Norway. She holds a Ph.D. in Conflict and Development Studies (Political Science) from Ghent University, Belgium. Her research focuses on resource extraction, forest and wildlife conservation and climate adaptation in the Indian Himalayas and South Asia. She is currently working on a project on natural resource management and corruption with a focus on Kenya. She has published on issues around militarised conservation, informal settlements within reserved forests, frontline forest staff and park-people relationships.

Ruth Gamble is a lecturer in the Department of Archaeology and History at La Trobe University. She is an environmental and cultural historian of Tibet and the Himalaya. Her first book, *Reincarnation in Tibetan Buddhism: the Third Karmapa and the Invention of a Tradition* (Oxford University Press, New York, 2018), traces the links between Tibet's reincarnation lineages and its sacred geography. She has also written a biography of the third Karmapa, Rangjung Dorje, for the *Lives of the Masters Series, Rangjung Dorje, Master of Mahamudra* (Shambhala 2020). Her upcoming book *Flowing Concrete: the transformation of a Himalayan River*, is an environmental history of the upper Brahmaputra River. Ruth has produced and co-authored an eTextbook Series, *Introduction to the Tibetan Language* (Australian National University Press, 2018), and numerous articles and book chapters on Himalayan sacred, secular and geopolitical environments.

Kinley Choki graduated from the Royal Thimphu College (RTC) with a Bachelor's degree in Political science and Sociology in 2018. Shortly after her graduation, she joined the Department of Political Science and Sociology at RTC as a research associate responsible for ethnographic fieldwork assignments. Her research areas are indigenous Knowledge, climate change, modern transformations and sustainability in the Bhutan highlands.

John Grim is cofounder and co-director of the Yale Forum on Religion and Ecology with his wife, Mary Evelyn Tucker. With Tucker, Grim directed a 10-conference series and book project at Harvard on "World Religions and Ecology." Grim is the author of *The Shaman: Patterns of Religious Healing among the Ojibway Indians* (University of Oklahoma Press, 1983) and the editor of *Indigenous Traditions and Ecology: The Interbeing of Cosmology and Community* (Harvard, 2001). His co-authored and co-edited books in recent years are *Ecology and Religion* (Island Press, 2014), *Thomas Berry: A Biography* (Columbia University Press, 2019), *Thomas Berry: Selected Writings on the Earth Community* (Orbis, 2014) and *The Routledge Handbook on*

Religion and Ecology (Routledge, 2016). John is the co-executive producer of the Emmy award-winning film, *Journey of the Universe*. He is the past president of the American Teilhard Association.

Thinley Dema graduated from Royal Thimphu College with a bachelor's degree in English and Environmental Studies in the year 2018. She recently received an "Indian Ambassadors' scholarship award" to pursue Masters in Ecology and Environmental studies in Nalanda University. She has been part of three research projects to date. She is the co-author of *Ecotourism and Social Cohesion: Contrasting Phobjikha and Laya Experiences* (Rig Tshoel, 2019), *Territory, relationality and the labour of deities: Importing Raffestin on the Bhutanese spiritual landscape* (Rig Tshoel, 2020) and (forthcoming) *Cities by Women: urban space, livelihoods and challenges of women street vendors in Thimphu city* (*Cities,* an Elsevier journal, forthcoming).

Zainab Khalid is a doctoral candidate in College of Earth and Environmental Sciences at Lanzhou University, China. She is also a program officer of The High Mountain Research Center at COMSATS University in Islamabad, Pakistan, specialising in physical and human geographies of Karakoram. Her dissertation research focuses on geohazard vulnerability assessment and risk perceptions in Western Himalayan Communities of Pakistan.

Madan Koirala is a professor in the Central Department of Environmental Science at Tribhuvan University and a member of Climate Change Council at Government of Nepal. His recent publications are the co-authored chapters published by Springer, entitled "Indigenous Practice in Agro-Pastoralism and Carbon Management from a Gender Perspective: A Case from Nepal" (2020), "Climate Change and Gender Policy" (2020) and "Designing Water Resource Use for Poverty Reduction in the HKH Region" (2020).

Charisma K. Lepcha teaches anthropology in Sikkim University, India. She completed her doctoral degree at the Department of Anthropology, North-Eastern Hill University in 2013. Her dissertation research focused on religion, culture and identity of the Lepcha in Eastern Himalayan borderlands. She was a fellow at the Indian Institute of Advanced Study (IIAS), Shimla. Her recent co-edited publications are *The Cultural Heritage of Sikkim* (Routledge 2019) and *India's Northeast: People, Power* and *Perspectives* (2020).

Mongfing Lepcha is a doctoral candidate at the Department of Anthropology, Sikkim University, Gangtok. Her thesis is titled "Comparative Study on the Fertility and Mortality Status among the Lepchas of Sikkim." She is interested in looking at the bio-cultural factors that influences the Lepcha women in both urban and rural areas. She is the General Secretary of the Research Scholars' Forum and has also presented papers, attended workshops and conferences in different parts of the country. Her research interests include public health, demography, tribal communities, bio-cultural anthropology.

Rongnyoo Lepcha is a Ph.D. candidate in the Department of Anthropology, Sikkim University, Gangtok. Her doctoral thesis is titled "Dynamics of Material Cultural of Lepchas in Sikkim, Kalimpong and Darjeeling." She completed her M.Phil. in 2017 from the same department and is an active member of the Research Scholar's Forum. She has published articles in local magazines and contributed to several edited volumes. She has also participated in workshops and presented papers in regional and national conferences. She is a self-taught artist who is using her artwork as part of her research work. She is interested in material culture, museum studies, tangible, intangible heritage, indigenous knowledge and art and ethnography.

Sarina Maharjan is a graduate student of the Central Department of Environmental Science, Tribhuvan University. She has recently completed her Master's thesis entitled "Flood vulnerability assessment in upstream and downstream regions of Mahakali River Basin in Nepal."

Dibas Shrestha is an assistant professor at Central Department of Hydrology and Meteorology, Tribhvan University in Nepal. He received his doctoral degree in Atmospheric Sciences from Nagoya University, Japan with specialisation in satellite meteorology. His present research areas are Climate Science, Climate Change and Mountain Hazards. His recent publications are "Elevation-dependent warming of maximum air temperature in Nepal during 1976–2015" (Elsevier, 2019) and "Ice cliff contribution to the tongue-wide ablation of Changri Nup Glacier, Nepal, Central Himalaya" (The cryosphere, European Geo-science Union 2019).

Dan Smyer Yü is Kuige Professor of Ethnology, School of Ethnology and Sociology and the National Centre for Borderlands Ethnic Studies in Southwest China at Yunnan University. He received his Ph.D. in anthropology from the University of California at Davis in 2006. He is the author of *The Spread of Tibetan Buddhism in China: Charisma, Money, Enlightenment* (Routledge 2011) and *Mindscaping the Landscape of Tibet: Place, Memorability, Eco-aesthetics* (De Gruyter 2015), and the co-editor of *Religion and Ecological Sustainability in China* (Routledge 2014) and *Trans-Himalayan Borderlands: Livelihoods, Territorialities, Modernities* (Amsterdam University Press 2017). His research interests are religion and ecology, environmental humanities, transboundary state effects, hydraulic politics, climate change and heritage preservation, Buddhism and peace-building, and comparative studies of Eurasian secularisms.

Jelle J. P. Wouters is a social anthropologist and teaches in the Department of Social Sciences at Royal Thimphu College (RTC), Bhutan. He holds an M.Phil. (distinction) in social anthropology from the University of Oxford and a Ph.D. in anthropology from the North-Eastern Hill University, Shillong, where he was also a Wenner-Gren grantee. Prior to joining RTC in 2015, he taught at Sikkim Central University, India, and was a visiting faculty at

Eberhard Karls University of Tubingen, Germany, under the "Excellence Initiative" of the German Research Foundation. He is the author of *In the Shadows of Naga Insurgency* (OUP 2018) and *Nagas as a Society Against voting and Other Essay* (Highlander Books 2019) and the co-editor of *Nagas in the 21st Century* (Highlander Books 2017) and *Democracy in Nagaland: Tribes, Traditions, and Tensions* (Highlander Books 2018).

Shailendra Yashwant is an independent documentary photographer and an environmental journalist based in India. He began his career as a news photographer and worked with several newspapers and magazines in India and South Asia, including *The Hindu* and *Outlook* magazine. Between 1999 and 2011, he worked as Campaigns Director with Greenpeace in India and Southeast Asia. Since 2012, he has been working on a long-term project documenting communities adapting to climate change and on photo projects for Save the Children, Sanctuary Asia, Oxfam India, UNICEF India, Infochangeindia and Climate Action Network South Asia.

Acknowledgements

This book project was primarily funded by Federal Department of Foreign Affairs (FDFA), Swiss Agency for Development and Cooperation (SDC) and Global Programme Climate Change and Environment (GPCCE) through a focus grant from the Himalayan University Consortium (HUC). The final book workshop held in Kunming, China, was partially supported by the National Centre for Borderlands Ethnic Studies in Southwest China (NaCBES) at Yunnan University and the First-Class Discipline Construction for Ethnology Project of Yunnan University. The research components of five individual chapters were also funded respectively by the British Academy Sustainable Development Program, the Irish Research Council, the Indian Council of Social Science Research, the Dutch Research Council, the HUC-TROSA Water Research Grant, the David Myer Fellowship at La Trobe University, the National Natural Science Foundation of China, the National Social Science Foundation of China, the Swedish Research Council, SSHRC Partnership Development Grant, and the Asian Modernities and Traditions Research Fund of Leiden University. Our heartfelt thanks go to their generosity and support for innovative interdisciplinary studies of the Himalayan environment with public engagements and policy implications.

HUC of ICIMOD, as the principal administering instrument of the FDFA, SDC and GPCCE grants, shows its superb leadership in supporting the project research executed by its Thematic Working Group on Trans-Himalayan Environmental Humanities. We are indebted to Chi Huyen Truong (Shachi), Bhawana Syangden, Anu Kumari Lama, Santosh Raj Pathak, Achala Sharma and Gao Liang at HUC. Led by Shachi, the HUC administrative team has demonstrated its care and professionalism throughout the field research phase and the bookmaking process through its ability to work with both policy and academic worlds, its fluency in international and intercultural communication, its familiarity with the different operating systems of the collaborative universities in the Himalayan region, and its willingness to take initiative and engage in creative approaches to coordinating multinational meetings and facilitating the timely disbursements of research funds. Without HUC's dedication, it would not have been possible to fulfil the goals of this project. Their team spirit for sustainable international collaboration has touched us all. Thank you so much, Shachi and colleagues!

For the success of our book workshop in Kunming in November 2019, we would like to express our gratitude to He Ming, the Director of NaCBES and the Leader of the First-Class Discipline Construction for Ethnology Project at Yunnan University (YNU), to Guan Kai, Li Zhinong and Zhao Chunsheng —the leadership of the School of Ethnology and Sociology at YNU—and to Zhu Min, Secretary of International Affairs at NaCBES. The generous support from YNU afforded our contributors and participants not only the conference venues and other logistical support but also, more importantly, the opportunity to meet many of its fine scholars and to know YNU's nearly a century-old campus, beautifully built and well-cared for. As an integral part of the founding history of China's ethnology, YNU is one of the most respected universities in China. The book workshop immensely benefitted from the participation of its scholars and graduate students. We look forward to more meaningful collaborative projects between YNU and HUC in the years to come!

We also would like to give our special thanks to John Grim, Co-Director of Yale Forum on Religion and Ecology, whose willingness to offer a conclusion to the book makes an immense difference for our chapters to meet a broader interdisciplinary audience. As the field findings of most of our contributors indicate the clear correlation between indigenous religious and ecological practices, this book is also characteristically a work of religion and ecology besides its environmental humanities orientation. Grim's conclusion dovetails the two fields and pushes their symbiotic interface to a new horizon in the field of Himalayan environmental studies. It lends more strength to our efforts to identify the indigenous environmental ethics and wisdom that are invaluable to the global environmental discourse and endeavour for a sustainable future.

During the time of our individual chapters' write-up and revision, many of our contributors presented their drafts and shared concept notes at different conference venues and thematic lecture events at the International Institute for Asian Studies (IIAS) and the Centre for Global Heritage and Development (CGHD) at Leiden University, Linnaeus University Centre for Concurrences in Colonial and Postcolonial Studies, Faculty of Humanities at University of Oslo, the Institute of Asian Research at University of British Columbia, the Centre for East Asian Studies at Jawaharlal Nehru University, India Institute of Technology Guwahati, the Institute for South Asian Studies at University of California at Berkeley, the Centre for Asia-Pacific Studies at Academic Sinica, the Center for International and Global Studies at Duke University, East Asian Studies and the Global Tea Initiative at the University of California Davis, the School of Humanities at Tallinn University, the School of Life Science at Peking University and the Institute for International and Area Studies at Tsinghua University. We are grateful to these globally vital venues of thought exchange and would like to thank these individual scholars in the forefront of innovating the fields of Asian studies, borderland studies and environmental humanities—Willem van Schendel, David Ludden, Gunnel Cederlöf, Mandy Sadan, Gunlög Fur, Joy Pachuau, C. Patterson Giersch, Bengt G. Karlsson, Thom van Dooren, Emily O'Gorman, Martin Saxer, Arupjyoti Saikia, Janet Sturgeon, Mary Evelyn Tucker, Prasenjit Duara, Hanna Havnevik, Wen-Chin Chang, Mett Hansen, Koen Wellens, Li Zhang, Lü Zhi, Halvor Eifring, Katharine Burnett, Tsering Shakya, Stéphane Gros, Lorence Cohen, Sara Shneiderman, Mark Turin, Zhang Jing, Iftekhar Iqbal, Emily Yeh,

Ralph Litzinger, Dawa, Jianxiong Ma, Nadine Salzmann and Zhang Haiyun for their kind feedback and/or organising of talks and discussions for our authors. Their support reaffirms our conviction that scholarship is an ever-growing colourful tapestry waiting for new threads to be woven into it.

As bookmaking involves so many minute but critical details, we would like to thank Wendy Smyer Yü for volunteering to copyedit our nonnative English authors' manuscripts, and the ICIMOD's cartographer Gauri Dangol for making the book map and environmental scientist Philippus Wester for climate change data. We owe our many thanks to them for the smooth publication of this book.

Dan Smyer Yü and Erik de Maaker

Figures and maps

Map—Geographical coverage of the chapters

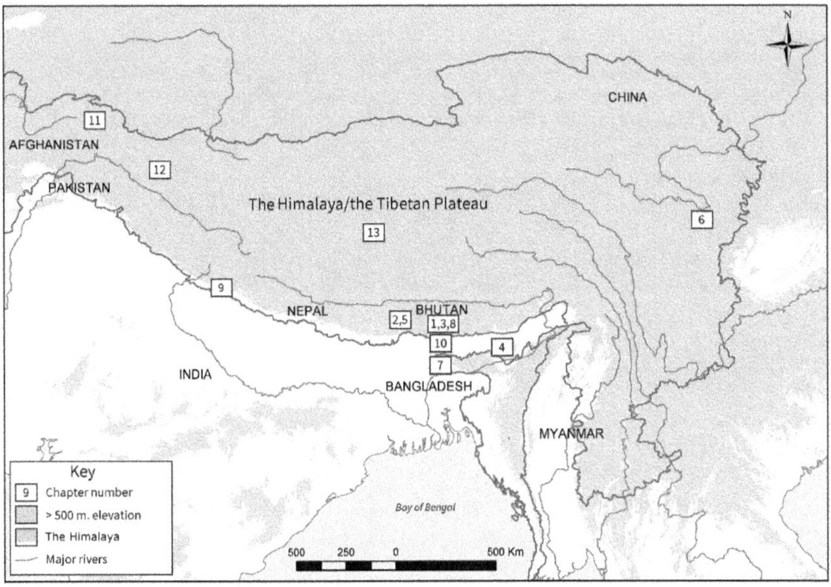

Situating environmental humanities in the New Himalayas

An introduction

Dan Smyer Yü

Context

The advent of the twenty-first century marks the rapid emergence of environmental humanities, an interdisciplinary field of environmental studies addressing locally manifested global environmental endangerment to "the maternal principle of the earth" (Berry 2009, 75), and exploring deeper, diverse meanings and values of humanity (Rose et al. 2012, 1–5) for a sustainable future. Its subject matters are not entirely new (Heise 2017, 1); however, its approaches are new in the sense of its capacity for a wide-ranging interdisciplinary alliance of natural and social scientists, scholars of humanities and the public. It finds its roots in a variety of modern environmentalisms and environmental studies worldwide (Guha and Martinez-Alier 1997). Community-based environmental movements and conservation efforts in the Indian Himalaya are a bundle of roots from which environmental humanities draws intellectual nutrients. In particular, the Chipko Movement in the 1970s spawned diverse understandings of environmental crises and enthused humanist approaches to ecological conservation not merely in the Indian Himalaya but also elsewhere in the world. Besides the causal force from the monsoon of the year, the detrimental 1970 flood in Uttarakhand, pinpointed the anthropogenic origins of natural disasters. The extractive tendency of modern development was identified as what the Chipko movement leader Sunderlal Bahuguna called "the butcher of Earth" (Guha 1989, 178). The emblematic image of the village women's hugging of trees as their resolute act of protecting the forest from industrial logging landmarked an inchoate era of humanist environmental studies from eco-spiritual, feminist, cultural and socio-political perspectives (Tewari 1995; Arora and Murthy 2015).

However, for the last 50 years, humanist approaches to environmental studies of the Himalayas have remained on the margin. The overarching trend in both academic and policy arenas has largely revolved around the conception, propagation and rejection of the "Theory of Himalayan Environmental Degradation" that originated in Europe and North America (Eckholm 1975; Thompson et al. 1986; Guthman 1997)—a scientific effort at guesswork about a Himalayan environmental super-crisis (Ives 2004, 23) and "a failed prediction" (Aase 2017, 1). The theory suggested that the environmental deterioration and natural disasters in the Himalayas were mostly caused by population growth and the expansion of native communities' subsistence farming. Regardless of its lack of scientific substantiation, the theory was adopted as "scientific" by policymakers and development specialists. Relocation of native populations from higher altitude areas of the Himalayas to lowland areas was suggested. Himalayan natives were thus blamed for environmental degradation.

Externally conceived assessment of local environmental conditions is not unique to the Indian Himalaya. It takes place in the adjacent Tibetan Plateau, too, in the case of nomads being pinned down as contributors to grassland degradation and, therefore, suggests that they have to be relocated to towns and cities as "ecological migrants" (Qi 2015, 47) as if nature could only recover itself by excluding its native human dwellers. While new outcomes of scientific studies are exonerating the Indian Himalayans, though not yet the Plateau nomads, the critical trend of Himalayan environmental studies continues to find its gravity in what Jack Ives identifies as "water and aesthetic amenities" (Ives 2004, 204–209), a set of natural heritages understood through their utilitarian qualities. The Himalayas thus continue to be perceived as a site of natural resources and the Himalayan poor are seen as passive recipients of specialists' technical and statesmen's technocratic calculations of how the natives can best utilise the gifts of the earth to bail themselves out of poverty and earn a surplus by offering their mountainous landscapes to global nature-tourist consumption. This environmental perception of the Himalayas is what Ramachandra Guha calls "the paradox of global environmentalism" (Guha 2000, 367), in which environmental conservation is mostly advocated by specialists, activists and nature lovers from affluent nations, all with track records of environmental destruction accrued from their imperial and colonial histories and modern development projects. Humanist approaches to Himalayan environmental studies are still at a prolonged budding phase, waiting to flourish.

This book is the first fruit of the Thematic Working Group on Trans-Himalayan Environmental Humanities conceived in fall 2017 under the auspices of the Himalayan University Consortium (HUC) housed at the International Center for Integrated Mountain Development (ICIMOD), a Himalaya-based, intergovernmental organisation. Co-led by Dan Smyer Yü and Erik de Maaker, the Group is designated to provide a formal introduction of environmental humanities to universities and research institutions based in and associated with the Himalayas. It supports multinational collaborative experiments in environmental humanities in both Himalayan and global contexts. It endeavours to identify knowledge-based synergies of local communities, scientists, scholars and policymakers by re-exploring, re-examining and re-articulating the fundamental meanings of human–earth relations, climate change, environmental challenges, resilience, sustainability and multispecies flourishing. The members of this book project come from twelve countries and are situated in the disciplines of anthropology, conflict studies, development studies, earth science, environmental science, ethnology, international relations, history and Tibetan studies. The geographical areas of our research cover the highlands of Bhutan, the Lepcha communities of Sikkim, the Hunza Valley of Pakistan, Sind Basin in Kashmir, Garo Hills in Meghalaya and Barpeta District in Assam, the lower region of Mahakali River in

Nepal, Southwest China, the Tibetan Plateau, Yunnan–Myanmar borderlands and Sikkim–Bhutan–Tibet borderlands. It is thus an international, interdisciplinary and interscientific group of scholars and scientists endeavouring to access environmental humanities to make sense in our individual disciplines and interdisciplinary conceptualisations of place-based environmental practices.

The initial texts of environmental humanities read among the Group members were mostly the foundational writings by leading environmental humanists, such as "Thinking Through the Environment, Unsettling the Humanities" (Rose et al. 2012), *Nourishing Terrains* (Rose 1996), "Lively Ethography" (van Dooren and Rose 2016), *The Mushroom at the End of the World* (Tsing 2015), *Arts, Religion, and the Environment* (Bergmann and Clingerman 2018), *The Routledge Companion to the Environmental Humanities* (Heise, Christensen and Niemann 2017), *Global Ecologies and the Environmental Humanities* (DeLoughrey, Didur and Carrigan 2015) and *The Environmental Humanities* (Emmett and Nye 2017). This body of literature gave us a deeply resonant sense of the inter-related conceptual momentums of "rethinking the ontological exceptionality of the human" (Rose et al. 2012, 2), "rethinking time, memory, and narrative" (Heise 2017, 4), understanding "the human as a geological force" (DeLoughrey et al. 2015, 12) and "thinking-feeling with the Earth" (Escobar 2016). What resonated with us most strongly are the conceptual works and case studies of Australian aboriginals' lived experiences of Country (Rose 2000; Gibson et al. 2015; Bawaka et al. 2013; Tebrakunna Country et al. 2019).

This body of literature indicates that indigenous environmental knowledge and practices play a large role in inspiring the works of the foundational environmental humanists. It also signifies that the global geographical spread of environmental humanities is concentrated in neo-Europe (Crosby 2004, 2), that is, places where Europeans established themselves outside of Europe, such as Australia and North America. Among the initial total 20 team members, three-fourths of us are native respectively to Assam, Bhutan, Kashmir, northern Myanmar, northern Pakistan, Sikkim, Tibet and Yunnan. Four non-Himalayan native project members had prior work in the field of environmental humanities in Europe, Australia and North America; whereas none of the Himalayan native scholars and scientists had exposure to this emerging interdiscipline prior to their participation in this project. These micro-statistics attests to an uneven global spread of environmental humanities resulting in fewer published contributions from the Himalayan region. For instance, between 2012 and 2019, the *Journal of Environmental Humanities* featured only one piece, by Sarah Besky (2017) on a postcolonial study of the Nepalese in Darjeeling, about the greater region. Overall, very few monographs and edited volumes are specifically environmental humanist approaches to the studies of the Himalayas. We take this scarcity of scholarly output as our opportunity to fill in gaps and build bridges by connecting our Himalayan-based findings and conceptual works with the larger community of environmental humanists around the world.

Situated in this context, for the purpose of comparing notes with the global community of environmental humanists, this introductory chapter shares with our readers the interdisciplinary and interscientific challenges our team members have met to co-create this book as an initiative of environmental humanities in what we call "the New Himalayas" by which we mean the anthropogenically impacted Himalayas since the early modern era. This work lays out three characteristics of the New Himalayas—the transboundary, the indigenous and the animist—as the formative factors of our environmental humanist approach to the study of Himalayan anthropogenic conditions. In the process, we propose that a planetary sustainable future finds its roots not merely in modern scientific and utilitarian calculations of the finite natural resources of the earth but also in our sentient relations with the earth as a planetary living being. Expressed diversely in our cultural memories, folklores, sacred geographies and our varied perceptions of the ontological union of the biotic and abiotic worlds, these sentient relations afford us a planetary common ground for building new environmental ethics for a sustainable future of the Himalayas as well as of the world. The earth's intrinsic value is entwined with and embodied in the affective bonds of places, humans, animals, plants and their associated spiritual worlds.

An experiment of undisciplining—challenges and dividends

The charisma of environmental humanities across the disciplines of the sciences and humanities comes from its new and re-visioned calls for "a thicker notion of humanity" (Rose et al. 2012, 2), "an unprecedented mix of science and conscience" (Rolston 1998, 125), "a seismic shift in environmental thought" (UCLA), "being and becoming with others" (van Dooren and Rose 2016, 80), "exploring nature's texture" (Bergmann and Clingerman 2018), spotlighting "the ghost of the climate refugee" (Armiero and Tucker 2017, 2), "a planetary ethic" (Bauman 2014) and more. It is innovatively seismic in the sense that the environment is being further set free from the enclosures of specific sciences and disciplines that have had the exclusive privilege to produce environmental knowledge and make distinctions between scientific and nonscientific knowledge of the environment. The implications of unidisciplinary environmental knowledge have woven together an epistemological reflexive consent normalising the hyperseparation of humans and nature, nature and culture, and the global and the indigenous. This scientific reflexivity has turned out to be "a radical exclusion" (Plumwood 1993, 48) that conceived the modern planetary worldview with multiple dualisms and is thus accountable for the rapid anthropogenic transformation of the earth's diverse environments. This is where environmental humanities comes into the scene and re-visions an integral terrain of environmental studies in which radical inclusion is advocated to become an interdisciplinary norm among environmental humanists from different fields of the sciences and humanities. The inclusive and democratic attributes of environmental humanities are thus inevitably magnetising more and more scholars, scientists and policymakers into more enlightened understandings of the earth's environmental crisis and exploring new humanist foundations for a sustainable future.

The notion of the environment is reshaping itself towards more lively, animated and empathetic understandings of "a more-than-human world" (Abram 1996, 24). Humankind is no longer in-but-not-of the environment. In practice, environmental humanities challenges its participants' interdisciplinary strengths and limits. For this research-based book project, in order to have more effective interdisciplinary outcomes, we advocated "undisciplining ourselves" from five epistemological reflexivities prone to the dualistic tendency mentioned above: (1) the notion of nature that excludes human presence, (2) indigeneity only understood in political terms, (3) scientific practices that treat the earth only as a physical and mechanical entity, (4) sustainable development promoted with the requirement of fostering economic growth and (5) area studies exclusively based on the cartographies of modern nations as analytical units. These five areas targeted for our experiment of undisciplining correspond respectively to five facts of Himalayan environmental studies: (1) nature in the Himalayas is often perceived to be extracted as material and aesthetic resources for human consumption (Ives 2004), (2) Himalayan indigenous studies largely focus on the political struggles of indigenous communities for rights and justice with little consideration of how their ecological worldviews and environmental ethics can be a valuable contribution to global environmental discourses (Guha 2000), (3) the studies of climate change and environmental flows in the Himalayas are highly quantified in the natural sciences with scant local human experiential perspectives and narratives included (Saikia and Thapa 2019), (4) sustainable development in the Himalayas continues to prioritise development over sustainability (Ives 2004; Ives and Messerli 1989) and (5) the integrality of the Himalayan region is cartographically fragmented by modern nation states (Davis in this book; Shneiderman 2010; Smyer Yü and Michaud 2017).

In the concept note of our project, we, therefore, invited team members:

- to tune into the conceptual foundations and ongoing trends of environmental humanities,
- to sense and articulate the intelligence and the agency of the earth in Himalayan indigenous contexts,
- to address planetary environmental questions from our place-based case studies,
- to not limit ourselves with our voices echoing only within the walls and fortresses of our own disciplines (Smyer Yü and de Maaker 2019).

Our aims were high and, expectedly, met challenging situations. In the process of revising our individual manuscripts, two environmental scientists, one ethnologist and one development specialist bowed out of the project. One of them apologetically expressed that environmental humanities is "out of our [co-authors] league." To be self-critical, the phrase—"the walls and fortresses of our own disciplines"—did convey a pejorative sense of intellectual incarceration if viewed from a Foucaultian perspective. Ontologically speaking, walls and fortresses safeguard our comfort zones and existential niches. Does being interdisciplinary mean tearing down disciplinary walls or mutually helping one another build stronger and more intelligible disciplines? Is the interdisciplinarity of environmental humanities a discipline to conform to or a destructively creative intelligence meant to tear down old walls for new beginnings? We have surely learned that the actual practice of environmental humanities is an art of working concurrently and symbiotically with differences, dichotomies and paradoxes, for example, one and many, the local and the planetary, the particular and the universal, and disciplinarity and interdisciplinarity. These dynamics suggest that the practice of environmental humanities is an open-ended, generative rite of passage in and beyond this bookmaking project. Environmental humanities has no ready-made framework of a traditional discipline. It is rather always in the process of being animated, co-shaping and co-becoming with whoever wishes to innovate his or her own interdisciplinary niche in its vast field.

Our experiment of undisciplining, on the one hand, reconfirms that environmental studies/sciences will continue their discipline-based approaches. On the other hand, it shows that increasing public awareness and policy demands for achievable visions for restoring the health of the earth towards a sustainable future reverberate across the sciences and humanities. This global awareness compels many of us to come together as a collaborative "critical mass" and reach out to one another situated in different disciplines, environmental contexts and cultural worlds. Undisciplining in this sense does not mean a simplistic act of breaking down walls and shifting our disciplinary grounds as if the united sense of the interdisciplinary world comes only from abolishing disciplinary distinctions and boundaries. This goes against the principle of evolutionary diversity. Likened to ecological dynamics, undisciplining in our interdisciplinary experiment reaffirms the uniqueness of each scholar's discipline as an intellectual ecological niche that connects itself to its neighbouring peers via ecotones or transitional terrains. What needs to be undisciplined is the artificial walls and borders that block or frustrate the transitional terrains connecting different disciplines. Seen from an ecological perspective, academic disciplines are thus the interconnected niches of our livelihoods. There is no single environmental humanist who does not have a specific disciplinary basis for his or her interdisciplinary work. For us, a large part of our environmental humanist work in this book project is concerned with how we effectively communicate our specialised findings, analyses and interpretations in an interdisciplinary language among academics and in what Rose calls an "earth language" (Rose 2013, 107) to engage the greater world. Our effort is being made and challenges are obviously ongoing. What we hope to communicate with our peer environmental humanists in the Himalayan region and elsewhere in the world are the outcomes from our individual but interrelated case studies of a transboundary mountainous and riverine earth that is sentient, animist and symbiotic but that is also anthropogenically impacted. Well-met challenges bring us rewards.

Himalayan humanities for the New Himalayas

Modern human–earth interactive dynamics point to an eco-geological state of the earth in which the Himalayas is encountering anthropogenic impacts but is forcefully able enough to overthrow excessively extractive human development schemes and modernisation projects through the course of natural disasters. The combined forces of humankind and the earth reflect the Anthropocene hypothesis

(Crutzen and Stoermer 2000) that the earth has transitioned into a human-induced geological age. In it, humankind is being considered a geological force that is altering the physical structure of the earth and bringing artificial climate change to every "corner." Whether or not humankind is a geological force is up worth debate; however, the Anthropocene hypothesis supports the globally recognised anthropogenic contributions to local and global environmental challenges shown in the environmental footprints of industrialisation, modernisation and economic globalisation activities. The Himalayan range, as we speak, is a recipient of such anthropogenic "gifts" in the forms of global human-induced climate change and regional geo-economic activities undertaken in Himalayan nations.

Based on the environmental issues and conditions discussed in our chapters, we propose that the anthropogenically impacted Himalayas is a "New Himalayas" to be approached with what we call "Himalayan humanities"—a wide range of existing and yet-to-be explored interdisciplinary and interscientific perspectives for the studies of the anthropogenic Himalayas. Our coinage of "New Himalayas" and "Himalayan humanities" was inspired by arctic environmental humanists' conceptualisation of "the New Arctic" (Evengård, Larsen and Paasche 2015) as the interscientific and interdisciplinary subject matter of the "Arctic humanities" (Sörlin 2015, 327–339, 2017, 433–442). In the spirit of conversing with the global community of environmental humanists, we find that the New Himalayas and the New Arctic are comparable in many environmental and cultural arenas. Both are highlands. The former is an altitudinal highland and the latter is a latitudinal highland. Both altitudinal and latitudinal highlands are homes of global water in both solid and liquid forms. Both host most of the planet's glaciers. The commonly shared newness of each in the Anthropocene consists of their vulnerability to human induced global climate change, susceptibility to the melting of ice and the modern endangerment and radical transformation of diverse indigenous peoples under the intrusion of external forces of change. Both are undoubtedly under anthropogenic duress.

The newness unique to the New Himalayas is a set of human-induced qualitative changes initiated in early modern era when old and new empires encountered one another and when modern forces of change began transforming the physical landscapes, modes of subsistence and native cultural heritages. Bhargabi Das's chapter recounts the British colonial effort to reinforce the water–land dichotomy for the sake of increasing the agricultural productivity of the riverine landscape in northeast India. This effort permanently changed the ecological system of the charlands. Likewise, British India's deforestation of Darjeeling for tea plantations around the same time period transformed the Lepcha from forest dwellers to farmworkers and urban labourers and, subsequently, led to the loss of Lepcha indigenous botanical knowledge (C. Lepcha in this book). This cumulatively understood newness has been compounded since the mid-twentieth century when newly independent modern republics began their modernisation efforts that assertively tapped deeper into modern agricultural potentials and hydraulic power to meet their nation-building objectives and geopolitical ambitions. In Ruth Gamble's chapter, a half century ago, the Chinese state's environmental perceptions of Tibet's wetlands as unproductive "wasteland" and its use of "exclusionary zones" of conservation in contemporary Tibet led to an attempt to disprove the traditional stewardship of native pastoralists and farmers for the environmental health of the wetlands.

Anthropogenic impacts in the New Himalayas are ample. Similar to the "New Arctic," what stands out in the New Himalayas is water, especially water in the form of transboundary rivers and their riparian environments transformed by dams and embankments. Inter-state contentions over the territoriality and hydraulic potentials of the transboundary rivers contribute more anthropogenic pressures to the mountainous environments and life communities in the Himalayan region. In this regard, the newness of the New Himalayas is understood as a range of anthropogenic preconditions and realities to be correlated and researched in the international sense as extensively discussed in Alexander Davis's chapter. The geopolitical demarcation of national borders in the region since the mid-twentieth century is not merely a boundary-drawing matter of individual nation-states, but is also a unilaterally decided legal basis for development projects with heavy transboundary environmental impacts. Notable projects include the constructions of hydraulic dams in Bhutan, Nepal, Pakistan, China and India. Each country has its own development reason to harness the hydroelectric potential from the Himalayan rivers. According to the recent studies by International Rivers, Bhutan's hydraulic goal of 1,000 MW by 2020, prioritises the financial gain of exporting 75% of the generated electricity to India regardless of the fact that Bhutan itself has not yet achieved 100% electricity coverage of its urban and rural areas. Similarly, over half of Nepal's rural population does not have electricity. The rationale behind Pakistan's dam building is the effort to give support to the nation's agricultural sector, which employs up to 66% of the nation's workers and generates 80% of its export economy. India and China, the two most populated trans-Himalayan countries and emerging global economic superpowers, each justify their ongoing pursuit of dam building agendas in a fierce competition over Himalayan transboundary hydraulic power, in an attempt to harness it to serve their growth-led economies (IR 2008; Gamble 2019). By 2019, there were over "550 hydropower projects in existence or under construction" (Xu et al. 2019, 144) in the Himalayan region. Human demand for energy is relentlessly stressing the New Himalayas.

The high environmental cost of hydroelectricity is an ongoing transboundary anthropogenic transformation of the New Himalayas. National economies can perhaps be boosted by the hydraulic energy but the same cannot necessarily be said of the sustainability of the environment, local human communities and non-human habitats. It is common environmental knowledge that the impoundment of megatons of water and sediment generated by a cascade dam increases the likelihood of imminent natural disasters such as earthquakes and floods to the already seismically active Himalayan region full of geologic faults and thrusts. Each dam creates an artificial lacustrine ecosystem impacting both upstream and downstream environments through deforestation, loss of flora and fauna diversity, habitat fragmentation and endangerment of indigenous cultural heritages (Grumbine and Pandit 2013, 36; Valdiya 2014, 485). However, national priorities dependent on transboundary rivers are rapidly becoming the source of international tensions and conflicts. The talk of water wars among geostrategists, policy advisors and state-sanctioned scientists is reverberating throughout the Himalayas and beyond, especially in the context of Sino-Indian relations. Based on his global understanding of "water as a weapon" (2013, xiii), Brahma Chellaney, a scholar-geostrategist, propagates the concept of water wars in the context of Sino-Indian "dam racing" (2012, 1) as a bilateral hyper-geopolitical stress. In this bilateral race for hydropower, transboundary rivers are perceived as "a bone of contention" (Mahapatra and Ratha 2016) between the two states.

While the New Himalaya region is being shaken by human geo-economic and geopolitical forces, the region is also undergoing human-centred environmental violence "connected to larger processes of material transformation and power relations" (Peluso and Watts 2001, 5), geared towards "the transformation of nature" into "resource environments" (Peluso and Watts 2001, 5) on both national and international scales. While the water-war scenario is hotly debated, the notion of water security is being further developed to prevent the potential for such competition. The regional rhetoric of water security (Chellaney 2012; Engelke and Michel 2019; Christopher 2004) and the global discourse of environmental security, more often than not, are exceptionally state-centric in their exclusive emphasis on the environment as a "pie" of resources to be sliced among nations or to be monopolised by a single nation. The Himalayan environment thus continues to be distilled down and reduced to a natural resource and energy source, and is not seen as a home of diverse ecosystems, habitats, culturally cherished natural heritages, and source of indigenous ecological worldviews and environmental ethics. In this regard, environmental security is not and cannot be equated with environmental conservation and sustainability. The latter demands more humanist understandings of the environment as an integral whole of livingness besides simply a physical, inorganic entity separate from yet heavily exploited by humankind.

While the Arctic environmental humanists are invoking the "cryo-historical moment" (Gk: cryo = ice and frost) to highlight "the fate of ice as a sign of the fate of our societies" (Sörlin 2015, 334; 2017, 437) on a global scale, we would like to propound the Himalayas as "a terrestrial ocean," an immensity of water in both solid and liquid forms on the surface and entwined in the mountainous landscape. The notion of terrestrial ocean is not a syntactic variation of the term terrestrial ice used in the context of the New Arctic (Sörlin 2015, 327), where it stands above the land but is separated from it. Neither is it a variation of the "water tower of Asia" concept metaphorically emphasising the liquid form of water. Instead, the Himalayas as a terrestrial ocean hosts water not merely in the form of terrestrial ice, like in the Arctic, but also in the form of what we can call "aquatic land" in which water fully saturates land and land entwines with water. From the perspective of earth sciences, the entire Critical Zone, the earth surface "between the top of the tree canopy and the base of weathered rock" (Dybas 2013, 1), could be said to be an aquatic land; however, like in the Arctic, the amount of water concentrated in the Himalayas and the greater environmental implications of this concentration compel us to recognise how water and land are entwined as one environmental body. In addition to its susceptibility to global climate change, the Himalaya, as a terrestrial ocean, has an anthropogenic history tracing back to British India and to the modern nation building processes of China and India. This terrestrial ocean, as a water–land body, has been subject to an artificial partition process that attempts to disembody water from land so that it may be harnessed for industrial use and the maximising of agricultural productivity. The flood-prone charlands in Bengal are also a legacy of the British India's agricultural endeavour to separate land from water (Lahiri-Dutt and Samanata 2013, 53; van Schendel 1991, 66; Das in this book). Other regions have seen the terrestrial ocean's disembodiment as well. In the modern human transformation of the vast wetland-area in Zoigé, eastern Tibet, the late-1950s expansion pastoral and agricultural lands led to a 60% decrease in area and a 13% increase in the desertification of the wetland (Wei 2018, 359). With the mindset of *kenhuang* or the transforming of "wasteland" into agricultural land (Gamblel in this book; An, Naiyun and Chang-e 2008, 824), the drained marshes and swamps did not become the desired grasslands and barley fields; instead, they were transformed into deserts and semi-deserts. Dam building across Himalayan nations and the Tibetan Plateau is another human method of changing the body of water-in-land into electricity and irrigation. Human material desires and comfort come with a high environmental cost.

These human-induced environmental conditions are demonstrative of this vast terrestrial ocean as an anthropogenically impacted inter-Asian region with global impacts. They demand integral perspectives to understand the entwinement of water and land in the context of the New Himalayas. Gamble's understanding and analysis of Tibetan wetlands from the native perspective as "muddy lands" and "wet meadows" recalls Lahiri-Dutt's "wet theory" (2014, 506–507) signifying the hybrid environments of water and land. Our notion of terrestrial ocean emphasises not only the mutual saturation of Himalayan water and land in the physical sense but also the affective intertwining of humans and the earth. Like their Arctic counterparts, the glaciers of the Himalayas are also "remnants of the major glaciation on a par with geological and paleontological features in the landscape providing witness of a deep time" (Sörlin 2015, 329) and they continue to serve as the ancient sources of the transregional rivers. The story of glaciers and spectacular rivers is both a story of the geological formation of the earth and a story of a humanised earth. The moving water from the Himalayas, embodied as transboundary rivers, is not merely a universal sustainer of life and a resource to be exploited but is also a transcultural thread networking together diverse civilisations and ethnolinguistic traditions as an integral part of the earth's spectacular diversity (Johnston 2012). Nature and culture are entwined and mutually embodied. While the earth naturalises humankind, humans humanise the earth. The humanity of the earth never stops speaking through our diverse mythologies, folklores, histories, literatures, philosophies, religions and spiritual senses of physical landscapes situated in different ecological niches and biospheres. Like the Arctic, the fate of the Himalayas is an inherent part of the fate of humankind. In this regard, the language of sustainability requires affective approaches, too.

Pluriverse of Himalayan humanities

The Himalayas can be considered a pluriverse or a "world of many worlds" (Blaser and De la Cadena 2018, 1–22; Escobar 2018, xvi). In the context of environmental geopolitics, this pluriverse is largely perceived as a human-centric universe with diversity centred upon human nation-states as the units into which the earth's various environments are divided. Human-designated national borders justify unilaterally decided hydraulic projects, slice the mountain bodies into different national territories and sever terrestrial animals' migratory paths. The geological contiguity, ecological continuity and ecotonal dynamics of the Himalayas are de facto secondary to the environmental vision of individual nation-states. Initiatives relating to transboundary environmental governance are often the aftereffects of unilateral implementations of modernisation projects with transboundary impacts. If geopolitics is set aside, the human-devised national units within the Himalayas are environmentally interdependent worlds that are undergoing a wide range of anthropogenic changes in the name of economic development. Like everywhere else in the world, the national practice of the universal ideology of growth-based development, on the one hand, engenders the state-corporate "ontological occupation" (Escobar 2016, 14) of land, flora and fauna, and human mindscapes and cultural behaviours that, therefore, produces anthropogenic effects. On the other hand, it compels humans to rethink the environment from more-than-human perspectives, to re-invoke locally rooted environmental knowledge and wisdom to address the global environmental crisis and to forge more integrated local–global conversations on the global values of that locally generated ecological knowledge and wisdom. This is what our authors strive to do.

Within the limits of our specialisations and geographical coverage, the chapters of this book are intended as an initiative for diverse humanist approaches to environmental studies of the New Himalayas. As they attest to the diversity of Himalayan habitats and environmental ethos under anthropogenic conditions, we group them in five subthemes—(I) Kindred of the Earth, Deities, Humans and Animals, (II) Aqueous Earth, (III) Evolving Cosmovisions, Climate Change and Community Resilience, and (IV) Transboundary Environmentality and Indigenous Commoning.

Part I Kindred of the Earth, Deities, Humans and Animals features chapters by Jelle Wouters, Thinley Dema and Charisma Lepcha (hereafter C. Lepcha). This thematic section invites readers into our project members' studies of the environment as a multitude of more-than-human worlds. It expands the current conceptualisation of the more-than-human world or the other-than-human world from a set of human–animal relations to human–earth relations and human–deity relations. The physical earth, in our studies of Himalayan indigenous environments, has a life of its own. Its biography is not limited to its physical–chemical evolution but also includes its ensoulment and its being animated with invisible spirits, supernatural beings and deities embodied in diverse landforms and environmental flows. The livingness of the earth is diversely expressed in humans' felt kindredness with the creative power of the earth personified in local deities and speaking through animals.

Jelle Wouters' chapter is exemplary in highlighting the ontological and affective relations of herders, yaks, deities and the physical environment of Bhutan's highlands. In this relational world, personhood is not limited to humankind but is found in yaks, place-based deities and animated mountainous landscapes. The chapter presents a more-than-human world speaking through Wouters' Bhutanese nomadic interlocutors, one in which the personhood of the earth, the yak and local deities embodies a long co-dwelling and co-becoming journey from a mythical time to contemporary-era Bhutan. Here, the interspecies relationship between humans and yaks is not merely shown as a horizontal intersubjectivity but as a relationship that is terrestrially rooted with the living earth. By engaging in deep conversation with Descola's animist discourse of nature and culture, and Haraway's and Tsing's studies of interspecies kin making, Wouter's chapter offers a fresh understanding of the trans-species personhood between the yaks and their human companions, and, in many ways, questions and blurs the artificial boundaries between animality and humanity. Like Descola's study of Amazonian animism, Wouters finds that the interspecies kinship between the herders and the yaks is found in their co-becoming process rooted in the highland environment and their mutual attentiveness mediated through their shared dwelling place and the local pantheon embodied in the mountains.

As a member of our project's research team in Bhutan led by Wouters, Thinley Dema presents a complex case of human–earth relations in her chapter with an ethnographic focus on the Three Brothers Mountains in Haa District, Bhutan. The eco-spirituality of the three deity mountains is expressed in the kindred relations between the deities and the earth, and between the deities and humans. The triadic relationship of the earth, humans and the deities is symbiotically woven together with religious, spiritual and economic factors. Like in Tibet, the entanglement of pre-Buddhist and Buddhist cosmologies is also embodied in the nomenclature and local people's perception of the Three Brother Mountains. While they are (re)named after three Bodhisattvas, their pre-Buddhist cosmic forces continue to sustain their supernatural position with their human companions. They are feared and revered. At the same time, as modern economic practice is becoming a norm in Bhutan, they are encountering humans' expanding utilitarian desires for the earth's physical resources. In this context, Thinley Dema offers two arguments concerning the relational ontology of the deities, the earth and humans. On the one hand, humans' eco-theological relations with the deity mountains have positive implications for environmental conservation. On the other hand, the mountains/deities are being ontologically occupied by humans' economic development. Local Bhutanese are being challenged to find a new equilibrium between the environment and humans' search for sustenance and wealth.

C. Lepcha's chapter invites readers into the world of the Lepcha in Dzongu, the Lepcha reserve where water, entwined with rocks, soil, streams, rivers and snows, is a foundational fabric of Lepcha country. C. Lepcha's ethnographic narrative resonates with the Australian aboriginals' idea of Country as "a living entity" (Rose 1996) that gives and receives life. It reclaims the Lepcha as indigenous botanists, destigmatising the colonial racialising condescension of the Lepcha as "born naturalists" and salvaging their indigenous ecological knowledge and practice from the postcolonial politically centred critique of the colonial essentialisation of indigenous peoples. Her argument is eloquently made through her mythical, religious, ecological and political narratives and discussions. Like the chapters by Wouters and Thinley Dema, C. Lepcha also regards the environment as both physical and spiritual. It is thus saturated with geo-spiritual forces. Water bodies in her case study are thus both environmental flows and living entities as human ancestors and relatives. They possess their own intrinsic agencies when they are in relation with humans. The kinship of water bodies and humans is identified in the sacred landscapes in Dzongu. C. Lepcha contextualises this human–nature kinship in colonial history as well as in contemporary local indigenous people's anti-dam-building movement. In both colonial and postcolonial contexts, the environmental history of the Lepcha shows a clear pattern of destructive transformation by the extractive economic practices of British India and the independent Indian state through deforestation, the introduction of cash crops such as cardamom, and the building of hydropower plants. As a work of environmental humanities, the chapter generates many comparative implications particularly in its conversation with Rose's documentation of the Australian aboriginal sense of their Country as "nourishing terrains" (1996).

In Part II Aqueous Earth, we have three case studies respectively from Bhargabi Das, Rongnyoo Lepcha and Mongfing Lepcha (thereafter R. Lepcha and M. Lepcha), and Ruth Gamble, all focusing on water but in different regions of the greater Himalayas. Although over half of our chapters touch on the topic of water, their chapters offer concentrated narratives of waters as an integral part of land and vice versa. Waterscapes are landscapes and, thus, both form an aqueous earth in their respective regions. Aqueous earth is not merely a physical manifestation, but also resides in the human mindscape as memories of how the earth and life began to take physical and often animated shape out of geo-spiritual creative forces and revealed in human affective expressions. Contextualised in the current anthropogenic conditions of the Himalayas and the Tibetan Plateau, and in the state-dominated development schemes intended to separate land from water or harness energy from water in Assam, Sikkim and eastern Tibet, the three chapters are intertextually comparable with each other regarding the politics of the environmental flows embodied in the land in the twenty-first century and rooted in the regionally varied, pre-colonial and colonial cultural and political histories, and in the modern nation-state building processes.

The chapter by Bhargabi Das, an experimental text of the lively ethography proposed by van Dooren and Rose in 1996, presents an anthropogenic aqueous earth in the form of shifting river islands, known as chars, in Assam. Das's humanist approach is built upon chardwellers' folk narratives, literary accounts and ontological resistance to the economically productive, environmentally exacerbating practices of the state. In her historical findings and ethnographic work, chardwellers in Assam, especially Bengali Muslims, have adapted to livelihoods that shift with the changing nature of the chars and the rivers that form and erode them. Their adaptation is in part driven by environment and subsistence. It is also an outcome of the systemic political and economic marginalisation that can be traced back to the colonial era of British India. As Das's research shows, the politics of the charlands finds a patterned continuity from the colonial era to postcolonial Assam, or, in Das's term, "neocolonial Assam." Assam's neocolonial pattern continues the same water–land hyperseparation initiated by the British for the purpose of maximising the agricultural productivity of the region's riverine landscape. Chardwellers' livelihoods and political representations, as in the time of British India, are severely marginalised. The Miya Poetry Movement in Das's study is a form chardwellers' political protest spoken through poems that crystallise the sufferings and hopes found in their everyday life. The aqueous earth in the charland of Assam is not merely a hybrid environment of water and land but is also an entwinement of the hardships faced by the riverine earth and its human dwellers.

R. Lepcha and M. Lepcha are indigenous scholar-artists based in Dzongu, Sikkim whose art is dedicated to the visual narratives of the Lepcha landscape as a storied world of ancestral spirits, supernatural beings, birds, animals and trees. Like Das, they make an attempt to write their chapter as a lively ethography. In it, they illustrate with their narratives and paintings the eco-spiritual ethos of their homeland in the context of the modern transformations and challenges that the contemporary Lepcha are facing. The environmental motif of the origin story in their narrative shares a commonality with other indigenous creation stories around the world. Creation never materialises itself ex nihilo but always starts with pre-existing, foundational environmental elements. In the case of the Lepcha, these elements are snow, water and land (mountains), all of which are both physical and spiritual and full of sentient expressions. The two rivers, Rongnyoo (Teesta) and Rungnyit (Rangit), are the primary spiritual–material threads interlacing both the mythical and historical memories and the social ethos of the Lepcha regarding their origin, and the eco-geological events experienced by their ancestors, such as earthquakes and an epic deluge. These past events all point to the region's geological history as an earth story of tectonic collisions leading to earthquakes and floods, both of which continue to be common events. In their conceptual discussions, R. Lepcha and M. Lepcha converse with case studies of the Australian Aboriginal conception of the earth as a living entity in the texts of Rose and van Dooren. Their lively ethographical approach is indeed a geo-poetic key to both tangible and intangible aspects of the Lepcha homeland as a more-than-human world.

Ruth Gamble's chapter enhances the environmental study of Tibet with her global perspective on water–land entwinement, delving into Tibet's wetlands as a critical part of the earth system. The ecological functions of wetlands across the world are evidenced in their capacities for storing and filtering water, recycling nutrients, controlling sediments and fostering biodiversity (Gamble in this book; Nyman 2011, 116). For the human utilitarian mindset, the value of wetlands is little and variable (LePage 2011, 12; Nyman 2011, 115). Gamble's careful study of the statist approach to the environmental management of Tibet's wetlands demonstrates the Chinese state's narrow, shifting valuations from "wasteland" to the "Plateau's kidney" (with its water purification function). In between, the "wasteland" underwent a state-initiated physical alteration from "muddy land" to dried, productive agro-pastoral land. Gamble's case study of the marshland in Zoigé, currently a Tibetan autonomous prefecture in Sichuan Province, confirms many natural scientists' assessments of human activities, such as the building of 700 aqueducts, as the primary cause for the marshland's 50% reduction in size between 1965 and 2000 (Li et al. 2014, 176; Bai et al. 2008, 2249). The marshland had been drained for four decades before the state took a conservation turn. However, this 180° turn to increase the size of the marshland has not been successful. Gamble argues that the ineffectiveness of this conservation attempt is largely due to what she calls "the statist approach," which treats the marshland as "sealed zones" excluding local communities' traditional interactions with the wetlands based on their lived ecological knowledge. Zoigé's marshland is thus an anthropogenic aqueous earth waiting to be re-embraced by its native human residents as its stewards.

Part III Evolving Cosmovisions, Climate Change and Community Resilience continues on to address human–earth relations in the arenas of cultivated landscapes, animated landforms and the environmental consequences of modern economic development and climate change with an emphasis on the agencies of humans, spiritual entities and the earth itself manifest in diverse forms and expressions. The chapters by Erik de Maaker, Kinley Choki and Rashila Deshar and her team share common ground, particularly in their ethnographic studies and conceptualisations of the complex relationships between indigenous cosmologies, local environmental knowledge, modern livelihoods and the concurrent but asymmetrical presence of tradition and modernity. With their fresh findings, the three chapters illustrate the current environmental conditions and human adaptive responses in the modern Himalayas.

Based on his long years of research in Garo Hills, in Northeast India, de Maaker's chapter narrates a case of what he calls the "Garo ontology" of agentive beings, inclusive of humans, ancestral spirits, animals and landforms that embody the spirits. With his ethnographic thread revolving around the *kusi*-boulders in Emangre Village that are inspirited with supernatural beings, de Maaker tells the changing story of the large boulders as a vibrant centre of the village's animist past and an abandoned but environmentally preserved historical site of the present. Between the past and the present, Christianity has played a role in changing the concurrently revered and feared human–spirit relations and in transforming the indigenous subsistence economy to monocultural cash-cropping. Both Christianity and modern monocultural livelihoods are the catalysts for the villagers to leave behind their animist tradition in the forests. From the perspective of new-animism studies as found in works by Graham Harvey, Philippe Descola and Eduardo Viveiros de Castro, de Maaker builds his argument concerning the environmental and social connections between the traditional animism-based relational ontology and the extractivist ontology of modern economic practices.

Similar to de Maaker's human–earth-deity theme, Kinley Choki's chapter addresses the fast changing traditional conception of the environment in the highlands of Bhutan. The keywords of the chapter—cordyceps, climate change and cosmological imbalance—reflect her focus on the role Bhutan's Buddhism-based indigenous cosmology plays in environmental conservation in the context of changing pastoral livelihoods, and the new wealth generated from digging and selling cordyceps, popularly known as "caterpillar fungus," from the highlands to the global market. The globally connected cordyceps economy is generating what Kinley Choki calls "unimaginable wealth" for Bhutanese highlanders, inevitably changing their pastoral lifestyle and the traditional cosmovision of the environment. The power of material wealth and the power of the spiritual understanding of the Bhutanese landscape are entering into a fraught relationship. Kinley Choki argues that the new wealth from the cordyceps economy is engendering a "cosmological imbalance." This imbalance is spiritually and materially manifest in a decline of the local communities' customary ritual observances by which they honour their mountain deities, and in the mountain deities' anger because of the environmental degradation caused by digging cordyceps as well as nonnative tourists' irreverent attitude towards Bhutanese native customs and deified landscapes.

As environmental scientists, Deshar and her co-authors take the challenge in this book to bridge natural sciences with social sciences. Their chapter documents how the indigenous groups, the Tharu and the Sunar, preserve their villages and homes in the flood-prone Mahakali River Basin. Aside from floods caused by monsoons and climate change, Deshar's team finds that most floods in the basin are human induced. The primary source of the flooding is the dams built by British India in 1928, and Tanakpur Barrage built by postcolonial India in the 1980s. When the barrages are opened, the water often inundates villages in the lower reaches of the rivers in Nepal. These human-induced floods are annual events in Nepal. The Tharu and the Sunar villagers have developed flood mitigation methods based on their indigenous knowledge and experience with tube wells used for water storage, and with large bamboo baskets for storing grains and seeds. Deshar's team argues that the successful coping strategies of the indigenous communities are the outcomes of the integration of their indigenous knowledge with modern technology and, that indigenous knowledge has a valuable role to play in sustainable living. Their chapter is a substantial interscientific contribution to the field of environmental studies on account of their willingness to translate their natural scientific findings into a humanist language so as to reach a broader audience. By adding a qualitative approach to their research, the chapter effectively reconstructs the primary causes of why the Tharu and the Sunar villages are susceptible to floods—namely climate change as a human-induced environmental change—the effective transboundary governance of the Himalayan rivers flowing through India and Nepal, and the deep-rooted Hindu caste system in Nepalese society.

Part IV Transboundary Environmentality and Indigenous Commoning features the chapters by Anwesha Dutta and Shailendra Yashwant, Zainab Khalid, Alexander Davis and Dan Smyer Yü. While all three chapters reaffirm the Himalayas as a world of many worlds and highlight the transboundary environmental connectivity between them, they reflect a geopolitical fact that the "many worlds" of the modern Himalayas are often understood as the bordered human nation-states rather than naturally formed environmental units. National border-making and this human-centric mindset inevitably generate anthropogenic contributions to ecosystems by severely disrupting natural regions with political borders. Borders delineate the territorial sovereignty of a given nation-state and defend it from external invasions. Meanwhile, they permit the nation's unilateral implementation of resource extraction projects that affect transboundary habitats within and beyond its sovereign territory.

In Dutta and Yashwant's chapter, the Dooars region of Assam that borders Bhutan attests to the anthropogenic impacts of the unilateral environmental decisions of neighbouring governments. Bhutanese and Indian states each approach their shared borderlands' environmental resources differently, but they produce the same anthropogenic effects as found elsewhere in the Himalayas: Bhutan's dam-building contributes to droughts and floods on the Indian side. While the Indian side is deforested to serve its national economic development, Bhutan keeps their side forested. Earlier studies of the Himalayan environment focused more on what Arun Agrawal calls "environmentality" or the combined form of a Foucaultian notion of governmentality and governmental rationality that refers to how governments make local environmental subjects serve the state's national development planning (Agrawal 2005, 216). The humanist approach of Dutta and Yashwant addresses such environmentality but goes well beyond it to highlight the efficacy of indigenous environmental knowledge as practiced by the Bodo people in the Bhutan–Assam borderlands. Their chapter presents a fascinating study of *jamfwi*, an indigenous irrigation system, as a combined case of political ecology and indigenous transboundary ecological practice. On the one hand, Dutta and Yashwant revisit the early modern territorial disputes induced by British colonial intervention and offer critical discussions of these disputes' legacy in contemporary Assam. On the other hand, they demonstrate the relevant role of indigenous environmental knowledge in local sustainable living as well as in the current planetary endeavours for deeper transboundary understandings of human–nature relations. The highpoint of their chapter is their conceptual dialogues with the existing studies of commons and commoning shown in the texts of Elinor Ostrom and Michelle Miller. As commons in their case study is specifically a transboundary water commons lying between India and Bhutan, they take the steps to expand the existing notion of it by interfacing local knowledge, modern science and cross-border community agencies.

Khalid's chapter is another case study of indigenous commons and commoning. It documents the practice of *rajaki* among residents in contemporary Hunza Valley. Culturally and etymologically rooted in the Burushaski language, *rajaki* refers to a mutual aid system overseeing the use of common pool resources available to a given community. In her findings, the enactment of *rajaki* is closely associated with maintaining *kuhl* or water channels that transport glacier water to communities for drinking and irrigation. The documented narratives from community members show that *rajaki* is a vital indigenous social mechanism helping balance the paradox of the abundance of Himalayan water and its low accessibility for humans. Khalid offers a twofold argument. On the one hand, *rajaki*, as an indigenous practice, plays a critical role in community sustainability. On the other hand, it encounters younger generations' reluctance to participate in it. Khalid's case of water commons is situated in a region bordering with China. Her conceptual emphasis is on the commoning as a set of actions taken within a single community for the purpose of illustrating *rajaki* as a system of commoning that strengthens the human–nature bond in both utilitarian and affective terms.

Alexander Davis' chapter opens a new arena of environmental humanities in the field of international relations. In many ways, his chapter recapitulates the transboundary nature of diverse environmental issues in the modern Himalayas, even as it reconfirms the modern Himalayas as a borderland of many nation-states with the existing conceptualisations "multiple-state space" (Shneiderman 2013, 28) and "multistate margins" (Smyer Yü 2017, 25). The pre-modern integrality of the Himalayas as a world region has been cartographically divided into strictly bordered individual nation-states. Human-induced environmental issues are part and parcel of the consequences of this modern cartographical process and of the unilaterally determined extraction of the natural resources of the Himalayas. By interfacing international relations studies and environmental humanities, Davis offers two correlated and convincing arguments that the geopolitics of Himalayan nations have engendered environmentally destructive feedback loops involving the environment, people and nonhuman species, and that the state-centric notion of environmental security, particularly in the international relations of China, India and Pakistan, is fixated upon territorial security instead of on treating the transboundary nature of the Himalayan environment as a priority for international environmental governance. Both arguments illustrate the anthropocentric policymaking process common among Himalayan nation-states and its anthropogenic consequences across the Himalayas. Davis's initiative to put international relations and environmental humanities in dialogue is generative of interdisciplinary implications and of policy meanings for a sustainable future of the Himalayas.

As a conceptual work, Smyer Yü's chapter takes a more-than-human perspective to offers his understanding of human–earth relations in the Tibetan-Himalayas through the discussion of what he calls "symbiotic indigeneity," "new-animist earth," and "transboundary water commoning." The chapter converses with its peer contributions in this book as well as with the current studies of river commons in Asia. Inspired by the ongoing studies of the Arctic as a multinational commons facing the challenges of climate change and human migration (Evengård et al. 2015) identical to those seen in the Himalayas, Smyer Yü illustrates the Himalayas as a natural transboundary commons through the Tibetan-Himalayan sourced water in both solid and liquid forms. Like the Arctic, the Himalayas is one of the largest homes of solid water, namely terrestrial ice, which is susceptible to climate change and the environmental impacts of regional modernisation projects. Unlike the Arctic, the Himalayan moving water is manifested in over ten large rivers and their tributaries, which rhizomically thread together the Tibetan Plateau, the Himalayas, and many parts of South, Southeast and East Asia. It is not exaggeration to say that the Tibetan-Himalayan water commons is planetary in nature and, thus deserves a multifaceted understanding with the combined efforts of earth sciences and environmental humanities. Based on his understanding of nature as a Deep Time concept, Smyer Yü argues for what he calls "nonhuman environmental commoning" and "environmental freedom" as a gift of nature and a maternal principle of the earth.

References

Aase, Tor H. 2017. "Are doomsday scenarios best seen as failed predictions or political detonators? The Case of the 'Theory of Himalayan Environmental Degradation'." *The Geographical Journal of Nepal*, 10, 1–14.

Abram, David. 1996. *The Spell of the Sensuous: Perception and Language in A More-Than-Human World*. New York: Vintage Books.

Agrawal, Arun. 2005. *Environmentality: Technology of Government and the Making of Subjects*. Durham: Duke University Press.

Allison, Elizabeth. 2017. "Spirits and Nature: The Intertwining of Sacred Cosmologies and Environmental Conservation in Bhutan." *JSRNC*, 11(2), 197–226.

An, Na, Gao Naiyun, and Liu Chang-e. 2008. "Wetland Degradation in China: Causes, Evaluation, and Protection Measures." *Chinese Journal of Ecology*, 27(5): 821–828. [in Chinese]

Armiero, Marco, and Richard Tucker. 2017. "Introduction: Migrants in environmental history." In *Environmental History of Modern Migrations*, edited by Marco Armiero and Richard Tucker. London: Routledge, 1–16.

Arora, Vibha. 2007a. "Unheard Voices of Protest in Sikkim." *Economic and Political Weekly*, 42, 34: 3451–3454.

Arora, Vibha. 2007b. "Assertive Identities, Indigeneity, and the Politics of Recognition as a Tribe: The Bhutias, the Lepchas and the Limbus of Sikkim." *Sociological Bulletin*, 56(2): 195–220.

Arora, Vibha. 2009. "'They are All Set to Dam(n) Our Future': Contested Development through Hydel Power in Democratic Sikkim." *Sociological Bulletin*, 58(1), Special Issue on Development of Democratic Routes in the Himalayan "Borderlands" (January–April), 94–114

Arora, Vanicka, and Manas Murthy. 2015. "*Re-discovering Uttarkhand's Cultural Identity: Issues for Consideration during Post-disaster Reconstruction*." Paper presented at the *7th i-Rec Conference : Reconstruction and Recovery in Urban Contexts*, London, July 6–8, 2015.

Bai, Junhong, Ouyang Hua, Cui Baoshan, Qinggai Wang, and Chen Hui. 2008. "Changes in Landscape Pattern of Alpine Wetlands on the Zoige Plateau in the Past four Decades." *Acta Ecological Sinica*, 28(5): 2245–2252.

Balikci, Anna. 2008. *Lamas, Shamans and Ancestors: Village Religion in Sikkim*. Leiden: Brill.

Bauman, Whitney. 2014. *Religion and Ecology: Developing a Planetary Ethic*. New York: Columbia University Press.

Bawaka, Country, Sandie Suchet-Pearson, Sarah Wright, Kate Lloyd, and Laklak Burarrwanga. 2013. "Caring as Country: Towards an Ontology of Co-becoming in Natural Resource Management." *Asia Pacific Viewpoint*, 54(2): 185–197.

Bergmann, Sigurd, and Forrest Clingerman eds. 2018. *Arts, Religion, and the Environment*. Leiden: Brill.

Berry, Thomas. 2009. *The Sacred Universe: Earth, Spirituality, and Religion in the Twenty-first Century*. New York: Columbia University Press.

Besky, Sarah. 2017. "The Land in Gorkhaland: On the Edges of Belonging in Darjeeling, India." *Environmental Humanities*, 9(1):18–39.

Blaser, Mario, and Marisol de la Cadena. 2018. "Introduction: Pluriverse, Proposals for a World of Many Worlds." In *A World of Many Worlds*, edited by Marisol de la Cadena and Mario Blaser. Durham: Duke University Press, 1–22.

Brantley, Susan L., William H. McDowell, William E. Dietrich, Timothy S. White, Praveen Kumar, Suzanne P. Anderson, Jon Chorover, Kathleen Ann Lohse, Roger C. Bales, Daniel D. Richter, Gordon Grant, and Jérôme Gaillardet. 2017. "Designing a Network of Critical Zone Observatories to Explore the Living Skin of the Terrestrial Earth." *Earth Surface Dynamics*, 5: 841–860.

Callicott, J. Baird. 2010. "The Conceptual Foundations of the Land Ethic." In *Environmental Ethics: The Big Questions*, edited by David R. Keller. West Sussex, UK: Blackwell Publishing, 201–210.

Chellaney, Brahma. 2012. *From Arms Racing to "Dam Racing" in Asia: How to Contain the Geopolitical Risks of the Dam-Building Competition*. Washington, DC: Transatlantic Academy.

Chellaney, Brahma. 2013. *Water, Peace, and War: Confronting the Global Water Crisis*. Lanham, Maryland: Rowman & Littlefield Publishers, Inc.

Christopher, Mark. 2004. "Water Wars: The Brahmaputra River and Sino-Indian Relations". *CIWAG Case Studies*. 7. https://digital-commons.usnwc.edu/ciwag-case-studies/7. Accessed October 26, 2020.

Crosby, Alfred W. 2004. *Ecological Imperialism: the Biological Expansion of Europe, 900–1900*. Cambridge: Cambridge University Press.

Crutzen, Paul J., and Eugene F. Stoermer. 2000 "The Anthropocene." *Global Change Newsletter*, 41: 17–18.

DeLoughrey, Elizabeth, Jill Didur and Anthony Carrigan. 2015. "Introduction: A Postcolonial Environmental Humanities." In *Global Ecologies and the Environmental Humanities: Postcolonial Approaches*, edited by Elizabeth, DeLoughrey, Jill Didur and Anthony Carrigan. London: Routledge, 1–32.

Drew, Georgina. 2017. *River Dialogues: Hindu Faith and the Political Ecology of Dams on the Sacred Ganga*. Tucson: The University of Arizona Press.

Dybas, Cheryl Lyn. 2013. *Discoveries in the Critical Zone: There Life Meets Rock*. Alexandria, Virginia: National Science Foundation.

Eckholm, Erik P. 1975. "The Deterioration of Mountain Environments." *Science*, 189: 764–770.

Emmett, Robert S. and David E. Nye. 2017. *The Environmental Humanities: A Critical Introduction*. Cambridge, MA: MIT Press.

Engelke, Peter, and David Michel. 2019. *Ecology Meets Geopolitics: Water Security in Himalayan Asia*. Atlantic Council: Washington, DC.

Escobar, Arturo. 2018. *Designs for the Pluriverse: Radical Interdependence, Autonomy, and the Making of Worlds*. Durham: Duke University Press.

Escobar, Arturo. 2016. "Thinking-Feeling with the Earth: Territorial Struggles and the Ontological Dimension of the Epistemologies of the South." *AIBR, Revista de Antropología Iberoamericana*, , 11(1): 11–32.

Evengård, Birgitta, Joan Nymand Larsen, and Øyvind Paasche, eds. 2015. *The New Arctic*. New York: Springer.

Gamble, Ruth. 2019. "How Dams Climb Mountains: China and India's State-Making Hydropower Contest in the Eastern-Himalaya Watershed." *Thesis Eleven*, 150(1): 42–67.

Gibson, Katherine, Deborah Bird Rose, and Ruth Fincher, eds. 2015. *Manifesto for Living in the Anthropocene*. Brooklyn, NY: Punctum Books.

Gorer, Geoffrey. 1938. *Himalayan Village: An Account of the Lepchas of Sikkim*. Melbourne: Nelson.

Grumbine, R. Edward and Maharaj K. Pandit. 2013. "Threats from India's Himalaya Dams." *Science*, 339: 36–37.

Guha, Ramachandra. 1989. *The Unquiet Woods: Ecological Change and Peasant Resistance in the Himalaya*. Berkeley: University of California Press.

Guha, Ramachandra, and Juan Martinez-Alier. 1997. *Varieties of Environmentalism: Essays North and South*. London: Earthscan.

Guha, Ramachandra. 2000. "The Paradox of Global Environmentalism." *Current History*, 2000, 367–370.

Guthman, Julie. 1997. "Representing Crisis: The Theory of Himalayan Environmental Degradation and the Project of Development in Post-Rana Nepal." *Development and Change*, 28(1): 45–69.

Haraway, Donna. 2016. *Staying with the Trouble: Making Kin in the Chthulucene*. Durham: Duke University Press.

Harvey, Graham. 1997. *Contemporary Paganism: Listening People, Speaking Earth*. New York: New York University Press.

Heise, Ursula K., Jon Christensen, and Michelle Niemann, eds. 2017. *The Routledge Companion to the Environmental Humanities*. London: Routledge.

Heise, Ursula K. 2017. "Introduction: planet, species, justice—and the stories we tell about them." In *The Routledge Companion to the Environmental Humanities*, edited by Ursula Heise, Jon Christensen, and Michelle Niemann. London: Routledge, 1–10.

Hodgson, Brian. 1874. *Essays on the Languages, Literature and Religion of Nepal and Tibet*. London: Trübner & Co.

Hogan, Linda. 2013. "We Call It Tradition." In *The Handbook of Contemporary Animism*, edited by Graham Harvey. London: Routledge, 17–26.

International Rivers. 2008. *Mountains of Concrete: Dam Building in the Himalayas*. Berkeley: International Rviers.

Ives, Jack D. 2004. *Himalayan Perceptions: Environmental change and the well-being of mountain peoples*. London: Routledge.

Ives, Jack, and Bruno Messerli. 1989. *The Himalayan Dilemma: Reconciling Developing and Conservation*. London: Routledge.

Johnston, Barbara Rose, ed. 2012. *Water, Cultural Diversity, and Global Environmental Change: Emerging Trends, Sustainable Futures?* Paris: UNESCO and Springer.

Khawas, Vimal. 2009. "Environmental Challenges and Human Security in the Himalayas." In *Environmental Concerns and Sustainable Development: Some Perspectives from India*, edited by Sakarama Somayaji and Gansesha Somayaji. The Energy and Resources Institute: New Delhi, 32–81.

Lahiri-Dutt, Kuntala. 2014. "Beyond the Water-Land Binary in Geography: Water/Lands of Bengal Re-visioning Hybridity." *ACME: An International E-Journal for Critical Geographies*, 13(3): 505–529.

Lahiri-Dutt, Kuntala, and Gopa Samanata. 2013. *Dancing with the River: People and Life on the Chars of South Asia*. New Haven: Yale University Press.

Latour, Bruno. 2014. "*How to make sure Gaia is not a God of Totality?*," *Written for the Rio de Janeiro meeting The Thousand Names of Gaia*, September 2014. http://www.bruno-latour.fr/sites/default/files/138–THOUSAND-NAMES_0.pdf. Accessed October 10, 2020.

Leopold, Aldo. 1949. "The Land Ethic." In *A Sand County Almanac: And Sketches Here and There, Aldo Leopold*. Oxford: Oxford University Press, 201-226.

LePage, Ben A. 2011. "Wetlands: A Multidisciplinary Perspective." In *Wetlands: Integrating Multidisciplinary Concept*, edited by Ben LePage. New York: Springer, 3–26.

Li, Zhiwei, Wang Zhaoin, Zhang Chendi, Han Lujie, and Zhao Na. 2014. "A Study on the Mechanism of Wetland Degradation in Ruoergai Swamp." *Advances in Water Science*, 25(2): 172–180. [in Chinese].

Little, Kerry. 2008. "Lepcha Narratives of Their Threatened Sacred Landscapes." *Transforming Cultures eJournal*, 3(1). http://epress.lib.uts.edu.au/journals/TfC.

Little, Kerry. 2010. "From the Villages to the Cities: the Battlegrounds for Lepcha Protests." *Transforming Cultures eJournal*, 5(1). https://epress.lib.uts.edu.au/index.php/TfC/article/view/1579, accessed October 24, 2020.

Little, Kerry. 2012. "Sanctity, Environment and Protest: A Lepcha Tale." In *Independent People's Tribunal on Dams, Environment and Displacement*, edited by Doma T. Bhutia. New Helhi: Human Rights Law Network, 83–94.

Little, Kerry. 2018. "Mythology as a Protest Narrative: A Lepcha Tale." In *Playing with Nature: History and Politics of Environment in North-East India*, edited by Sajal Nag. New York: Routledge, 341–352.

Lovelock, James. 2000. *Gaia: A New Look at Life on Earth*. Oxford: Oxford University Press.

Mahapatra, Sushana Kumar, and Keshab Chandra Ratha. 2016. "Brahmaputra River: A Bone of Contention between India and China." *Water Utility Journal*, 13: 91–99.

Margulis, Lynn. 1998. *Symbiotic Planet: A New Look at Evolution*. New York: Basic Books.

Margulis, Lynn. 1995. "Gaia is a Tough Bitch." In *The Third Culture*, edited by John Brockman. New York: Touchstone, 129.

Mullard, Saul. 2011. *Opening the Hidden Land: State Formation and the Construction of Sikkimese History*. Leiden: Brill.

Nyman, John. 2011. "Ecological Functions of Wetlands." In *Wetlands: Integrating Multidisciplinary Concept*, edited by Ben LePage. New York: Springer, 115–128.

Pandit, Maharaji K. and R. Edward Grumbine. 2012. "Potential effects of ongoing and proposed hydropower development on terrestrial biological diversity in the Indian Himalaya." *Conservation Biology*, 26(6): 1061–1071.

Peluso, Nancy, and Michael Watts. 2001. "Violent Environments." In *Violent Environments*, edited by N. Peluso and M. Watts. Ithaca: Cornell University Press, 3–38.

Plumwood, Val. 1993. *Feminism and the Mastery of Nature*. London: Routledge.

Pope Francis. 2015. *Laudato Si': On Care For Our Common Home*. Vatican. http://w2.vatican.va/content/dam/francesco/pdf/encyclicals/documents/papa-francesco_20150524_enciclica-laudato-si_en.pdf. Accessed Oct 6, 2020.

Qi, Jinyu. 2015. "Study on Social Adaptation and Community Cultural Reconstruction of Ecological Migration in Sanjiangyuan Area." *Journal of Minzu University of China*, 42(3): 47–53.

Richter, Daniel, and Sharon A. Billings. 2015. "One Physical System: Tansley's Ecosystem as Earth's Critical Zone." *New Phytologist*, 206: 900–912.

Rigby, Catherine. 2019. "Weaving the Environmental Humanities: Australian Strands, Configurations, and Provocations." *Green Letters*, 23(1): 5–18.

Risley, Herbert Hope. 1894. *The Gazetteer of Sikhim*. Calcutta: The Bengal Secretariat Press.

Rigzin, Ngodup Dokhampa. 1998. "The Hidden Valley of Sikkim." *The Bulletin of Tibetology*, 1: 75–86.

Rolston, Holms 1998 "Challenges in Environmental Ethics." In *Environmental Philosophy: From Animal Rights to Radical Ecology*, edited by Michael E. Zimmerman, J. Baird Callicott, George Sessions, Karen J. Warren and John Clark. Upper Saddle River, NJ: Prentice Hall, 124-144.

Rose, Deborah Bird. 1996. *Nourishing Terrains: Australian Aboriginal Views of Landscape and Wilderness*. Canberra: Australian Heritage Commission.

Rose, Deborah Bird. 2000. *Dingo Makes Us Human: Life and Land in an Australian Aboriginal Culture*. Cambridge: Cambridge University Press.

Rose, Deborah Bird. 2013. "Val Plumwood's Philosophical Animism: Attentive Interactions in the Sentient World." *Environmental Humanities*, 3: 93–109.

Rose, Deborah Bird, Thom van Dooren, Matthew Chrulew, Stuart Cooke, Matthew Kearnes, and Emily O'Gorman. 2012. "Thinking Through the Environment, Unsettling the Humanities." *Environmental Humanities*, 1: 1–5.

Rudwick, Martin J. S. 2008. *Worlds Before Adam: The Reconstruction of Geohistory in the Age of Reform*. Chicago: The University of Chicago Press.

Rudwick, Martin J.S. 2005 *Bursting the Limits of Time: The Reconstruciton of Geohistory in the Age of Revolution*. Chicago: The University of Chicago Press.

Saikia, Anup, and Pankaj Thapa, eds. 2019. *Environmental Change in the Himalayan Region: Twelve Case Studies*. Cham, Switzerland: Springer.

Samuel, Geoffrey. 1993. *Civilized Shamans: Buddhism in Tibetan Societies*. Washington, DC: Smithsonian Institution Press.

Shneiderman, Sara. 2013. "Himalayan border citizens: Sovereignty and mobility in the Nepal-Tibetan Autonomous Region (TAR) of China border zone." *Political Geography*, 35: 25–36.

Shneiderman, Sara. 2010. "Are the Central Himalayas in Zomia? Some Scholarly and Political Considerations across Time and Space." *Journal of Global History* 5(2): 289–312.

Smyer Yü, Dan. 2014. "Sentience of the Earth: Eco-Buddhist Mandalizing of Dwelling Place in Amdo, Tibet." *Journal for the Study of Religion, Nature and Culture* 8(4): 483–501.

Smyer Yü, Dan. 2015. *Mindscaping the Landscape of Tibet: Place, Memorability, Ecoaesthetics*. Boston: De Gruyter.

Smyer Yü, Dan. 2017. "Introduction: Trans-Himalayas as Multistate Margins." In *Trans-Himalayan Borderlands: Livelihoods, Territorialities, Modernities, edited by Dan Smyer Yü and Jean Michaud*. Amsterdam: Amsterdam University Press, 11–44.

Smyer Yü, Dan, and Jean Michaud. 2017. *Trans-Himalayan Borderlands: Livelihoods, Territorialities, Modernities*. Amsterdam: Amsterdam University Press.

Smyer Yü, Dan, and Erik de Maaker 2019. "Revision Guidelines" (Book project internal document).

Smyer Yü, Dan, and Sonam Wangmo. 2019. "A Historical Sketch of Buddhism in Tibet." In *Oxford Research Encyclopedia of Asian History*, edited by David Luden. Oxford: Oxford University Press.

Sörlin, Sverker. 2015. "Cryo-History: Narratives of Ice and the Emerging Arctic Humanities." In *The New Arctic*, edited by Birgitta Evengård, Joan Nymand Larsen, and Øyvind Paasche. New York: Springer, 327–339.

Sörlin, Sverker. 2017. "Scaling the planetary humanities: environmental globalization and the Arctic." In *The Routledge Companion to the Environmental Humanities*, edited by Ursula Heise, Jon Christensen, and Michelle Niemann. London: Routledge, 433–442.

Spoon, Jeremy. 2014. "Everyday Buddhism and Environmental Decisions in the World's Highest Ecosystem." *JSRNC*, 8(4): 429–459.

Tebrakunna Country, Emma Lee, and Robyn Eversole. 2019. "Rethinking the Regions: Indigenous Peoples and Regional Development." *Regional Studies*, 53(11): 1509–1519.

Tewari, D.D. 1995. "The Chipko: the Dialectics of Economics and Environment." *Dialectical Anthropology*, 20(2): 133–168.

Thompson, M., M. Warburton, and T. Hatley. 1986. *Uncertainty on a Himalayan Scale: An Institutional Theory of Environmental Perception and a Strategic Framework for the Sustainable Development of the Himalaya*. London: Milton Ash Editions.

Tsing, L. Anna. 2015. *The Mushroom at the End of the World: On the Possibility of Life in Capitalist Ruins*. Princeton: Princeton University Press.

UCLA. 2020. "What is the Environmental Humanities?" http://environmental.humanities.ucla.edu/?page_id=52. Accessed September 30, 2020

United Nations Human Rights Council (UNHRC). 2013. The United Nations Declaration on the Rights of Indigenous Peoples: A Manual for National Human Rights Institutions. Sydney, Australia: Asia Pacific Forum of National Human Rights Institutions, 2013.

Valdiya, K.S. 2014. "High Dams in Central Himalaya in Context of Active Faults, Seismicity and Societal Problems." *Journal of Geological Society of India* 49, 479–494.

Van Dooren, Thom, and Deborah Bird Rose. 2016. "Lively Ethography: Storying Animist Worlds." *Environmental Humanities*, 8(1): 77–94.

Van Schendel, Willem. 1991. *Three Deltas: Accumulation and Poverty in Rural Burma, Bengal and South India*. New Delhi: Sage Publications.

Viveiros de Castro, Eduardo. 1998. "Cosmological Deixis and Amerindian Perspectivism." *The Journal of the Royal Anthropological Institute*, 4(3): 469–488.

Wei, Huanxin. 2018. *On Value in Use: the Exploration of Human Sustainable Society*. Salt Lake City: American Academic Press.

Xu, Jianchu, Ram P. Chaudhary, and Pratikshya Kandel. 2019. "Sustaining Biodiversity and Ecosystem Services in the Hindu Kush Himalaya." In *The Hindu Kush Himalaya Assessment: Mountains, Climate Change, Sustainability and People*, edited by Philippus Wester, Arabinda Mishra, Aditi Mukherji, Arun Bhakta Shrestha. Cham: Switzerland: Springer, 127–165.

Part I

Kindred of the earth, deities, humans and animals

1 Relatedness, trans-species knots and yak personhood in the Bhutan highlands

Jelle J. P. Wouters

Folkloric entrances

I open this chapter with an assemblage of folkloric narratives and myths, each of which variously accounts for the origins, social significance, and spiritual, symbolic and ontological status of yaks and herders in Bhutan's and the surrounding highlands. The first narrative, or rather a variety of tales, concerns the early arrival of yaks in the highlands. The second narrative is told by Tibetan pastoralists, and with whom Bhutanese herders share a similar mythological complex, while the third comes from Lingzhi in the Bhutan highlands, which is the ethnographic locale from which this chapter emerges.

A well-known Bhutanese folktale tells that yak and buffalo once lived together in the lush meadows of southern Bhutan. Back then, both yaks and buffalos grew fur on their bodies. One day, buffalo told the yak about salt from the Tibetan area, which he heard was delicious. This made yak also curious to taste it. For that, however, they would have to travel a long distance and brave highlands and mountains on their way. Anxious about the cold weather there, the buffalo proposed he lend his fur to yak, who agreed. Once in the highlands, the yak found the salt delicious and the grasses fresh, so much so that he decided to remain there, in what today are the Bhutan highlands. Down in the meadows, the buffalo waited in vain for the yak's return. This explains, the folktale concludes, why buffalo have little hair on their bodies, whereas yaks are furry and thick-haired. It further explains why yaks live in the highlands and buffalo in the lowlands, as well as why buffalos are often seen raising their heads to gaze in the distance, from where they still anticipate the yak to appear, carrying with him salt from Tibet.

This folktale has several variations, for instance versions that replace salt with butter, or that elevate the tale into the realm of the sacred by narrating that Guru Rinpoche (Padmasambhava) cursed the buffalo for defying nature by voluntarily shedding his fur and, as a punishment, permanently separated buffalos and yaks and sowed eternal enmity between them. A different narrative traces the presence of yaks in the Bhutan highlands to the arrival of the great Tibetan Lama Zhabdrung Ngawang Namgyal who came to Bhutan in the seventeenth century (more below) and blessed the highlands with a bounty of yaks. Yet another story holds that yaks were initially the cattle of a *yeti* but that humans overpowered him, stole his yaks and escorted them to the Bhutan highlands. What these tales commonly purport is that yaks have not always been in the highlands, but arrived there by circumstance.

The second meta-narrative concerns the shifting terrain of power, standing, authority, rights and dominion between yaks and humans. It tells that in the deep past, yaks enjoyed a sacred or supernatural status and were part of an expansive pantheon of deities. Humans, in contrast, survived on a much lower and more subservient strata within the earthly and cosmic hierarchy. Yaks' initial descent on earth, in the Himalayas and the Tibetan Plateau, therefore lent them a superior position, akin to the gods of the earth, and they exerted control over humans. Gradually, however, and for a range of reasons, humans evolved and eventually elevated themselves above yaks, whom they subsequently subjected to human utilitarian and functional purposes such as transportation, fibre for making tents, ropes and whips, as well as for the occasional protein. After the introduction and spread of Buddhism in the region, which in parts replaced and in parts mingled with pre-existing Bon and animistic belief practices, yaks became fully reduced as exclusively belonging to the animal realm, one of the undesirable realms to be (re)born into, and were thus permanently positioned lower in hierarchy compared to humans. This explains why humans today herd yaks, instead of, as was a possibility in the mythical past, yaks herding humans.[1]

The third narrative is a story widely told in Lingzhi. It tells of an ancestor by the name of Madhagoki who herded thousands of yaks. Madhagoki was flourishing because he enjoyed the benevolence and protection of Jo-Gyem, a powerful deity within an all-encompassing and hierarchically constituted cosmic order that consisted of *lha* (enlightened beings), *lhu* (serpents), *dhe* (evil spirits), *tshen* (personal guardians), *nep* (spiritual proprietors of a territory) and other spirits that called the Lingzhi area their home. Each year, Madhagoki sacrificed his most prized yak to Jo-Gyem, who in return bestowed fortune on him. One year, however, he decided to offer Jo-Gyem a horse instead of a yak. In those days, horses were scarce, rare and therefore particularly precious in the highlands and Madhagoki, in good faith, thought that a horse might please Jo-Gyem even more than his best yak. As Madhagoki ritually sacrificed a horse and offered its meat and blood to Jo-Gyem, the deity's voice echoed through the mountains, asking Madhagoki: "Should I let it go?" Madhagoki interpreted this question as Jo-Gyem offering him even greater riches and he immediately responded: "Let it go!" No sooner, Jo-Gyem made a lake burst and which flooded Madhagoki's village, washing away lives and property.

This story carries multiple meanings, all of which are constitutive of the highland ontological matrix. First, that worldly well-being and fortune cannot be established and maintained without the explicit protection and care of the territorial deities, who demand reverence through ritual sacrifices and offerings in return. Second, that irrespective of material or monetary value attached to other nonhuman animals such as horses, only yaks are ascribed with the spiritual potency that quality for sacrificial offering. And third, that deities, when angered, can exert punishment and revenge through causing natural calamities and by inducing destruction and misfortune.

Introduction: co-dwelling, co-evolving and co-shaping

For centuries, yaks have co-dwelled, co-evolved and co-shaped the land with pastoralists and grassland ecosystems in the Bhutan highlands. Because of yaks, humans in the highlands became herders, and subsequently developed relevant social organisation and culture, patterns of land-usage, and migratory routes. It also reflects in a cosmology whose numinous beings profess a preferential association with yaks in terms of rituals, offerings, protection and sacrifice. The commands and the clout of these differently manifested deities constitute a major determining condition of social life. Through the presence of yaks, moreover, humans have been enabled to sustain themselves in high altitude grasslands. Yak dung is used for cooking in a landscape largely devoid of fuels. Their produce of dairy, fur, skin and meat allowed herders to construct tents, weave clothes and blankets, eat and drink, and to trade with neighbouring populations on the Tibetan Plateau and in Bhutan's Paro and Thimphu valleys to obtain rice and other foodstuffs. Jointly, yaks and herders form a landscape that is continuously modified, where grasses grow abundantly and larger bushes and trees are trimmed. As such, the Bhutan highlands do not constitute "pristine" nature, the way it is popularly presented; in actuality, the highland is the result of active processes of place-making, resulting in a deeply socialised and carefully managed landscape that has co-evolved and co-shaped with herders and yaks for centuries.

The implication? That, in understanding life in the Bhutan highlands, it is impossible to draw a clear ontological distinction between herders, on the one hand, and yaks, deities and the environment on the other; the epistemological cul-de-sac, that is, of apprehending the physical, social and cosmological environment as independently constituted and functioning according to their own autonomous laws, categories and registers of existence. In contrast, drawing on Descola (2005), they are part of the same social sphere, interconnected in an encompassing social continuum, and thus they conjure a complex, integrated and layered regime of sociability in which herders, yaks, deities and the environment are constituted relationally through a dense network of intersubjectivity.

From this, it follows that it is insufficient, in good-old ethno-ecological fashion, to conclude that herders possess a remarkable empirical understanding of the complex interrelations between yaks, other nonhuman animals and the environment, and march through life with a clear awareness of the cosmic order. Such a characterisation is insufficient because it suggests a conception of the world in which herders exist as a transcendental and separate domain largely autonomous from nonhuman animals and cosmic and natural beings who, in ethnographic reality, they conceive of, and act towards, as social partners and relatives, if not always on terms of harmony and conviviality. Any ethnographic theory of highland social life in Bhutan must take such intersubjective connections, or relational ontology, as the kernel of an elementary and integrated schema.

Within this ontological field, this chapter asks what it means for herders and yaks to live a life that is deeply intertwined, entangled, knotted and ultimately co-constituted with one another. To be sure: the deep, mutual affinity I will discuss as characterising the ties between herders and yaks is indissociable from a wider intersubjective ambience in the highlands that also draws in deities, other nonhuman animals and the physical environment and climate (change). Both for want of space and because the relational ontology between nature and deities in Bhutan will be dwelled on in ethnographic detail by two other chapters in this volume, I limit my discussion here to the intersubjective connectivity, relations and experiences that exist between herders and yaks. I am particularly interested in how these entanglements and knots produce a schema and sense of "relatedness" (Carsten 2000) that reveals itself through social practicing of kin-making between herders and yaks.[2] The interspecies kinship relations that thus emerge are not based on a "pre-given" state of being in terms of shared semen, blood or saliva but are cultivated and nourished through "performance," "forms of doing" (Schneider 1984), and a "mutuality of being" (Sahlins 2011) that results from lifelong and mutual processes of "becoming with" (Haraway 2008, 2).

As opposed to classical models of kinship that were long incarcerated in the realms of biology and the human, I focus on social processes of generating relatedness through kin-making that manifest themselves as a trans-species affair. In so doing, I build on fresh scholarship that confronts the universalistic edifice of a Cartesian dualism by revealing how human nature is often constituted as an "interspecies relationship" (Tsing 2012, 141); scholarship that so probes beyond the "ontological frontiers of humanity" (Descola 2005, 23); scholarship, in short, that ultimately decentralises the "anthropos" in anthropology.

The ethnographic locale of this chapter is Lingzhi, located in Bhutan's western highlands. Lingzhi is remote, both in the sense that it requires a 3 day walk from where the road terminates to reach it, and in terms of popular social imagination that long perceived yak-herders as living harsh, difficult and deprived lives. Lingzhi is the last settlement before the Bhutan highlands fold and unfold into Tibet. It is part of Bhutan's northern highland belt where yak-herding has been a common subsistence strategy for many generations, and continues to be practiced by the majority of highlanders (Gyamtsho 2001). To herders here, yaks are the "heart and soul" of their livelihoods and lifeworlds (Wangchuk et al. 2013) and function as the "camels of the snow"; without their presence "one cannot imagine how humans could survive in this beautiful but hostile region" (Gyamtsho 2001, 2).

Ethnographic research for this article was carried out in 2018 and 2019 by a research team consisting of my students Kinley Choki, Tshering Pelzom, Pema Gayltshen, Pema Namgyel and Thinley Dema, who, divided over several fieldtrips, spent roughly 3 months in the Bhutan highlands. Participant observation was complemented by interviews, of which well over a hundred were taken, recorded, translated and transcribed. What follows in this chapter is based on the analysis of the data thus collected, as well as being informed by three fieldtrips I was able to undertake into the highlands, and meetings and interviews I had with yak-herders when they happened to come down to Thimphu. In order to stay as close to the ethnography as possible, I will here and there insert ethnographic observations and insights from the field-notes by the research team and mention the name of the particular researcher who made them.

Yaks and yak-herding have generated considerable scholarship in Bhutan. This literature largely preoccupies itself with the utilitarian and functioning purposes yaks have for humans, with high altitude grassland management systems (both indigenous and by the state), yak health and diseases, and with policy-planning, intervention and evaluation more broadly (e.g. Ura 1993, 2007; Gyamtsho 2001; Roder et al. 2001; Wangda 2016; Wangdi and Wangchuk 2017). This chapter adds to this important scholarship another dimension by forefronting the inter-subjective networks of relatedness that connect herders and yaks.

Derrida's cat and Haraway's dog: extralinguistic communication and animal personhood

Social processes of kin-making between herders and yaks manifest themselves as a relation of reciprocity. Said otherwise, yaks are not the passive recipients of humans' attention, labour, care, love (and occasional dislike), or for that matter only the unwitting recipients of anthropomorphism with herders interpreting their behaviour using analogies with human conduct. Instead, this chapter sees yaks, as Lingzhips do, as active, agential and responsive beings who are cultivating trans-species knots of their own.

The ontology, agency and sensorial perception of nonhuman animals, and of "companion species" in particular, has been variously reflected on, arguably most prolifically through Derrida's (2002) cat and Haraway's (2016) dog. This scholarship blurs the boundaries between humans and animals, humanity and animality, structure and agency, nature and culture, and in the upshot demands us to think afresh about human and other-than-human worlds. In *What Is an Animal?*, Ingold (1998, 9) influentially postulates that nonhuman animals "embody attributes of personhood which in the West are popularly identified with the condition of 'Humanity.'" Without having read Derrida, Haraway or Ingold, herders in the Bhutan highlands know that yaks are persons, just as they themselves are, that there are different kind of yaks, just as there are different kind of humans, and that humans not only know yaks, but that yaks also know humans and respond to them through "extralinguistic communication" (Descola 2005, 15). All of this points to fragments of yak personhood.

Anthropology has long recognised that the human body is a necessary but not sufficient condition of personhood (Collins 1985). This scholarship situates the body as a canvas upon which social relations and values that culturally define "a person" are inscribed and through which the body becomes the owner of an ethical, moral and social identity, thus transforming it into a person in the Mauss (1985) sense of the term. However, the conceptions that bodies of nonhuman animals are also deeply socialised—the receptacle and embodiment, that is, of affective relations, social values and intersubjective belonging—has, until recently, escaped serious notice within academia. It hasn't escaped Lingzhips, who, as shall become clear, recognise the potential and performance of yak personhood and who perceive of the yak body, akin to the human body, as the socialised physical form of personality, agency, relations and morality.

Lingzhips are further aware that it is not just they, as herders, who are cultivating relationships with yaks, but that yaks apply their own agency, reflexive consciousness, intentionality and emotions in shaping trans-species relationships with particular humans. I will illustrate this through engaging the yak construction of sensory perception, which, in turn, animates the differential postures and responses yaks adopt towards particular humans, cultivating intimate trans-species relationships with some and seeking to evade others. The upshot of all of this is at least twofold: first, elements of personhood are not reserved to the condition of humanity but, in the Bhutan highlands, extend to yaks and, secondly, that it is not just humans who are composed relationally but also yaks. As Donna Haraway (2016, 102) writes more broadly and in pursuit of rethinking the ties of genealogy and kin, and of kin and species, "kin making is making persons, not necessarily as individuals or humans," then emphasising that "it matters how kin generate kin" (Haraway 2016, 3).

Trans-species relatedness, kin-making and knots

The sun was sinking lower in the sky above Cheybesa, draining away the light of day and replacing it with the chill of twilight. A caravan of two herders and roughly twenty horses, each one fully loaded with supplies from the Paro valley, from where they had started their journey three days prior, entered the narrow valley, just in time to dodge the now quickly approaching darkness.

Cheybesa is a small hamlet surrounded by undulating hills covered with a rug of grasses. From behind rises a great line of mountains, scarfed with snow, that reach for the sky. In the middle of the valley runs a stream; the water is freshly melted, cold, slow-flowing and languid in pace, and a crucial life source for all sentient beings in Cheybesa. Winter has begun to creep in and most herders have returned from their faraway *tsamdros* (pastures). Soon Cheybesa will be snowed in and will likely remain so for a couple of months. During this period yaks will have to fend mostly for themselves.

For some time, I have been talking with Tshering and Tenzin, both herders of many autumns. Looking at the sky, and noticing the imminent sundown, they now each call out to their yaks, who, however, seem to be nowhere in sight. "They will come soon," Tenzin says as I inquire. He is right. A few minutes later, about a dozen yaks appear, walking in single file as they cross the narrow, makeshift bridge that leads across the stream. As the yaks make their way past several houses, they separate themselves into two groups: half of them saunter into Tshering's fenced front yard while the others enter Tenzin's. These are the yaks that don't reproduce or provide milk and who are now exempted from migratory life in the highlands. After a day's gazing, the yaks have now come in for the night, for an evening snack of grasses that were dried and stored during the summer, and "karma feed" (a manufactured food-supplement brought in, at subsidised rates, from Thimphu) and for protection against wildlife attacks, especially by hungry bears,—who, both Tshering and Tenzin insist, have grown more audacious and intrepid in recent years: "more like humans," as they put it—foxes, and the occasional tiger and leopard. I ask Tenzin whether all his yaks, of which he herds well over a 100, recognise his voice and are always as obedient. This first question seems incongruous to Tenzin. He shrugs, "Of course, they recognize my voice. They are like my children. Have you ever seen a child returning home to the wrong house? A child knows where he is loved and belongs. So does a yak." Tenzin extends his analogy between yaks and his children to answer my second query:

> None of my children are the same. My eldest daughter, for instance, is calm, obedient, and hard-working and I don't have to worry about her much. My son is very different, however. He is naughty and careless and I always worry about his future. Similarly, each of my yaks is different. Some are gentle and obedient. Others are mischievous, stubborn, and always cause trouble. There are all kinds of yaks and I must treat each yak according to her habits, moods, and personality.

He then points to his yaks, one by one, and mentions their names (Gama, Gayu, Chari, Wanghu and Dromdung) and attributes idiosyncratic personality traits to each: cheerful, gentle, solitary, playful and obstinate. "We name each yak when they turn three years of age," Tenzin says, "just as we name our children. We usually name them after their distinctive features and behavior [become apparent]." "I miss them," he adds as he looks into the distance towards the hill behind which he has left his other yaks to spend the winter months. "Winter is a difficult time for them, as soon the snow will cover up most of the grasses. I hope they will be alright."

In conversing about their yaks, Lingzhips often emphasise similarities between raising children and raising yaks as both rely on an intense and prolonged process of caring, nourishing, sacrificing and loving, as well as cause worry, anxiety and the occasional exasperation. Says the herder Dorji:

> As much as I worry about my children, I worry about my yaks. If one of them is left behind in the forest, I can't sleep during the night. Instead I must go out and search for her. When my yaks are not near me, I constantly worry that they will be attacked by a bear or tiger. I also worry that they'll fall sick in my absence, or stumble of a cliff. And I worry that there is not enough grass for them to eat. Not a single day goes by in which I don't worry about them.

Akin to Tenzin and Dorji, Lingzhips habitually invoke an idiom of relatedness to describe their connections to their yaks. Most herders refer to their yaks as "like their children," although some insist that they are simultaneously "like their parents" because while they nurture yaks, yaks equally nurture them by providing and provisioning. As states Chimphu: "My yaks are both like my children and parents. I have seen them grow right in front of my eyes. I have taken care of them, fed them, and worried for them. I cannot imagine my life without them. At the same time yaks are like our parents too because although we raise them, they also raise us by providing food and income."

This sense of relatedness between herders and yaks emerges from a dense network that includes co-dwelling, co-movement and co-dependency. While much has changed in recent years, including the emergence of a lucrative fungus economy (see the chapter by Kinley Choki, this volume), and while much will continue to change, it remains hard to overestimate the extent to which herders' livelihoods depend on their yaks' produce, a dependency that is reflected in the centrality yaks assume in social, cultural and religious lifeworlds. At least one Lingzhip, in an act of apparent defiance, decided to sell his entire herd in order to fully concentrate on collecting *yartsa gunbu* (*Ophiocordyceps sinensis*, an endemic caterpillar fungus) and other lucrative medicinal plants, but only to bear the brunt of the entire community who condemned his decision as foolish and short-sighted. This included his wife, who confessed the shame she felt about no longer being able to offer homemade butter and cheese to visitors. Consider also rituals. Most rituals that are conducted implore the protection and health of yaks, ranging from a large and lavish annual ritual participated in by the entire community and directed at the superior deity, to the more low-key, routine offerings, backed up by a prayer, to lower—yet known to be vengeful—territorial deities of the places herders migrate back and forth to, and which serve to acknowledge that the herders and yaks are mere guests of the local territorial deity.

In all of this, it must be stressed that herders don't refer to yaks as a general category but speak about their connections to individual yaks with whom they cultivate a sense of relatedness. The particular form this relation takes is again expressed and experienced differently in connection to different yaks. Herders' description of yaks "like their children" or "like their parents," which is complemented by feelings and expressions of joy at a yak's birth and loss at a yak's demise akin to gaining or losing a family member, is symbolic of the deep affinity and reciprocal relations that exist between them. It is of course not evocative of a consanguineous family in any literal sense, and for that matter glosses over the degree of obvious difference, separation and hierarchy that also characterises herders' relations with their yaks.

Like reciprocity, hierarchy and power differentials are of course everywhere intrinsic to the social world, both human and nonhuman. But if Lingzhips assume measures of control over their yaks, this is not exercised on the basis of absolute subordination, even less on forceful domination, but on an integrated set of practices in which all entities have a role and place to fulfil, and in which one's status and position can change from one rebirth to the next, and which, Lingzhips tell, often sees herders reincarnated as yaks and yaks as herders. This hierarchy and difference does not undermine the real possibility of cultivating relatedness and practicing kin-making, which establishes yaks as relatives from a different species. This is not altogether surprising because, as Eduardo Viveiros de Castro (2012, 38) notes, "kin can only be made out of others," and these "others" need not always be fellow humans.

Indeed, it is clearly evident that yaks in the Bhutan highlands are seen and treated as persons, as social, embodied and sentient beings with whom it is possible to establish relations of attachment, intimacy, care, love and worry. The following observation comes from the field-diary of Tshering Pelzom:

> Each yak has its own personality and a herder talks about his or her yaks in the same ways as they talk about their children. Some yaks are talked of as

> stubborn while others are described as gentle, for instance. When we were in Juphu [a small yak-herding settlement], Acho [brother] Tandin talked about a particular yak who refused to accept her calf. I asked him the reason for it. He responded: "This yak is always like this. She never accepts her calf. Usually it is the young mother yaks who now and then do that, but she is already an old yak, yet still she behaves like this." Acho Tandin then took the calf into his room and said he would sleep next to her that night, as she was too young to sleep alone.

Tandin's criticism of the mother's apparent refusal to take responsibility for her calf, as well as his refusal to leave the calf alone during the cold of the night, taking it instead into his bedroom and sharing his blankets and body warmth with it, is further illustrative of the close relationships of interspecies intimacy and care herders cultivate with their yaks. This intimacy is no doubt fostered by daily proximities and everyday practices of labour and care. It is further strengthened by a lifelong process of *co*-movement between herders and yaks. I emphasise "co-movement," rather than "movement" because it is not always clear who is leading whom. At times, herders initiate the movement, especially when yaks are asked to carry luggage, loads or the occasional human. At other times, however, yaks take the lead with Lingzhips following their tails, particularly when it comes to annual migratory routes and from which yaks refuse to deviate, making, as well, Lingzhips explain, the decision regarding when it is time to migrate.

Regardless of who initiates the movement, it is while migrating and "staying with" that herders cultivate and experience a deep connection with their yaks, not in the least also because during this period, contact with other humans, except for sometimes immediate family members, is extremely limited. When alone with the yaks, herders speak and sing to them, share their personal trials and tribulations, and eat, sleep and defecate where the yaks also do. Here again, Lingzhips cherish no pretension that yaks will respond or counsel them through words, they are after all not human relatives, but they nevertheless experience and express an affective connection of interspecies relatedness, solidarity and trust, and allegiance.

Of course, these relations are not always amicable, now and then break down, and also involve displeasure and disappointment, just as human relations, especially those of close relatives, have the occasional misunderstandings, arguments and fights. But quarrelling relatives are still relatives, and likewise herders' occasional dissatisfaction with an individual yak does not result in the disseverance of the relationship; rather, it is part of and indicative of the very existence of deep affinity. At least so in most cases, as some yaks, Lingzhips confess, are difficult to love and forgive as they seem permanently bad-tempered, aggressive and ill-disposed. We have recorded several stories of yaks taking this hostility up a notch by actually exercising violence against humans. "Some yaks are just bad, just like there are human thieves, murders, and molesters," says Phuntsho. He then narrates a personal experience:

> Once, when my granddaughter was still very young, my son was holding her as he went near to one of his yaks. After a while, I heard his cry. I rushed to see what had happened, and when I reached them I saw that he was near a tree covered in blood. A yak had hit him many times and he could not run because he was carrying his daughter. I had a hard time in chasing the yak away.

Discussing animal intimacies in the Indian Himalaya, Radhika Govindrajan (2018) shows ethnographically how the rhythm of *paharis* (hill/mountain peoples) is closely entangled with their domestic animals, mostly cattle and goats, and how "daily proximities and entanglements produced a deeply felt sense of mutuality between villagers and livestock animals that was predicated on their imbrications in another's lives" (2018, 7). So, as the ethnography above suggests, is broadly the case in the Bhutan highlands, where the social alterity is shaped by, to apply Donna Haraway's (2008, 35) apt phrases, "multispecies knots" and social processes of "becoming with" that are not based on biological or genealogical associations but are nourished through herders' daily practices and sensibilities that are deeply knotted with their yaks; knots that cannot be disentangled without fragmenting the very flow, essence and meaning of highland life.

This section has been written from the herders' point of view. That, in the establishing and nourishing of these trans-species knots, yaks themselves are active, agential and responsive beings is the focus of the next section.

Intentionality, agency and affective lives among yaks

Ingold's question "What is an animal?," discussed above, has now been variously reframed as well as narrowed down to engage particular ethnographic locales, as Piers Vitebsky and Anatoly Alekseyev (2015) have done for Siberia by asking: "What is a reindeer?" They address this question by showing how scientific, indigenous, political and popular perspectives of reindeers vastly vary. However, rather than asking the analytically probing question "what is a yak?," this section continuous our exploration into the dense network of relatedness and intersubjectivity between herders and yaks. Here, however, by adopting, to the extent possible, the yak's point of view, who may well probe the question "what is a human?" I hope to show that yaks in the Bhutan highlands are not the passive recipients of herders' labour, love and care, but that they are self-conscious and intentional subjects that lead affective lives, and gaze and act back towards humans.

Whereas the previous section showed how, in their kin-making practices towards yaks, Lingzhips guide themselves by the distinctive moods and personalities of individual yaks, this section uses ethnographic fragments to suggest that yaks, in their kin-making with humans, too, do not perceive of herders (and humans more broadly) as a singular, abstract category. Instead, they apply their agency and subjectivities to foster distinct relationships with particular herders. First an obvious caveat, namely—and as yak herders themselves lament, especially during the bouts of inevitable loneliness they experience while migrating with their yaks—yaks do not talk, at least not in a language intelligible to the human species. From this follows that approaching yaks' agency, capacity for affective living, and intentionality runs the risk of interpreting these in terms of human representations of yaks' signs and symbols. Be this as it may, careful ethnographic attentiveness and listening to the stories and experiences of herders, especially to those whose entire lives have been spent in intimate proximity and entanglement with yaks, offer glimpses into the ways yaks themselves are agential and purposive beings that cultivate trans-species relations with particular herders on their own terms.

To discuss this, I borrow insights from the anthropology of the senses that show how just as much as the body is socially constructed, so too are our senses and that thence the sensorium, and the meanings and practices associated with it, is culturally variegated with different peoples enumerating, classifying, experiencing and acting upon the senses in discrepant ways (see Classen 1997; Goody 2002; Le Breton 2017). Expanding such insights beyond the human realm (see Abram 1996), I suggest that the framework of sensory perception can be usefully applied to approach the agency, intentionality and capacity for emotions of yaks; a preliminary exploration, that is, of the yak construction of sensory perception. I then show how this sensory perception animates the clearly differential postures and responses yaks assume towards different humans. The current state of this ethnographic research allows me to engage with three sensory perceptions: sight, touch and sound.

Sight—or the (in)famous yak's gaze. No outsider visiting the highlands, this author included, can escape the feeling of intimidation as he is being stared at probingly, suspiciously, somewhat threateningly by one or multiple yaks along the way. Momentarily ceasing their grazing, the yaks refuse to move, and enter in an intense, prolonged and piercing gaze at the human unfamiliar to them, as though expecting to be given a justification for the unexpected, unannounced arrival of the stranger. This deep, intimidating gaze impressed all members of the research team. Kinley Choki writes:

> When we walk by yaks, they would just stare at us for a long time without averting their eyes even for a second. It feels quite intimidating. The herders explained that they were suspicious of us since they have not met us before. Their stare was them studying us and trying to figure what our intentions were. Even after we walked passed them and we turned out heads a little, we found that they were still staring at us. But once when we spent some days in a remote yak-herding camp, the yaks who had all stared at us for the first day, from the second day onward began to realize that our intentions were good and they stopped staring at us.

This discontinuation of the stare after establishing familiarity is crucial. Lingzhips themselves, indeed, are not subject to this gaze, at least not by the yaks they herd, or by the yaks of fellow-herders they have established a sense of mutual familiarity with. "Yaks recognize our faces as much as we recognize theirs," says the yak-herder Tandin. He continues:

> When my yaks see me approaching, they make soft noises of recognition and immediately come forward, especially when we haven't met each other for some time. I know they feel joyful to see me. They know I care for them deeply. But when a stranger approaches them, they immediately become suspicious and want to find out that person's intentions. Yaks don't trust all humans. First, they want to find out what kind of human is approaching them. If they don't trust him, they will simply run away.

The yak's sight, for all its precision in distinguishing one human from another, can be deceived or confused, however. A story told to us on multiple occasions details how one Lingzhip had gone down to Thimphu, some years prior, and had brought back with him colourful raincoats, which he distributed among his fellow-herders. They replaced the earlier fur coats which, while unrivalled in the warmth they provide, became easily soaked, heavy and uncomfortable to wear in the rain. The herders' new and colourful appearance, however, startled the yaks, who reacted with apprehension and suspicion towards their own herders, and unable to identify them, initially galloped away. It took the yaks some time to familiarise themselves with the new raincoats and to associate a particular colour with a particular herder, after which the status quo of them recognising and acknowledging their herder resumed.

Touch—and its gendered variation. In detailing their relations with yaks, Lingzhips regularly emphasised how their yaks adopt different postures and behaviour with male and female herders, a variance that appears reflective of gendered differences in labour and the related "touching" of their yaks. Among Lingzhips, the conventional divisions of labour prescribes (though not exclusively) it as a female duty to milk the yaks; a daily act that exceeds the biological and also manifests itself as a socio-physiological act that connects female herders and yaks in intimate ways. The gentle squeezing and pulling motion performed by female herders, then, is experienced by yaks as a recurrent movement of affectionate contact, and for yaks it turns into a sensory touch of familiarity, touch and care.

Yaks grow attached to a particular herder's touch, which they easily differentiate from an unfamiliar touch that cannot necessarily be trusted. This is to the extent that yaks express visible reluctance, and sometimes outright aggression and hostility, to have their teats touched by anyone except for the hands and touch they have grown familiar with. Relates Tshering Pelzom, who repeatedly attempted to milk one of the yaks: "the yak just wouldn't allow me. Whenever I came close, she would show her anger by stamping on the ground. Even when she was tied down to a pole, she still found ways to resist my coming near. I tried several times, but the yak just didn't allow me to touch her. When Pema [the yak's female herder] came the yak was at peace and allowed Pema to milk her."

The characteristic touch of male herders tends to be different. Their primary duty is herding, engaging them as pack animals and keeping them together, and this requires the occasional discipline and punishment, necessary acts Lingzhips again compare with raising children. Like a child, Lingzhips compare, a yak that is taught no boundaries and is never punished becomes spoiled and entitled. Disciplining is part of parenting, and this applies equally to children and yaks.

The disciplining of yaks comes in the form of "the stick," which male herders always carry with them and now and then feel obligated to unleash on yaks that behave stubbornly, for instance, by refusing to walk in line while carrying loads or by acting aggressively towards another yak. Such occasional beatings do not only work to set the particular yak straight, but also, as Tandin explains, "to set an example to the other yaks. It warns them against copying bad behavior." In disciplining their yaks, male herders, however, need to thread a fine balance: those who refuse to use the stick at all experience continuous hardship in controlling their yaks with the latter readily taking advantage of the herder's meekness. Those herders "abusing" the stick, however, see their yaks galloping away whenever they approach them.

What these instances reveal is that yaks profess the practical consciousness and agency to reciprocate the differential touch they receive from female and male herders (and from different herders within both genders). All Lingzhips know this. "My yaks act more calm and obedient toward my wife and daughter. This is because they never receive beatings from them," says the yak-herder Sonam. He explains further: "Toward me they behave a little more distant. They know I care for them. They recognize and acknowledge my presence, but they also know that I don't tolerate any wrong behavior from them. That's why they are a little more careful in their interaction with me." Or as another herder narrated, again invoking a kin-like relationship:

> The relation we share is like that between parents and children. The only thing is that they cannot speak [in human language]. Other than that, their behavior is no different than the behavior of any human being. They recognize the faces of herders who beat them and who do not beat them. If they are walking in front and see a yak herder who often beats yaks walking behind them, the yaks will become restless and may even try to run away. If the herder never beats yaks, the yaks will act stubborn in front of this herder. Yaks are like humans as we too don't want to be around someone who is aggressive and abusive."

Sound—the herder's voice. As already illustrated by the vignette in which the herders Tenzin and Tshering called out to their yaks, yaks have the faculty to recognise and differentiate human voices. They readily respond to voices they are familiar with, but tend to ignore the commands made in voices unfamiliar to them. Dorji explains: "my yaks always recognize my voice. Even if I am on the other side of a hillock and they cannot see me, they come if I call out to them. My voice is enough for them to know it is me approaching." Yaks not only have the faculty to associate a particular voice with a particular person, but also to associate the voice with the kind of herder that person is. Continues Dorji: "Even if I shout at my yaks and tell them to go away, for instance when I am angry with them, they will not go because they know me all too well. They just think: 'Ah, it is this person. He will not do anything against us even if he is angry, so we will not go away.'" Voices unfamiliar, however, fall on deaf yak ears, as Dorji explains further: "If a different persons shouts at my yaks angrily, my yaks will immediately vacate their place. They recognize that is not a familiar voice, as well as the anger in the voice. So they feel it is safer to just leave." Dorji concludes by saying what we heard, in various versions and expressions, over and over in the Bhutan highlands: "Yaks recognize who we are. They know us very well and they know how to deal with different kinds of humans."

Conclusion

Through ethnographic insights and fragments, this chapter discussed the dense networks and practices of relatedness, kin-making and intersubjectivity that connect herders and yaks. In so doing, it emphasised how herders and yaks intermingle and affect one another, and mutually apprehend each other as persons with distinctive subjectivities who are, however, engaged in the same social sphere of relations and affective living. What prolonged social processes of co-dwelling, co-movement and co-dependence nourished locally, in the upshot, is a "mutuality of being" between herders and yaks, along the same lines that Marshall Sahlins (2011, 2) projects as characteristic of human kinship: "persons who are members of one another, who participate intrinsically in each other's existence," and who share "intersubjective belonging."

While yaks are rather indisputably the most significant and sacred nonhuman animal that dwell in the Bhutan highlands, the multilayered sociabilities, sensibilities and subjectivities that bind the ties between herders and yaks are certainly not the only trans-species knots that establish social life in the highlands. These also extend to horses, whose numbers are rapidly rising locally, to vultures, who ultimately eat the herders' flesh and bones in sky-burials, and to bears, who ever more boisterously break into herders' houses and steal rice and vegetables. They also extend into the densely populated cosmic polity that hierarchically encompasses Lingzhi society, and whose deities perform distinctive roles as agents of history, governors of fortune and resident protectors. Each of these knots requires further ethnographic unravelling.

As such, this chapter is just one chapter looking into much broader and more complicated social webs of relatedness and kin-making that bind humans, nonhuman animals, numinous beings, as well as the physical landscape itself, and that jointly engender an encompassing, relational ontology that refuses to break up highland life into a set of autonomous social, natural and cosmic domains.

Notes

1 I thank Dan Smyer Yü for relating this Tibetan folktale from the Golok area in Amdo, currently in China's Qinghai Province.

2 By proposing the term "relatedness," Janet Carsten subscribed to the critique voiced evocatively by Schneider (1984) that then conventional conceptual models of kinship were incarcerated in projections of 'biological essence' that relied on a clear-cut analytical distinction between biology and culture with clear antecedents in Euro-American epistemologies. By offering the concept of relatedness, Carsten proposes to foreground 'indigenous statements and practices' (2000, 3) that may often focus on the daily doing, performing and nurturing of what it means to be related in particular ethnographic locales. What Carsten and her contributors convincingly showed was that the actual form and substance relatedness takes may well "fall quite outside what anthropologists have conventionally understood as kinship" (2000, 3). Whereas

earlier studies of relatedness, for all their innovativeness, continued to limit social processes of 'kin-making' to the realm of the human, this chapter builds upon scholarship that recognizes the possibility that regimes of relatedness may involve both human and nonhuman agents, be they animals (Haraway 2008, 2016, Govindrajan 2018), gods and spirits (Ramberg 2014) or plants (Galvin 2018)

References

Abram, David. 1996. *The Spell of the Chapter 1Sensuous. Perception and Language in the More-than-Human Worlds*. New York: Pantheon.

Collins, Stevens 1985. 'Categories, Concepts or Predicaments? Remarks on Mauss's use of Philosphical Terminology.' In: *The Category of the Person: Anthropology, Philosophy, History*, edited by Carrithers, Michael, Steven Collins, Steven Lukes, 46-84. Cambridge: Cambridge University Press

Viveiros de Castro, Eduardo 2012. 'Immanence and Fear: Stranger-Events and Subjects in Amazonia', translated by David Rodgers. *Hau: Journal of Ethnographic Theory* 2(1): 27-43

Carrithers, Michael, Steven Collins, Steven Lukes, eds. 1985. *The Category of the Person: Anthropology, Philosophy, History*. Cambridge: Cambridge University Press.

Carsten, Janet, ed. 2000. *Cultures of Relatedness: New Approaches to the Study of Kinship*. Cambridge: Cambridge University Press.

Classen, Constance. 1997. "Foundations for an Anthropology of the Senses." *International Social Science Journal* 49 (153): 401–412.

Derrida, Jacques. 2002. "The Animal That Therefore I AM (More to Follow). Translated by David Wills." *Critical Inquiry* 28: 369–418.

Descola, Philippe. 2005. *Beyond Nature and Culture*. Translated by Janet Lloyd. Chicago: University of Chicago Press.

Galvin, Shaila Seshia. 2018. "Interspecies Relations and Agrarian Worlds." *Annual Review of Anthropology* 47: 233–249.

Goody, Jack. 2002. "The Anthropology of the Senses and Sensations." *La Ricerca Folklorica* 45: 17–28.

Govindrajan, Radhika. 2018. *Animal Intimacies: Interspecies Relatedness in India's Central Himalayas*. Chicago and London: University of Chicago Press.

Gyamtsho, Pema. 2001. "Economy of Yak Herders." *Journal of Bhutan Studies* 2 (1): 1–45.

Haraway, Donna Jeane. 2016. *Staying with the Trouble: Making Kin in the Chthulucene*. Durham and London: Duke University Press.

Haraway, Donna Jeanne. 2008. *When Species Meet*. Minneapolis: University of Minnesota Press.

Ingold, Tim, ed. 1998. *What is an Animal?* London: Routledge.

Le Breton, David. 2017. *Sensing the World: An Anthropology of the Senses*. Translated by Carmen Ruschiensky. Bloomsbury: New York.

Mauss, Marcel. 1985 [1939]. "A Category of the Human Mind: The Notion of Person; The Notion of self." In *The Category of the Person: Anthropology, Philosophy, History*, edited by Michael Carrithers, Steven Collins and Steven Lukes, 1–45. Cambridge: Cambridge University Press.

Ramberg, Lucinda. 2014. *Given to the Goddess: South Indian Devadasis and the Sexuality of Religion*. Durham and London: Duke University Press.

Roder, Walter, Kinzang Wangdi, Pema Gyamtsho, and Karma Dorji. 2001. *Feeding the Herds: Improving Fodder Resources in Bhutan*. Kathmandu: International Centre for Integrated Mountain Development.

Sahlins, Marshall. 2011. "What Kinship Is (Part One)." *Journal of the Royal Anthropological Institute, NS*, 17 (1): 2–19.

Schneider, David Murry. 1984. *A Critique of the Study of Kinship*. Ann Arbor: University of Michigan Press.

Tsing, Anna. 2012. "Unruly Edges: Mushroom as Companion Species: For Donna Haraway." *Environmental Humanities* 1 (1): 141–154.

Ura, Karma. 1993. "The Nomads' Gamble: Pastoralists of Northern Bhutan." *Journal of South Asia Research* 13 (2): 81–100.

Ura, Karma. 2007. "The Herdsmen Dilemma." *Journal of Bhutan Studies* 7 (1): 1–129.

Vitebsky, Piers and Anatoly Alekseyev. 2015. "What is a Reindeer? Indigenous Perspectives from Northeast Siberia." *Polar Record* 51 (4): 413–421.

Wangchuk, Dorji, Wanchai Dhammasaccakarn, Punya Tepsing, and Thongphon Promsaka Na Sakolnakarn. 2013. "The Yaks: Heart and Soul of the Himalayan Tribes of Bhutan." *Journal of Environmental Research and Management* 4 (2): 189–196.

Wangda, Pema. 2016. "Yaks and Yak Herding: Challenges and Opportunities in the Bhutanese Himalayas." In *Yaks on the Move: Transboundary Challenges and Opportunities for Yak Raising in a Chang HKH Region*, edited by Wu Ning, Yi Shaoling, Srijana Joshi and Neha Bisht, 65–76. Kathmandu: International Centre for Integrated Mountain Development.

Wangdi, Jigme and Kesang Wangchuk. 2017. "Productive and Reproductive Traits of Yak Types in Bhutan: Characteristics and Comparisons." *Journal of Applied Animal Research* 46 (1): 893–897.

2 Lepcha water view and climate change in Sikkim Himalaya

Charisma K. Lepcha

The children were drenched despite wearing raincoats when they arrived at the monastery that afternoon. They carried with them Buddhist scriptures wrapped in plastic (to avoid getting the books wet) and had walked in single file behind monks and elders holding prayer flags, cymbals and trumpets as part of the annual *bumkor* procession. Buddhist scriptures were taken out of the monastery and carried on people's shoulders as they travelled around the village houses to bless villagers as part of an important community event in Dzongu. People lined up near their houses to receive blessings from the contingent carrying the holy books as the procession made a full circle of the village before returning to the monastery. *Bumkor* is a Buddhist ritual practiced by the indigenous Lepchas in north Sikkim every February. Though well ahead of the monsoon, one could observe the continued drizzle throughout the day that overshadowed the event. But the villagers seemed prepared as those who participated in the promenade carried umbrellas and raincoats to withstand the weather. It was a commonly held belief that rains during *bumkor* meant blessings from above. Recent years, however, have seen more rains along with landslides, slope failures, road blockades and other transformations of the physical landscape in Dzongu (Figure 2.1).

Located in the heart of the tiny Himalayan state of Sikkim, Dzongu is home to the indigenous Lepchas whose understanding of their environment has often been ignored and discounted in Sikkim's "State Green Mission" to build an "organic Sikkim."[1] Only a 3-hour journey from the state capital, Dzongu was set aside as a "Lepcha reserve" since the time of the king[2] with a royal proclamation as early as 1958 stating that "outsiders (non-Dzongu natives) may remain in North Sikkim area only under a permit issued by Sikkim Darbar" (Government of Sikkim, Law Department 1984, 38) thus barring entry and residence of non-Dzongu residents into the reserve area. It is a culturally homogenous Lepcha terrain where legends boast of how the first creation was sent to live here, and is known as *fokram takram* or the source of Lepcha origin and life. It is a "place that gives and receives life" (Rose 1996, 7) as the mythological stories come alive and the land becomes sacred and inescapable to a Lepcha mind. It has always been the homeland for Lepchas, no matter where they are in the world, as it continues to "shape their interactions and understandings—their co-becomings" (Country et al. 2016, 457) in a larger society where Dzongu is often projected as "remote" and the people "backward." Borrowing from Rose (1996), Dzongu country is however, not

Figure 2.1 Bumkor procession arrives in Tingvong monastery.
All photos by the author, unless specified.

just imagined, "it is lived in and lived with" (Rose, 7). It holds narratives similar to those of indigenous peoples across the world who have been historically marginalised and silenced by the state as their land, their water and their environment have been snatched from their grip by the hands of the powerful who have used it for their advantage. But Dzongu is "a living entity" (Rose) holding a past, present and future, with a "consciousness, and a will toward life" (Rose) as this paper explores the indigenous understanding of water among Lepchas of Dzongu country and how climate change is experienced through the various interactions in their landscape.

From free-flowing rivers to forest streams, water spirits and rituals to prevent hailstorms, the meaning and usage of water is embedded in the everyday lives of the indigenous population. Water runs deep underneath the rocks and crevices but is also visible and accessible in the form of rivers, streams and waterfalls as the snow melts descend from the dynamic Himalaya. These water bodies are arteries of interconnectedness that connect both human and nonhuman, tangible and intangible (Rose 1996, Country et al. 2016) in a "water based" (Yazzie and Baldy 2018, 2) relationship. Indigenous peoples' fight for water rights, in recognition of it as a living being, is further explored as water is also seen as "as an ancestor and as a relative with agency within this network of life, one who deserves respect, care and protection" (Yazzie and Baldy, 1). Along similar lines, an attempt has been made to examine the "water view" (Yazzie and Baldy) of the indigenous population of Dzongu.

This chapter pays heed to water-based knowledge that is found in the Lepcha religious beliefs and livelihood practices that are often ultimately discarded or play second fiddle within the scientific framework. Based on an ethnographic team research with Rongynoo Lepcha and Mongfing Lepcha, it is written as what Thom van Dooren and Deborah Rose call a "lively ethography" or "a mode of knowing, engaging, and storytelling that recognizes the meaningful lives of others and that, in so doing, enlivens our capacity to respond to them by singing up their character or ethos" (van Dooren and Rose 2016, 77). The stories and the ethos I am about to tell do not concern others but are our Lepcha stories and ethos. As an indigenous scholar, I write this chapter in response to the increasing acknowledgement of indigenous knowledge, especially in relation to the environment and natural resources because of their "deep connection to land—is gaining support with the concept of "indigenous environmental knowledge" (Dove 2006, 196). That kind of knowledge is supposed to assume a "harmonious engagement with nature" (de Maaker 2018, 31) thereby creating an "organic linkage" (Baviskar (2006, 38) between indigenous culture and ecology. It also presumes all indigenous communities know their land and environment, a concept that has been criticised for essentialising native people (Berkes 1999). However, colonial ethnographers' connecting of indigenous Lepchas to nature has been carried forth by post-colonial anthropologists and deserves some attention. MacDonald (1930, 8) wrote,

> Every Lepcha is a born naturalist, and living as they do in the forests, they know the habits of every beast and bird, and make first rate collectors. They have a name for every living thing, including plants, which is found in their country.

Likewise, Waddell (1998) also writes that "his [the Lepchas'] close companionship with nature has made him a naturalist, a tender lover of flowers, and something of a philosopher" (Waddell, 93). These writings stigmatised the natives as "born botanists," but it also confirmed their environmental knowledge and their ability to live in their natural surroundings. They stereotyped the Lepcha ethos but held the position that they were first stewards of the earth, of their land. Arora (2006) called them "primordial environmentalists" (Arora 2006, 72) when Lepchas went on a hunger strike against dams in Dzongu, when water was under attack by capitalist attempts to "harness, obstruct, pollute, and discipline water through infrastructure projects" (Yazzie and Baldy 2018, 3).

These developments connected Lepchas with the shared struggles of indigenous people across the world in a bid to safeguard their sacred waters. They were committed to reclaiming their relationship to water while being accountable to their "water view." In the wake of these developments and the growing call to give space to an indigenous "water view," this chapter sets off looking at how water is embedded in the Lepcha landscape, stories, beliefs and everyday lives. It explores the different types of water—its usages and meanings in different contexts—while paying attention to the production of meanings about climate change in the Lepcha landscape. It also makes an attempt to tell the story of water as understood by the people through the popular folk tales and myths that are still told around kitchen hearths and on riverbanks. These stories reveal "ways of constituting, of responding to (and) in a living world" (van Dooren and Rose 2016, 85), and actually become powerful ways of thinking of and understanding water not only in terms of conservation and management but also in terms of holding "possibilities and interpretations" (Smith 2005) of water's own story of "interdependency and respect" (Yazzie and Baldy 2018, 3). This chapter also highlights the "extreme events" (Sherpa 2014) like the earthquake in 2011 and slope failure in 2016 that have directly affected Dzongu residents whose landscape has changed as erratic climatic conditions have increased in the last decade or so. Thus, the overall sub-themes of this paper overlap between water and mythology, water and religion, water and gender, water and livelihood, and water conflict and dams.

Lepcha homestay in *Himalayan Village*

In 1938, Geoffrey Gorer published *Himalayan Village: An Account of the Lepchas of Sikkim* after his stay in Lingthem, lower Dzongu for 3 months along with John Morris who went on to write *Living with the Lepchas: A Book about Sikkim Himalayas* (1938). These monographs were the first detailed study of Lepcha people, still referred to by scholars studying this community. Gorer's book was restudied by Rip Roshina Gowloog in *Lingthem Revisited* (2005), and occasional research on different topics has researchers flocking to this village. In recent days, Lingthem is a somewhat-popular homestay destination for offbeat travellers looking to spend a night or two in a Lepcha village. Early ethnographers would be baffled to hear about Lingthem as a tourist destination but there are some researchers today who have spent time here studying and researching the same. Lepcha Homestay opened in 2009 as part of an alternate means for Dzongu youth to earn a livelihood as they protested hydro projects in their land. They were seen to be anti-state and, as a result, government jobs were difficult to get. The owner of Lepcha Homestay was and is an activist and a victim of state bureaucracy: eventually he was transferred to a different district during the anti-dam movement. Now, he takes tourists for short hikes around his village and indulges them with stories that allow them to travel in various worlds, if they are keen to hear. When asked about the source of water in his village, he promises to take me there, but not before first telling me about how water came to his village (Figure 2.2):

> Long ago, there were eight sisters who were traveling nearby when Lingthem Anue Pandi decided to make way to Lingthem from the confluence of Rongnyoo and Rongnyong. As she walked uphill, she felt hungry and thirsty so she started eating the soft cheese locally known as *churpee* that she had carried. It made her thirstier but she could not find any water. She saw that it was dry land so she took her sickle and hit the rock and water gushed out. The source seemed modest and the dripping water made an attempt to try fill the tank (that has now been built in that place); and the good news was that the trickling water never dried up.

This water source was a constant in villagers' lives and they continue to use water from this tank for irrigation. In memory of Lingthem Anue Pandi, villagers have

Figure 2.2 Lingthem Anue Pandi's water source.

also built a resting shed for tired travellers who walk this path. My friend who had been telling me the story and I rested for a bit after tracing Anue Pandi's steps. We were walking uphill to the water source of his village in upper Lingthem where the forest was dense and gushing water could be heard but not seen. The forest streams hidden underneath the thick jungle gave indication of Lingthem's water source. The running water echoed off the mountain top as the leaves and the wind played games we could not understand. I was told that people are scared to come here because of the spirits, and indeed, there was an absence of human-centred existence that made way for the "more-than human construction of knowledge" (Country et al. 2016, 457) in regards to the relationality of the place to the people. Not wanting to disturb the spirits, we descended towards the village. Along the path was a small stream that had only unveiled itself from the forest canopy when it was pointed out to me. "In case of a cut or a wound, wash in this stream and the wound would be healed" was a commonly held belief that pointed to injured villagers' relationship to water.

Nobody questioned these beliefs nor dared pollute these streams as their accountability to their water view remained intact. Across the valley, the next hill looked bruised and I was reminded of the massive landslide of August 2016 that caused the entire mountainside to collapse, leaving a wound on the landscape. The ecology was disturbed as the river transformed to a lake 150 feet wide. Roads disappeared, houses were submerged and the debris would end up blanketing the area for quite some time. New buildings were made old. The local school closed down and there were other disruptions in the everyday way of life, while those on the other side of the lake were stuck on their hill without access to hospital, police station, or the capital city. Life had come to a standstill. If this was one disaster, another disaster was lying in wait as flash floods were a threat that would eventually drown many newly built houses. It so happened that many villagers had left their ancestral Lingthem and moved southwards to the up-and-coming Passingdang block as it was closer to the road. There, concrete houses were built, but the recent developments of landslides and flash floods have since made them think twice before moving to newer homes as most of the families who had settled in Passingdang in the last decade were heard to be making their way back to their original homes.

Lingthem is just one of the 14 villages in Dzongu country in which the people are experiencing a drastic change in their relationship with land and water. The youth of Lingthem have been active in taking back their waters through different modes of resistance; perhaps unsurprisingly political influences have not been on their side. The goal of this project was to visit all 14 villages to map the water bodies in Dzongu. But due to landslides, road blockades and time constraints, only 10 of the 14 villages were visited: Gor, Leven, Lingdem, Lingthem, Passingdang, Safo, Salim Pakel, Shipgyer and Tingvong, as illustrated in Figure 2.3. There are two main rivers Teesta and Rongnyong that runs through these villages although villagers are mostly dependent on forest streams for their daily water source.

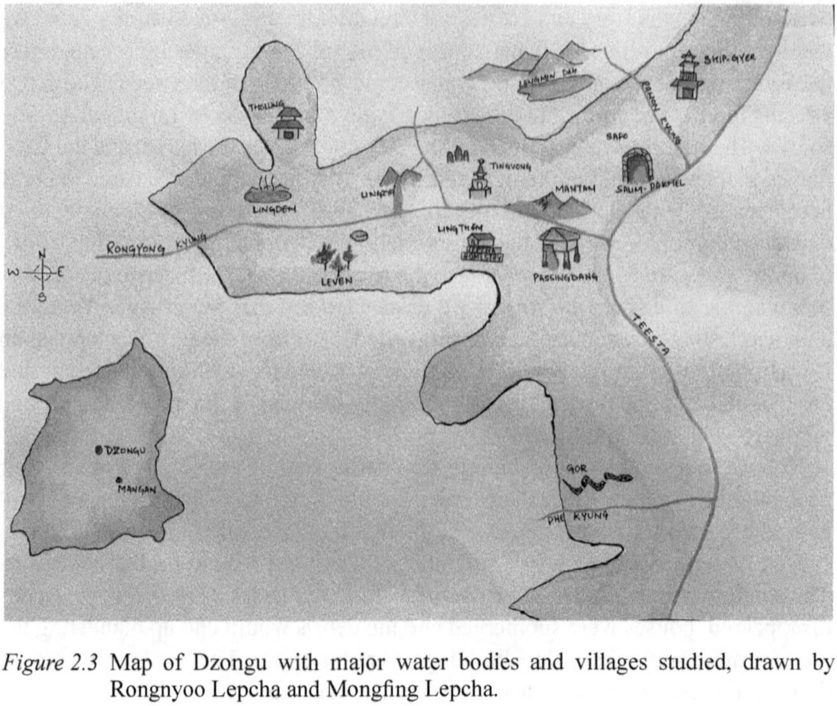

Figure 2.3 Map of Dzongu with major water bodies and villages studied, drawn by Rongnyoo Lepcha and Mongfing Lepcha.

Villages in Dzongu vary in terms of temperature, vegetation and the number of water bodies present in each village. There are rivers, waterfalls (both large and small), hot springs, lakes, ponds and streams found inside Dzongu, though not in every village. Hence, ecologically, Dzongu is not a heterogeneous entity as it is often made out to be. There are places in upper Dzongu where yaks used to graze and snowfall was part of the seasonal change. Today, villagers have only heard that "it used to snow in their village" and they reminisce of the time when there used to be snowfall in Dzongu. It is likely that snowfall was not something foreign to the indigenous population since their creation narrative points to Lepchas being created from the snow of Mt. Kanchenjunga and that it was "rare to find a Lepcha village from where Khanchendzonga [*sic*] cannot be sighted" (Wangchuk and Zulca 2007, 32). But the myth, the stories and the people have also undergone changes due to climatic, ecological, migratory and religious factors, just as the story of Lepchas and Mt. Kanchenjunga has undergone drastic changes in the course of time. The following section will examine the Lepcha creation myth, as it represents people's basic outlook "concerning natural or historical phenomena" (Peyton 1969, 31) at that time. Taking a cue from the origin story, the significance of snow—a different form of water in Lepcha cosmology—is evident in the creation myth itself. Additional myths related to water will be discussed and illustrated below.

Water and myths

The story of Kanchenjunga is about the sacred origin of the indigenous Lepchas. It was their "sacred history: relating an event that took place in primordial time, a fabled time of the 'beginnings'" (Eliade 1974, 60). In the beginning, the Himalayas were created and Kanchenjunga was the first creation of their mother creator Itbudebu Rum and is revered as the eldest brother. These Himalayan mountains are the elder brother and sister to every other landform and river created thereafter. These mountains were living beings, siblings and kin to the country. Itbudebu Rum then created everything— "the sun, the moon, the stars, the earth, the forests, the plants and the animals" (Kotturan 1983, 20). Tradition has it that Itbudebu Rum picked balls of snow from the top of Kanchenjunga and created the first humans Fudongthing and Nazongnyu from the fresh snows of Mt. Kanchenjunga.

> With the snow in the right hand (s)he created a man and called him Phadong Thing which meant "the most powerful." With the snow in his left hand (s)he created a woman whom (s)he called Nazyonguyu which meant "ever fortunate." The Lepchas believed that these two were the first parents of the Lepcha people.
>
> (Kotturan 1983, 20)

Created by God herself, the Lepchas call themselves *mutanchi rongcup rumkup* meaning mother's loved ones/children of snowy peaks/children of God. "The snowy peaks refer to Kanchenjunga because it was from its snows that the first couple was created"

(Wangchuk and Zulca 2007, 31). Therefore, Lepchas consider Mt. Kanchenjunga to be their place of origin and have always lived in the shadows of Kanchenjunga. But if we analyse the residential pattern of the Lepcha population, we know that they do not live in snowy areas and are not always comfortable with snow. Some Lepchas even die without having seen snow in their lifetime. In that, I propose looking into the snow melt of Mt. Kanchenjunga and venture towards the "relational theory of water culture to remind us that we are water based and that we have water memory" (Yazzie and Baldy 2018, 2) as it was the waters and the rivers formed by this snowmelt that spread into places where Lepchas resided. Lepchas depend on streams and rivers for their livelihood. While snow made for difficult-to-negotiate terrain, Lepcha people have their indigenous ways of navigating the rivers to sustain themselves. In that, there is also a systematic arrangement of how they have named their rivers, for example, with all river names starting with the sound represented by the English letter "R": Rongnyoo, Rangeet, Reshi, Relli, Rey, Rongchong and the list goes on.

Lingzya waterfall

This story tells how, many years ago, a hunter came traveling from the south and chanced on the Lingzya waterfall. He was thirsty from his wanderings and decided to refresh himself. He decided to take bath and as he was drinking the water, he found a long strand of hair. Seeing that, he thought there might be people living in the area so he decided to trace the source of the waterfall. Upon climbing up the cliff, he saw a beautiful girl taking bath. He was taken aback by her beauty and vowed to marry her. But the girl put forth a condition—he had to meet her parents, who were both snakes. Other suitors had also come to ask for her hand in marriage but were unable to face her parents. The hunter was not afraid and when he finally met them, the snakes wrapped themselves around his body. The hunter was a strong man and withstood their grip thus passing the test and marrying the most beautiful girl he had ever seen (Figures 2.4 and 2.5).

The story of Lingzya falls is not just limited to a water body but also about human and nonhuman interaction in the Dzongu landscape. The hunter can clearly be seen as an "outsider" who discovered the waterfall and the strand of hair which leads him to a beautiful girl. But the fact that the girl's parents are snakes also tells of the continuity between humans and nonhumans in the Lepcha worldview. The acceptance that snakes could be parents-in-law for a human and vice versa "dismantles the hierarchy of knowledge that would place human-centred" (Country et al. 2016, 457) understanding as the way of ordering one's world. It is an example of "co-becoming" (Country et al. 2016), where everything exists in a state of emergence and relationality.

Tale of the water spirit and the hunter

It is also told that one day, a hunter was fishing and caught a huge fish. It was not an ordinary fish, but a water spirit (*ung dokhi*). The spirit wanted to be free and offered anything the hunter wanted in return for his freedom. Grabbing this opportunity, the hunter wished to become very rich. After being granted his wish,

Figure 2.4 The story of Lingzya Waterfall.

Art by Mongfing Lepcha.

Figure 2.5 The hunter and the hair.

Art by Rongnyoo Lepcha.

he set the spirit free. The hunter had become a rich man and he was happy for the most part. But one day, the sun was shining so brightly as to blind everyone. The hunter was furious and rebuked the sun, “I have all the riches. How dare you shine brighter than my treasures!” This was the point when his pride took over and subsequently his treasure vanished and he went back to being a hunter just as he had been before meeting the water spirit (Figure 2.6).

In Lepcha cosmology, the fear of the spirits is rampant. There are both benevolent and malevolent spirits that can be seen as gods and demons respectively. The gods are called *rum* and the demons are called *mung.* “The *rum* is supposed to cause happiness while the *mung* causes sorrow” (de Stocks 1975, 59). There are numerous gods and an even wider array of spirits that cause illness and misfortune if unhappy and bring health and fortune if properly propitiated. Siiger explains how a powerful *rum* “turned into a primarily malicious being because the hunters have neglected proper ceremonial sacrifices” (Siiger 1978, 428). Hence, the idea of a water spirit is very much embedded in the psyche of the people. Below is a villager’s account of a water spirit:

> There was a stream near the village where everyone collected water and it was sufficient for everyone in the village. One evening my grandmother faced a frightening situation when she sensed cold hands grabbing her wrists

Figure 2.6 The water spirit and the hunter.

Art by Rongnyoo Lepcha.

> as she went to collect water. She was scared as she prayed but was finally released.

The presence of *mung* near water bodies is a shared fact as children are not allowed to go near the streams especially in the evening. People believe that the rivers can take away the *apil* (soul) of children and do not allow them to play near rivers or streams. On most occasions, people tend to avoid streams and if they are passing through they leave a small twig or a leaf on a stone while crossing the stream with an infant so that the infant's *apil* is not left behind or taken away by the *mung* in the vicinity. The fear of such water spirits or the *mung* in their environment reveals the "multispecies" (Rose et al. 2012, 4) existence of beings in their country. One can see people's response as reverence for their surroundings that allows them to be responsible stewards of their environment.

Water and religion

The Dzongu landscape is full of good and bad spirits, with the latter occupying a dominant position once Buddhism was introduced and easily appropriated into their everyday life. It was believed that Guru Rinpoche (Padmasambhava) who introduced Buddhism to Tibet passed through Sikkim and blessed many caves, mountains, rivers. Balikci (2008, 83) recounts Guru Rinpoche's acts that

> tamed all the supernatural beings of the land, including the mountain god Kanchendzonga, during his eighth century visit to Sikkim and … bound them through solemn oaths into being protectors of the faith and to refrain from causing harm to sentient beings.

Guru Rinpoche is also believed to have passed through Dzongu as legends of his antics fill the rocks and streams, especially on the route to Tholung monastery. Located at an altitude of 8,500 feet and situated within Dzongu, this monastery is considered to be a sacred monastery owing to its historical, political and geographical significance. Because of its remoteness and lack of connectivity, only Dzongu residents make their way uphill for the handful of celebrations that take place in Tholung monastery. The name "Tholung" is a derivation of two Lepcha words, *atho* and *long* which indicate a high and rocky place. The popular story tells of the monastery being built on a pond. In addition, there are about a dozen bodies of water in the Tholung landscape and each stream, brook, pond, waterfall or lake narrates a different story. For any visitor, these water bodies may appear all the same, but different water bodies differ even in the way they are referenced at different points in any given day. There were streams that were not to be touched after the sun sets and brooks from which water cannot be consumed before the sun rises. And if one does not observe these directives, illness, paralysis and accident are likely results. These narratives also establish the creation of holy and sacred water spaces, revealing the local understanding of water and its purposes and utility. For example, a hot spring nearby served as a medicinal bath for pilgrims, and there exists a clear understanding of which water is to be used for what purpose, although some water bodies are not restricted in their use and the water could be used at any time.

The coming of Buddhism meant monasteries opened up in all villages of Dzongu as a new way of life was established. There were various rituals, prayers, activities, do and don'ts and the act of coming together to a monastery for worship gave a sense of religious solidarity. Buddhism was more organised than the indigenous religion although Lepchas have been criticised for not performing religious rites "beyond perhaps turning a prayer wheel and prostrating themselves before Lamaist images" (Gorer 2005, 274). They have been blamed for creating a "double-layered religious system" (Torri 2010, 149) as they practice Buddhism and the indigenous religion side by side. But it is the everyday practice of Buddhism as embodied by the people in their given environment, "beyond the bounds and often without the approval of religious authorities" (Ammerman 2021) that tells of their religion. One of the most striking image in a Lepcha village are the Buddhist prayer wheels often painted in bright colours powered by water mills set in a local stream and rotating on their own. It is a home-grown technology that serves the purpose of turning an ordinary stream into a place of worship with the prayer wheel, the *khadas*, and other religious paraphernalia indicating it to be a holy place. On the other hand, the water that supports the prayer wheel is now holy water too so people refrain from polluting or dirtying it as it flows downwards (Figure 2.7).

There is a general belief that if the water bodies are dirtied, villagers will be affected with various illness and diseases because the deity would be upset and

Figure 2.7 Prayer wheel powered by water.

angry. In order to prevent such misfortunes, the *bongthing* (male shaman) would have to cleanse the water and the people too.

Another local observance is recognition of the presence of *lu*—snakes that reside near springs and that bring either prosperity to the village when properly propitiated, or illness and harm to villagers when ignored. For this reason, the spring which is the *lu*' s abode is routinely cleaned and looked after. No one is allowed to litter or dirty the place and rituals are performed to please the *lu*, while *lu sang*, a type of incense, is specially prepared and offered to the *lu*. The *lu* in Salim Pakel village is believed to be a green snake and the spring where it resides has a perennial flow of water. People are keen to observe its sanctity and refrain from dirtying the place to prevent any of them from falling sick.

According to a Buddhist tradition of releasing fish after a ritual or for a special occasion, these fishes are not supposed to be consumed by the people. But there are some who tend to ignore or who are not informed who end up eating the fish. In that case, the family conducts a ritual so the person concerned does not fall sick. Chak Tshering Lepcha recounts an event some 23 years ago when his family had released some fishes which were caught by people from the neighbouring village. As a result of this, his grandson fell sick and they had to perform another ritual for his recovery. The *bongthing* told them that the river spirits were angry as someone had eaten the fish offered to the spirit.

Lepcha *bongthing* also uses dried fish known as *ngu chi* during certain rituals. In Saffo and Lingthem villages, there is a specific clan that catches a specific fish from a specific river to use during a ritual called *Kabi Rum Faat.* A week before the ritual, the fish is caught and dried to be used during the ceremony (Figure 2.8).

Indigenous Lepcha rituals are also observed for different climatic conditions, although these rituals are performed on a smaller, household, level and are not often discussed when talking about religion. Some important rituals related to water are discussed below to explain how these practices have been embedded in the Lepcha belief system from the start.

Satap Rum Faat is an important ritual observed to prevent hailstorms that would damage crops. It is performed in March by a *bongthing* who prays by sacrificing a rooster to the god of hail. *Ungrum Faat* is another ritual, performed to appease the rain gods during the months of June and July and here too a rooster is offered for the appeasement of the gods.

Water and gender

The popular water related myths are not gender specific and we see both men and women participating in their roles associated with water. An elderly person in Passingdang also mentioned how water bodies can be categorised into both male and female genders. In early days, the source of water would be a little distant from people's homes so they had to fetch water in bamboo containers called *padam*. Villagers today no longer have to carry water as everyone has a private supply in their homes, transported from forest streams or rivers through bamboo or ordinary pipes (Figure 2.9).

Figure 2.8 *Ngu chi*, the dried fish used in rituals.
Photo by Rongnyoo Lepcha.

Figure 2.9 Water connection using plastic pipes and containers.

Both men and women are involved in the water collecting process. It so happens that excessive rain or fallen leaves can block the water source and people take turns to resolve such issues. During wet rice cultivation, both men and women prepare the field during the monsoon season. The men are engaged with ploughing and irrigating the field while the women plant seedlings into the field. When it comes to hunting and fishing, it is usually the men who go out and women are responsible for gathering and for processing game when it is brought back home. Women also engage in works related to water outside the home, but the majority of their water-related responsibilities lie within the confines of the home. Women water the vegetables in their kitchen gardens, wash clothes, cook and collect water for kitchen and other household uses.

Men use a bamboo basket trap (*phip*) with a wide mouth at the top that narrows at the bottom, inverted downwards and placed at an *ung taksot* (waterfall), the fish can get in but cannot get out once they are inside the basket. Another method of catching fish uses a wild plant known as *tuknyil* (*Artemisia vulgaris*) to intoxicate the fish which are then collected in a *tukshyor* (bamboo strainer).

Water and livelihood

Water is considered pure and central to life; each river carries significance and is linked with the lives of the people in their day-to-day life as well as through folktales and myths. Some rivers, like the Rongnyoo and Rangeet, have more religious significance than others and are greatly entwined with the Lepcha socio-religious life. For example, the Rongnyoo River is considered to be a provider and is seen in a motherly light. In everyday life, once water enters the home, it is used starting with early morning rituals in which it is offered to the gods and deities. Uses then extend to the household chores that require water and to agriculture for which it is central; the villagers mostly depend on rain water to irrigate their fields.

The villagers also deeply believe in keeping the sources of water clean since dirtying the source may lead to illnesses or sometimes even cause the sources of water to dry up completely. Local water sources have not dried up completely, though the water drawn from the source is used not only for drinking, but also for watering crops and animals, etc. In recent times, villagers have noticed decreasing rainfall patterns that have affected the seasonal crops that are planted before the rains come. One of these seasonal crops is cardamom. Still the main cash crop, a large area is devoted to it even though the harvest yield and the market price are not as good as they were before. At one time, North Sikkim was one of the largest cardamom producers of the country. Most Lepchas in Dzongu grew cardamom although it was not a native crop. The Lepcha folktale on cardamom is also related to a body of water.

> Once, the daughter of a blacksmith was playing near a *thyol* (muddy body of water) and she was eaten up by a crocodile. To avenge his daughter's death, the blacksmith made a sword and dived into the water. But before entering the water, he said that if he were to die while submerged, the water

would turn red, and a white and yellow flower would bloom and that the plant should be grown by the people. Indeed, the blacksmith died and the flower that was bloomed was the cardamom and it eventually spread widely.

With time, people are gradually planting crops for market, especially after Sikkim was declared an "Organic State" and the value of local vegetables subsequently increased. Vegetables such as cauliflower, cabbage, peas, beans and potatoes are now planted mainly as cash crops that are sent to the market. In recent days, newer cash crops like oranges and ground apple (*Smallanthus sonchifolius*, also known as yacon) have been introduced in the village as cash crops. Yacon showed itself to be suited to the region by yielding more than expected, and since the market price of the crop is good, more villagers are willing to plant the crop.

Living in relationship with the local bodies of water are aquatic species that are of interest too. The bigger rivers have fish, though the village streams do not have indigenous fish. Instead, they are populated with the fish that have been released by the people of the village. There are also various kinds of *taluk* (frogs, which are edible, and toads, which are inedible). Frogs and toads are used as medicines: *luk chuk taluk* has a reddish colour and *nampok taluk* is bigger than other frogs and has a darker colour (almost black) and is regarded as having more medicinal properties. The greenish *luk eng taluk* is hunted at night during the monsoon and is also consumed as medicine. Hunters use a flaming torch to hypnotise the frogs who are then easily speared with a sharp bamboo stick. The *taluk* are either eaten fresh or smoked, dried and stored for later use. Usually, the men hunt and collect the frogs and the women do the cooking.

Some of the birds, such as *ungsumketfo* (Oriental magpie robin [*Copsychus saularis*]) and *chyakmongfo* (whistling thrush [*Myophonus caeruleus*]), that are found in the environment are consumed by the villagers for medicinal purposes. People also fish, though mainly to the Rongyong and Rongnyoo Rivers. Many years back, traditional fishing methods included the use of the traditional rod made of bamboo, and fishing nets. Today, fishing rods are more popular and are widely used. The fish are caught and collected in a small bamboo basket called a *ngu tongdong*.

Apart from fishing in natural bodies of water, some of the households in the village are engaged in the raising of fish in fisheries, under the capacity of a government program. The selected participants are given training before starting the business and the government provides the materials to make a fish pond. The fish they rear are mainly trout, which need to be resupplied from Lachung every 2–3 years. Ultimately, this business has not flourished to the extent that the fish could be supplied to the market in bulk, as their growth process is slow, perhaps because they are kept captive and in a controlled environment, but the trout are sold within the village. The source of water for the fish pond is the same pipeline as the drinking water supply.

Lingdem village is also home to a hot spring known as *tato pani* (in Nepali) *Song Byung* or *Brong Ungthrum* in Lepcha. Local people would visit the hot spring in hopes of curing their illnesses mostly related to bones and joint pains. But with the opening of ecotourism, the hot spring is a huge draw for the tourists as there is also a small inn just below the hot spring. According to the locals, the original hot spring used to be somewhere above the present hot spring and the hot spring had actually shifted from its original place since people had dirtied the previous habitation. Apparently, the present water had sprouted as a result of drilling because of the construction of roads in the area.

Centrality of water

From the time of one's birth, water plays an important role in the lives of the Lepchas. Newborns are taken to a spring and given a bath. When a couple is married, a *padam* (bamboo water container) is used to store water known as *yum ung* (elixir) as sign of purity. The bride and the groom are also sprinkled with the water for purification. The groom is sprinkled as he heads out towards the bride's home and the bride is sprinkled when she enters the groom's house. The placement of the *padam* is symbolic as well. It is placed outside for the bride, signifying that she is going out of the house, while at the groom's house the *padam* is placed inside the gate to signify that the bride is coming into the house after marriage.

By and large, the significance of water in everyday life of this indigenous community is manifold. Villagers depend on water for their everyday lives but they also revere the water bodies as a way of preventing the spirits from causing them harm. Each lake, pond and stream carries stories about the environment, while aquatic animals like fish and tadpoles have their own function to remind people of the "ideas about management of water [that] are deeply influenced with cultural values" (Amrith 2018, 323).

Unmanageable waters

Of major concern for this study is the changing rainfall pattern because "it is these rains that bring the Himalayan river-systems to life" (Rahman 2020) and destruction. Earlier there was a relatively longer Spring and Autumn which meant clear blue sky and enough sunshine before and after monsoon. Now the rainy days have taken over as it rains every season; sometimes it rains every day, which means that rain has become incessant and unpredictable. In the past people were able to predict the weather, which would assist in their agricultural activities. In addition to the difficulty caused by unpredictable weather, and the challenges that increased precipitation brings to traditional agriculture, continuous downpours have also caused landslides, even during the once-dry winters, and causing a transformation in the landscape of Dzongu (Figure 2.10).

Landslides have become so common in Sikkim, nobody bothers much. But on August 13, 2016, a massive slope failure and its resulting landslide caused a whole mountainside to come crumbling down, eating up the road and creating an artificial lake. According to the National Remote Sensing Center, "Three hundred meters of road was washed away and nine villages were completely cut off from other villages." Houses were submerged and villages were separated on the two sides of the lake. Likewise, the Mantam village disappeared and while there was no casualty, much of agricultural land was also damaged (Figures 2.11 and 2.12).

Figure 2.10 Landslide during February.

Three years later, villages are still divided by the lake while the only bridge connecting the two sides of the river is only semifunctional. People walk across this broken bridge, but during monsoon season, when the water level rises, one is better off staying on one side of the river. The water level has decreased and the lake has turned to a river but the flow and its current is strong. There was another temporary hanging bridge that was constructed for pedestrians but that is nonfunctional today and parts of the road that were submerged are still visible (Figures 2.13 and 2.14).

People fear the river and their apprehensions are embedded in the way they look at the world. After the landslide, a Dzongu native posted in social media, "our deities have started responding now" clearly indicating that the gods are angry. And rightly so, because the damming of the rivers has definitely changed the landscape of Dzongu. Villages like Shipgyer and Salim Pakel are witnessing tunnelling of their mountains first hand as hydro power projects have invaded their villages (Figure 2.15).

Of the 26 dams proposed in Sikkim, 6 were slated to be built in Dzongu which has been met with protests especially from the local youths. They opposed the dam with religious-cultural and ecological reasons, and rightly so. They were labelled "anti-Sikkimese" for resisting "development" as their sacred waters are being harnessed to create electricity. These dams do not benefit the indigenous population in many levels. Firstly, it distances them from their water sources and the sacred connection they have with the water creatures and spirits of Dzongu

Figure 2.11 Mantam landslide.

Figure 2.12 Submerged houses.

Figure 2.13 Semifunctional Mantam Bridge.

Figure 2.14 Road submerged by the river.

Figure 2.15 Tunnelling set up in Salim Pakel.

Photo by Mongfing Lepcha.

country. Secondly, the electricity produced by these dams is not for the Lepchas; so the Lepchas cannot even take indirect advantage of their water resources. Thirdly, the creation of these dams has been detrimental to the ecology of the fragile Himalayas as the blasting and tunnelling of the mountains have led to extreme events in the landscape of Dzongu. As politicians have not done anything for the people of Dzongu, the local people are managing among themselves even to the extent of accepting the altered landscape. Natural disasters like the Mantam landslide are reminding us that climate change is not only experienced or measured by the quantitative data but also how

> they mark the seasons as they used to be, or of epochal storms that now seem portents of the future. But they also mark it through traces on the landscapes, through memories of old houses and old neighbors. Traces of those earlier times lie embedded as debris at the water's edge (Amrith 2018, 330).

The lively ethography I narrate thus far shows that the Lepcha dwelling place is being heavily impacted by global climate change and regional modernisation projects such as the dam building. Natural disasters in Sikkim Himalaya have the anthropogenic signature besides the ecological, meteorological and geological forces behind them. I share the same sentiment with what Rongynoo Lepcha and Mongfing Lepcha write in their chapter in this book—while our traditional, land-based cultural and spiritual ethos is a source of inspiration for the global environmental discourse, it has undergone many involuntary, anthropogenic changes since the advent of modernity over a century ago. Like Linda Hogan, I also recognise "what once victimized us is now a special area of religious studies [indigenous animism]" (Hogan 2014, 21) and cherish the fact that our indigenous ethos and environmental wisdoms are being gradually included in the global endeavour for a sustainable future. At the same time, I also hear our land is crying out loud for healing, restoration and wholesomeness like anywhere else in the world.

Notes

1 The Government of Sikkim launched its State Green Mission in 2006. Sikkim was later certified as a "100 percent organic" state by the Food and Agriculture Organization for promoting agroecological and sustainable food systems.

2 Prior to merging with India in 1975, Sikkim was ruled by the Namgyal dynasty who came from Tibet and established their kingdom in Sikkim.

References

Ammerman, Nancy T. (2021). "Lived Religion as an Emerging Field: An Assessment of its Contours and Frontiers." *Nordic Journal of Religion and Society* 29 (2): 83–99.

Amrith, Sunil. 2018. *Unruly Waters: How Mountain Rivers and Monsoons have shaped South Asia's History*. New Delhi: Allen Lane.

Arora, Vibha. 2006. "The Forest of Symbols embodied by the Tholung Sacred Landscape of North Sikkim, India." *Conservation and Society*, 4(1): 55-83.

Balikci, Anna. 2008. *Lamas, Shamans and Ancestors: Village Religion in Sikkim*. Leiden: Brill.

Baviskar, Amita. 2006. "The Politics of Being 'Indigenous'." In Bengt Karlsson and Tanka Bahadur Subba (eds.) *Indigeneity in India*. London: Kegan Paul.

Berkes, Fikret. 1999. *Sacred Ecology: Traditional Ecological Knowledge and Resource Management*. Philadelphia: Taylor & Francis.

Country, Bawaka, Sarah Wright, Sandie Suchet-Pearson, Kate Lloyd, Laklak Burarrwanga, Ritjilili Ganambarr, Merrkiyawuy Ganambarr-Stubbs, Banbapuy Ganambarr, Djawundil Maymuru, and Jill Sweeney. 2016. "Co-becoming Bawaka: Towards a Relational Understanding of Place/Space." *Progress in Human Geography* 40 (4): 455–475.

De Maaker, Erik. 2018. "On the Nature of Indigenous Land: Ownership, Access, and Farming in Upland Northeast India." In Eva Gerharz, Nasir Uddin, and Pradeep Chakkarath, eds. *Indigeneity on the Move: Varying Manifestations of a Contested Concept*. New York: Berghahn Books.

Dove, Michael R. 2006. "Indigenous People and Environmental Politics." *Annual Review of Anthropology* 35: 191–208.

Eliade, Mircea. 1974. "Archaic Myths and Historical Man." *Philosophy of Religion* ed. Norbert O. Schedler. New York: Macmillan. 60–70.

Gorer, Geoffrey. 2005 [1938]. *Himalayan Village: An Account of the Lepchas of Sikkim*. Kathmandu: Pilgrims Publishing.

Government of Sikkim, Law Department. 1984. *Sikkim Code Vol. III. Old Laws of Sikkim*. Gangtok: Law Department.

Gowloog, Rip Roshina. 1995. *Lingthem Revisited: Social Changes in a Lepcha Village of North Sikkim*. New Delhi: Har Anand Publications.

Hogan, Linda. 2014. "We Call It Tradition." In Graham Harvey, ed. *The Handbook of Contemporary Animism*. New York: Routledge, 17–26.

Kotturan, George. 1983. *The Himalayan Gateway: History and Culture of Sikkim*. New Delhi: Sterling Publishers Pvt. Ltd.

MacDonald, David. 1930. *Touring in Sikkim and Tibet*. Delhi: Asian Educational Services.

Morris, John. 1938. *Living with the Lepchas: A Book about Sikkim Himalayas*. London: Heinemann.

Peyton, Henry H. 1969. "Myths and Legends." *Interpretations* 2 (1): 31–36.

Rahman, Mirza Zulfiqur. 2020. "The Languages and Imaginations of Monsoon Rains below the Eastern Himalaya" *Australia Himalaya Research Network blog*. https://aushimalaya.net/2020/10/29/the-languages-and-imaginations-of-monsoon-rains-below-the-eastern-himalaya/ Date of Access: October 31, 2020.

Rose, Deborah Bird. 1996. *Nourishing Terrains: Australian Aboriginal Views of Landscape and Wilderness*. Canberra: Australian Heritage Commission.

Rose, Deborah Bird, Thom van Dooren, Matthew Chrulew, Stuart Cooke, Matthew Kearnes, and Emily O'Gorman. 2012. "Thinking through the Environment, Unsettling the Humanities." *Environmental Humanities* 1: 1–5.

Sherpa, Pasang Y. 2014. "Climate Change Impacts among Sherpas: An Anthropological Study in the Everest Region, Nepal." *Habitat Himalaya* 8 (1). http://resourceshimalaya.org/wp-content/uploads/2017/08/Final-HABITAT-HIMALAYA-2014.pdf

Siiger, Halfdan. 1978. "The Abominable Snowman: Himalayan Religion and Folklore from the Lepchas of Sikkim." In James F. Fisher, ed. *Himalayan Anthropology. The Indo-Tibetan Interface*. Paris: The Hague, 421–430.

Smith, Mick. 2005. "Hermeneutics and the Culture of Birds: The Environmental Allegory of Easter Island." *Ethics, Place, and Environment* 8 (1): 21–38.

de Stocks, Beauvoir Cheridah. 1975 [1925]. *Sikkim: Customs and Folklore*. Delhi: Cosmo Publications.

Torri, Davide. 2010. "In the Shadow of the Devil. Traditional Patterns of Lepcha Culture Reinterpreted." In Fabrizio Ferrari ed. *Health and Religious Rituals in South Asia, Disease, Possession and Healing*. Routledge (South Asian Religion Series). 148–165.

van Dooren, Thom and Deborah Bird Rose. 2016. "Lively Ethography: Storying Animist Worlds." *Environmental Humanities* 8 (1): 77–92.

Waddell, Lawrence. 1998 [1899]. *Among the Himalayas*. Kathmandu: Ratna Pustak Bhandar.

Wangchuk, Pema and Mita Zulca. 2007. *Khangchendzonga: Sacred Summit*. Kathmandu: Pema Wangchuk/Hillside Press.

Yazzie, Melanie K. and Cutcha Risling Baldy. 2018. "Introduction: Indigenous Peoples and the Politics of Water." *Decolonization: Indigeneity, Education & Society* 7 (1): 1–18.

3 Eco-spiritual and economic perspectives in Bhutan's Haa district

Thinley Dema

Introduction

Haa, a small district located in the west of Bhutan, is presided over by three mountains, collectively referred to as the "Three Brothers," and individually named *Jampelyang, Chana-Dorji* and *Chenrezig.* From the perspective of my native reverence, they are actually addressed as the Three Deities Mountains. For non-Bhutanese, it may be quite incomprehensible to see mountains as gods and spirits, but for us they are interchangeably mountain deities. These three mountains are not merely three unique physical landforms, but are also the spiritual heart of the *Haaps*, the human inhabitants of the Haa valley. There is hardly any written history about the mountains' origins; however, oral accounts in their variety are the closest, most intimate history we have of these mountains. As this chapter progresses, I trust my readers will soon have the same sense I do regarding these mountains. As terrestrial beings, we are all capable of sensing the livingness of the earth.

I now turn to a widely told folktale about these mountains. Long before the dawn of the Buddhist civilisation in Bhutan, these three mountains were believed to be the physical manifestation of three gods. It is said that the three male entities mysteriously rose from the earth. They were further said to be glowing and their appearance radiated power and authority. One of them held a scroll and was the god of wisdom, *Jampelyang*. This script is the book of knowledge. The second mountain was the god of power given that *Chana-Dorji* held a sword and is today well known for his wrath. The third mountain, *Chenrezig*, had an eye embedded in both his hands and was the god of compassion. Together they are a trinity; later, in the eighth century, *Haaps* considered these deities to be the gods of Vajrayana Buddhism and, in reference to their male forms, *Haaps* referred to the mountains as *meri puensum*: in the native language, Dzongkha, *me* means human, *ri* means mountains, and *puensum* means three siblings.

Besides the folktale mentioned above, there are numerous stories that indicate the spiritual essence of these three mountains. One such story suggests that in the eighth century Guru Rinpoche blessed the mountains and named them after the lords of three families of Vajrayana Buddhism: *Jampelyang*, *Chana-Dorji* and *Chenrezig.*[1] The mountains are also called *Rigsum Gompa,* which can be interpreted as the three temples where each of the three gods resides. As Rose (1996, 7) suggests, "...country is a living entity with a yesterday, today and tomorrow with a consciousness and a will towards life," and *Haaps* consider these three identical mountains with their magnificent heights and replicative physical appearance to be abodes of the Gods and an important part of people's lives (Figure 3.1).

Situated within another larger ethnographic research project,[2] my engagements and interviews with *Haaps* about these three mountains in Haa valley constitute the heart of my investigation for this chapter, in which I explore how *Haaps* see the environment and interact with it accordingly. Eco-spiritual perceptions display the spiritual connections between environment and deities. This chapter initially delves into an eco-spiritual perception of these mountains. In adopting this view, I recognise the presence of spiritual beings and sacredness of nature within Haa. As a corollary, I assess how such an understanding of the environment motivates strategies are appropriate for the environment. While this viewpoint has prevailed for generations, increasingly another perspective has come to compete with this, particularly among younger generations. This I broadly conceive of as the economic perspective. The perceptions of both eco-spiritual and economic realities have their respective ontological manifestations and are also entangled in a relational ontology.

Relational ontology, in this chapter, has a twofold meaning. On the one hand, it pertains to the animist relations of humans, deities and the earth because mountains and deities are interchangeable, and humans and deities have a spiritual kindred bond. Overall, the earth, from the perceptions of both humans and deities, is alive with a strong animist characteristic. As my case study, in principle, is similar to Eduardo Viveiros de Castro's Amazonian case, the relational ontology in Haa Valley in essence is an animist ontology, which "postulates the social character of relations between humans and non-humans: the space between nature and society is itself social" (Viveiros de Castro 2020, 473). On the other hand, relational ontology in my work also signifies the ontological force of economic globalisation, which is pushing our traditional Himalayan lifestyles and values

Figure 3.1 Image of the Three Brothers Mountains.

towards the global standardisation of human economic life or the "one-world world: a world that has granted itself the right to assimilate all other worlds" (de la Cadena and Blaser 2018, 3). In this sense, it is understood as what Arturo Escobar calls the "ontological occupation of the territories" (Escobar 2016, 14).

In the next section, I describe the differences between the eco-spiritual and the economic perceptions. I also draw the perspective of the Bhutanese government into my discussion. In the process, I engage with Escobar's "earth thinking" and "ontological occupation" (Escobar 2016) for the purpose of illustrating the tension between the two divergent perceptions aforementioned. Next, I relate the eco-spiritual stories and ontological meanings of the Three Brothers Mountains, particularly based on the narratives from the elders in the Haa Valley. In the last portion of this chapter, I discuss, through examples, how the local eco-spiritual perception is now simultaneously being challenged by the neoliberal logic of economic development and being transformed into a financial motivator for environmental conservation.

A spiritual landscape

Haaps conceive of the mountains as the abodes of *yul-lha* (local deities) and *kay-lha* (birth deities), of lakes as inhabited by *tshomem* (lake deities), of the surface land as belonging to *nep* (local deities) and *lu* (dragon-like water deities), and of cliff as belonging to *tshen* (personal guardians and cliff deities), while *dhe* (evil spirits) inhabit unholy areas. As such, deities reside in many natural elements (Pommaret 2004; Allison 2015). The mountains are named after enlightened gods but the region is also inhabited by deities who are unenlightened beings, like humans. They are immortal beings who have power to grant wishes like gods and, thus, they are between the human and enlightened worlds. Under appropriate conditions, these deities and spirits of the landscape protect the land, the people and their faith (Spoon 2010) but they can also harm people if they are offended.

Haaps believe the trees on the mountains have a strong religious meaning and are inhabited by deities. The trees are traditionally respected due to the presence of spiritual beings and such respect has led to high biodiversity within what otherwise runs the risk of turning into a degraded environment (Schaaf 2003). Trees are anthropomorphised and show a relational significance with humans. The story of the Lord Buddha's life suggests the importance of trees as all four major events associate with his life occurred underneath trees. The first event was when Lord Buddha was born, second, when he received enlightenment, third, his first teaching and fourth, his death. In 2004, a Royal decree was issued by Her Royal Highness *Ashi* Kesang Choden, the Queen Grandmother of Bhutan, to conserve trees in Haa Valley because, according to the Buddhist precepts, it is an unethical deed to cut down trees as they are sentient beings and, therefore, deserve moral consideration from humans.

Aup Chundu is the major local deity of Haa Valley. Historically, he was a demon who feasted on the hot blood of animals that was regularly sacrificed by the locals who acted out of fear.[3] This fear continued until *Chana-Dorji* and *Guru Rinpoche* subdued him with the introduction of Vajrayana Buddhism and ordered

Figure 3.2 The white temples in Haa.

Figure 3.3 The black temples in Haa.

him not to corrupt or harm any living being in the area.[4] He submitted and promised to the two enlightened beings that he would look after every being present in Haa. In the eighth century, when Guru Rinpoche visited Haa, he discouraged general animal sacrifice to *Aup Chundu*; however, *Haaps* continued their practice of animal sacrifice out of fear that *Aup Chundu* would punish the residents with diseases and natural calamities.

These narratives express a relationship between deities, traditional values and the natural environment. In the three mountains, there are deities and divinities that are present to guard the region. Because of that, these mountains are home to various ancestral temples and a place of residence for countless legendary deities. Haa is known for its grand stories about how the Black and White Temples were constructed, right at the foot of the Three Brothers Mountains. These two temples figure centrally, and are also constitutive of the spiritual landscape of the valley. A 39-year-old woman narrates this story:

> We believe that *Lhakhang Karpo* (White Temple) and *Lhakhang Nakpo* (Black Temple) was part of the 108 stupas which were constructed by King Songtsen Gampo in the sixth century, with a statue of Maitreya Buddha[5] placed in the main halls to subdue the demon that was bothering the people. Many such temples were constructed in Bhutan to subdue the huge demon and these temples were part of those 108 temples (Figures 3.2 and 3.3).

Another version states that a long time ago, a black bird and a white bird flew to Haa Valley from Tibet. The black bird landed on the upper side of the valley and the white bird on the lower side. In these respective locations, the black and white temples were constructed corresponding to the colour of each bird. Another story states that the local deity, *Aup Chundu*, became offended because the people did not worship him as a deity, and because of this, his face turned half white and half black. In order to appease the local deity the *Haaps* constructed the temples. These stories represent the history of Haa that has been passed on for generations and that culminates in stories that influence peoples' thinking and livelihood. The eco-spiritual perceptions of the people have been built upon these oral tradition that has been recited to them by their elders, even during the post-Buddhism era. A 34-year-old woman recited a story about a *lu*:

> There is a story about a *lu* in the Three Brothers Mountains, and a man. Once there was a man who had cut down trees and taken out all the rocks from the mountains. This disturbed the *lu* and the man ran away from Bhutan to Tibet because he had offended the *lu*; but the *lu* followed him to take revenge. The *lu* transformed itself into a leaf and was able to get a hold of the man. My mother narrated me this story and she said that the moral of the story was that we should never harm the *lu* and its environment because no matter where we go it will find us and take revenge.

This story tells us that *lu* can be offended when their surroundings are polluted or disturbed. They can also be revengeful and will punish the people who pollute the environment (Ura 2001). This story plays a major role in how *Haaps* relate to the deities during their daily life and it has also influenced how *Haaps* interact with the areas that have spiritual forces residing within them. The spiritual value of the Three Brothers Mountains have helped guide the consciousness of *Haaps*, which in turn has created particular perceptions of the environment.

Haaps' knowledge of the landscape and their capacity to co-become with their local deities in the shared environment is important for mutual survival. The knowledge that they have about their surrounding makes them capable of working and operating appropriately in the environment. The spiritual forces of the deities cannot be seen but if a person has the spiritual knowledge of the environment, he or she will be able to live in that area peacefully without fearing the deities. These stories influence the way the locals behave near areas surrounded by deities.

Local deities and stories

In this section, I investigate the relational ontology between the spiritually understood physical environment and the *Haaps*. In 2013, the majority of *Haaps* decided to end the age-old tradition of animal sacrifice after consulting with relevant authorities and the local deities. There were few who were against this decision so in order to persuade them a dice rolling ceremony was conducted to gain approval from *Aup Chundu*. *Haaps*' deities approved with doing away with animal sacrifice and his holiness *Je Khenpo*, the chief abbot of the Central Monastic Body of Bhutan, officially ended the age-old tradition. A 50-year-old man said:

> We have stopped sacrificing yaks while conducting rituals for *Aup Chundu*'s *selkha* (offering). However, we still offer meat during the rituals because the last time we released the sacrificial yak, that same yak jumped off a cliff and died during the ritual.

After this event, *Haaps* decided to conduct a *Bho-yak* ritual as a tribute to *Aup Chundu*. This ritual was meant to satiate his taste for meat and blood. During one of my visits to Haa, I attended this ritual celebration that is conducted in April. The ritual begins at two-thirty in the morning in *Lhakhang Karpo*. The head monk takes a sacred white scarf from the statue of *Aup Chundu* and places it on the saddle of his white horse. *Haaps* believe that *Aup Chundu* actually rides the horse from *Lhakhang Karpo*, and the sacred white scarf symbolises the deity himself. When the horse reaches the celebration ground, it makes three rounds of the grounds without anyone guiding it. It then goes to the middle of the grounds where it jumps and neighs loudly. This event strengthens the community members' belief in the deities. Afterward, meat brought from a shop is offered to the deities in the middle of the temple grounds.

The narratives of the abovementioned rituals thus reveal that there is a strong connection between the landscape and religion in Haa. They also show the shift in the eco-spiritual perceptions of the *Haaps*: in the past, animal sacrifice was considered to be mandatory to appease the deities but now rituals for the protection and the well-being of the community and the strengthening of the bond between *Haaps* and the deities are conducted without sacrificing animals. According to Pommaret (2004, 49) the term by which the deity is addressed shows that there is a sense of bond, even of kinship, between the people and the deities. *Haaps* addresses their local deity as *Aup Chundu*, with the term *Aup* translated as "father," showing the intimate relationship between the deities and locals. This familiarity connotes the idea of kinship with the deities and directly links *Haaps* with the deities. They believe that *Aup Chundu* is the father of the people of Haa (Pommaret 2004, 49) and say that he protects children born there. Though he was long regarded as a Dharma protector, *Haaps* to the requirement to sacrifice animals for him every year meant he was not taken seriously; after *Aup Chundu* accepted the cessation of animal sacrifices, *Haaps* really considered him as having become a true Dharma protector. The majority of *Haaps* said that they were ambivalent or uncomfortable with the practice of animal sacrifice, as they had to eat the meat of the freshly killed animal to honour the deity. Therefore, when the practice of animal sacrifice was stopped, they felt like *Aup Chundu* understood them and that he was willing to sacrifice his own desires to please his children.

A 29-year-old woman, referring to *Lhakhang Karpo*, narrated the importance of the temple and the rituals that are conducted to appease the deities and ask for their protection of the community. *Aup Chundu* may have accepted the discontinuation of animal sacrifice but the conducting of rituals is still mandatory. She narrated thus:

> I remember last time we could not conduct the annual community ritual because the temple was under reconstruction. At that time there were unusual weather conditions which shocked everyone. So this time we all decided to conduct the *Bonko* ritual to appease the deities.

In 2009, *Lhakhang Karpo* was reconstructed for the National Day celebration so the locals were unable to conduct the annual community rituals to please the deities who showed their displeasure in the form of natural calamities. In the past, people used to sacrifice yaks during the ceremony but now this tradition has been discontinued. Other rituals, such as *Lochoe*, that assures families' well-being, and *Gyalpo*, in honour of the King of the deities of the local area, are conducted at the household level and are performed by a spiritual master (*phajo*) and a practitioner who can communicate with the dead and with the gods (*pawo*), to honour families' household deities. These various rituals have been conducted by *Haaps* for generations and have become a tradition to ensure their well-being and to please local deities. This eco-spiritual perception of the environment, supported by religious belief, has been a major influence on human–environment interactions and is visible in the daily lives of the people.

Haaps' knowledge about the area and their capability to interact with the deities helps them in their survival. They conduct rituals, keep their surroundings clean, and they offer flowers as a symbol of peace so that the deities will grant wishes and protection. *Haaps* fear that if they stop conducting rituals, the deities will be offended and create manifold problems for them. Additionally, houses in Haa Valley have small white prayer flags on the rooftops to indicate that the house has given offerings to appease the local deities and has conducted their rituals. During the rituals, locals offer butter, cheese, flowers, milk, or clean water in the area inhabited by deities to please them.

Haaps are able to identify the personality and preference of each deity present in their surroundings. They believe that *lu* are easily offended if offered eggs, meat and alcohol, and that offense manifests as joint pain, skin diseases and dreams about snakes. Other deities such as *tshen* would be pleased to be offered alcohol, but if *tshen* are otherwise offended, people will have headaches, feel nauseous and will vomit. These characteristics were often mentioned during our fieldwork and pointed to the different personalities of the deities. A 48-year-old monk commented, "When we get sick, we know that the deities are offended, and when we have strange dreams of snakes it shows that *lu* are angered by our actions."

This comment shows that dreams are of interest to *Haaps* and they experience them to be a source of creativity and knowledge that is crucial to them as they can portray deities' teachings and judgments. As Bhutanese, we believe that dreams can indicate the past, present and future of a person depending on their situation and experiences. There is also a spiritual dimension to dreaming so dreams can heal, teach, warn and guide us (Holbeche 1998). In the highest form of Bhutanese Buddhist practices, the monks use the dream state for spiritual practices. Dreams reveal our own personal mythology, which is the base from where we perceive and experience the world around us. Dreams about snakes are a type of prodromic dream which gives *Haaps* a chance to act before they become sick. *Haaps* say that they dream of snakes when they fail to keep the surroundings clean or when they do not conduct rituals on an annual basis to appease the *lu*.

A well-known story mentions a young girl who kept on having nightmares about snakes over the course of a year and who became seriously sick. She was taken to the hospital but doctors were unable to cure her so they consulted a monk who conducted an astrological reading and performed a ritual known as *Lu-choe*, after which she was cured. *Haaps* experience dreams as a bridge between the human and spiritual worlds by which deities communicate with humans.

Deities and environmental conservation

Haaps suggests the hills, mountains and forests that surround them are the abodes and citadels of manifold deities that preside over and protect the area and its natural resources (Pommaret 2012; Ura 2001). In this section, I will discuss people's perceptions and experiences of the landscape that culminate in *Haaps*' motivation and behaviour that works to conserve the environment.

The Royal Government of Bhutan (1998a) has proclaimed that the practice of Vajrayana Buddhism is responsible for the pristine forests and mountains which are conserved and recognised as sacred areas, as it interlinks religion and traditional beliefs with nature conservation. Strong belief in rituals, religion, stories and fear of offending deities played a major role in moulding the way *Haaps* interact with the environment. According to Ura (2001), cultural and spiritual beliefs of the locals motivate them to protect certain landscapes as they are linked to the invocation of deities. Having knowledge of the environment can introduce new perceptions of that environment. A 57-year-old woman commented, "If we know that the area is inhabited by deities, then we feel the need to keep the surrounding clean out of fear and respect for them."

This statement shows why *Haaps* feels the need to keep their surroundings clean. The existence of deities in the environment and fear of angering them creates an ideology of spiritual essence. Locals fear that if they fail to keep the surrounding of the deities clean then the deities will get offended and will harm the community. Deities are like neighbours: if we throw garbage in our neighbour's house then the neighbour will get angry and fight with us. Similarly, in case of deities they show their displeasure in the form of causing sickness and natural calamities in the community. The fear of getting sick or having unpredictable climate prompts the locals to maintain their surroundings and conserve the environment. When they get sick, they conduct the rituals recommended by spiritual masters and they claim to be cured after the prescribed rituals. A few rituals are to be conducted annually, and failing to conduct them bring natural calamities.

The presence of deities, divinities and spirits in the mountains has influenced *Haaps* to worship them and it has resulted in strategies that conserve the forests and other so-called resources. As such, their recognition of the sacredness of the land has promoted conservation of the area as well. The use of the natural resources from the mountains is considered inauspicious. *Haaps* claims that not only do they respect their religion, but they also live in mythical areas along with spiritual beings. Their culture and beliefs manifest in shared values that make it possible for a conservation ethic and practice to emerge. Bhutan's Biodiversity Action Plan of 1998 states that "Traditional and local beliefs promoted the conservation of the environment, and key ecological areas were recognized as the abodes of gods, goddesses, protective deities and mountain, river, forest, and underworld spirits" (RGoB 1998b, 19).

Haaps believe that if someone harms the environment then they will fall sick because the deities protecting the environment become insulted and angry. Locals believe that they should not pollute the rivers, lakes and air, as there are deities inhabiting such areas. Behaviour and actions of people change when they are in the environmental force field of the deities. An eco-spiritual perception of the mountains influences eco-friendly strategies to conserve the environment. Therefore, the government does not have to implement strict policies to enforce conservation behaviour. People's beliefs and the fear of the unknown spiritual world influence the behaviour of people who dwell in spiritual surroundings.

The rise of instrumental orientations towards the environment

In 2002, the government started a road construction project from Haa Valley to Takchu Gompa, a temple located behind the Three Brothers Mountains. As soon as the road pavement work began the machines broke down for no logical reason. The contractor of the project was a religious person so he conducted an astrological reading, called *mo*, and found that a deity inhabited the areas surrounding the construction site. The government officials conducted rituals for the deity that allowed the machines to work so they could begin construction smoothly. People say that they cannot see deities, but they were able to witness the true power of the deities during this 2002 road construction project.

Development projects and economic growth in the valley have, however, affected the way *Haaps* perceive the natural environment, and are thus reshaping the way they relate themselves to the sacred areas. This narrative portrays tension between the deities and the human-centred economic growth in Haa. *Haaps* believe that the deities were interfering in the road construction by showing their resentment towards economic development. The presence of deities and people's belief in them has affected the way *Haaps* think about modern development (Allison 2004). Globalisation and economic development can undermine the value of sacred mountains and threaten to diminish religious and environmental ethics (Bernbaum 2010). Globalisation has negatively affected not only the physical environment of the mountains but also the way locals view the sacred mountains (Verschuuren 2010). Additionally, due to development people are exposed to different ideas which make them question the existence of deities thus degrading their experience of the sacredness of the mountains.

Rapid development has started to change the perception the people have of the environment. Areas that were once considered to be inhabited by deities are being exploited by locals because they claim that the deities are not harming them. The locals' strong faith in deities is slowly fading away as there is intrusion from economic activities. Due to minimal development efforts in the past, the people of the region have begun to target possibilities for improving local livelihoods, subsequently neglecting the presence of the deities in the area. When *Haaps* observe that the deities are not harming the people for their misdeeds the locals becomes cynical about the presence of the deities thus they start to exploit the natural resources. They have started seeing economic potential in the Three Brothers Mountains and started utilising the environment to promote their economic business in many ways, including logging timber.

Now *Haaps*' environmental perception is undergoing changes, as the ongoing economic development is progressively eroding the traditional belief in the environmentally embodied local deities. Increasing global demands for natural resources has put economic pressure on mountainous areas and their associated sacred sites (Bernbaum 2010). Interaction with economic sectors has led to a change in people's belief system and the way they perceive the environment. The older generations carry the knowledge of the local practice related to deities (Allison 2004) and to local ecosystems, but the youth are slowly deconstructing the traditional belief system and economic improvement is being given priority in a way that intensifies inter-generational differences in the Haa community. The younger generations are not familiar with the practices and requirements, thus losing touch with their ancestral traditions and beliefs (Allison 2004, 454). *Haaps*' knowledge and perception of the spiritual landscape has been replaced with an eye towards the marketable benefits of the landscape, thereby promoting the economic perception. This is shown by a comment from a 21-year-old, homestay-owner's son who was brought up in an ecotourism-operating family that granted him a different perception of the environment compared to the older generation. He commented:

> I feel that as a youth we have lost the connection with the spiritual environment … I can see the difference in how we youth and how our grandparents perceive the environment … our parents and grandparents lived in a different era with a different belief system and they had different living styles where they used to work in field and herd yaks.

This statement is the perfect way to represent the inter-generational differences as it shows that in the past, the elder *Haaps* were involved with yak herding and agriculture but now the younger generations are involved with a service-based lifestyle. The younger generations are at the peak of social transition where the community is shifting from farming and herding way of life to more luxuries life of tourism. *Haaps* in their thirties and forties are the last generation to work in the fields and herd yaks, and the introduction of new market advantages has shifted the community's perception. The younger generations have lost the deep natural connection that the older generations have with the landscape that was gained over a lifetime of experiences.

Due to ongoing development, there is a tension in Haa Valley between material wealth and spiritual wellbeing. Tourists mainly visit Haa Valley for pilgrimages and to know more about the place. Tourism has also encouraged the locals to keep the areas surrounding the houses clean and to maintain the trekking path to sacred lacks and temples. A 25-year-old woman who works in the homestay business said: "Tourism is increasing in Haa and I feel that in order to keep it up we need to advertise and utilize the available natural resources."

Haaps need to conserve the environment to attract tourist for their own benefit. As traditional belief systems fade away with modernisation, *Haaps* have started to perceive the environment as a potential source of income. The shift in the ideology related to the environment has led to a reconstruction of people's views on the environment. The desire to earn has influenced *Haaps* to utilise the natural environment in novel ways.

Among the various forms of economic pressure resulting from global change, tourists have the greatest and most widespread impacts on sacred mountains (Bernbaum 2010). The tourists who visit Haa Valley to explore the sacred places and to hike around the Three Brothers Mountains have disrespected the essence of the area by polluting it. In her work in eastern Bhutan, Karst (2016) reports that ecotourism fostered discord in human relations while nature was seen as a commodity through which the sector could be used to harness its economic value. A 24-year-old woman commented: "Tourists are not responsible tourists, they throw garbage everywhere, and they generally don't respect our natural environment; they do not have respect."

Tourists hike to lakes and mountains to see scenic views, but these activities impact the natural environment. When people travel, they produce waste along the way which tends to pollute the environment. *Haaps* say that when the tourists disrespect the local deities by throwing out their garbage and smoking in areas which are considered to be sacred, the deities get offended and show their displeasure in the form of weather events, or by affecting the community. Locals state that due to the bad deeds of the tourist they have to face the consequences.

Neoliberal logic

The neoliberal logic found in the Bhutanese case is a paradox of economic development with two seemingly opposing poles. On the one hand, it "proposes that human well-being can best be advanced by liberating individual entrepreneurial freedoms and skills within an institutional framework characterized by strong private property rights, free markets, and free trade" (Harvey 2005, 2). This aspect of neoliberalism produces an effect of what Harvey (2005, 3) calls "creative destruction, not only of prior institutional frameworks and powers but also of divisions of labour, social relations, welfare provisions, technological mixes, ways of life and thought, reproductive activities, attachments to the land and habits of the heart." This means that while the *Haaps* encounter the neoliberal economic changes erosive to their traditional values, they also embrace it as an opportunity for a cash income generated by ecotourism that depends on their maintaining the environmental integrity of the deities' mountains. A gradual increase in ecotourism and its resulting economic benefit appear to be reshaping perceptions of land-use among locals, which has influenced decision making regarding natural resources (Spoon 2010). The human–environment relationship has shifted after the establishment of new production relations. According to Montes and Kafley (2019), "Ecotourism, and its associated neoliberal logic, produce a redistribution of power in which humans are no longer seen as inferior to the land and associated deities, but are elevated to a place of dominance."

This economic progress exists at the cost of the spiritual well-being; however, there are a few economic sectors that coexist with the deities. Not all the economic perceptions have negative impact on Bhutan's spiritual landscape. According to Bernbaum (2010), ecotourism can be used as a tool to aid the conservation of sacred natural sites. *Haaps* have started seeing value in the instrumental evaluation of the environment that the tourism sector promotes, rather than granting the land an intrinsic evaluation. One of the homestay owners in her in her thirties commented: "We need to conserve our environment because most of the tourist comes for our village to site see and to experience the natural environment."

Now the mountains serve as a magnet to attract tourists, and locals have used these sacred mountains to propel their business ventures (Denkar 2018). These holy mountains are seen as a marketable product to promote ecotourism, with the resulting change in view affecting the behaviour of the society in Haa Valley. Most of the locals had indicated that they recognise the need to conserve the natural environment for its economic benefit since most tourists visit Haa Valley to see and experience the "pristine" or "untouched" environment. In the previous years, they conserved the environment due to societal values or ethics, but now the underlying motivation for conservation is based on the economic value granted to resources (Montes et al. 2019).

Ecotourism serves as a sustainable way of earning money, provided it is adequately managed. In Haa, ecotourism has served as a method to avoid extraction of resources without hampering the continuation economic growth. *Haaps* also use the stories of deities associated with the mountains to attract tourists and therefore indirectly protect the natural areas. A local woman in her twenties stated: "After we established homestays, there was increase in the number of tourist, so now there are lots of changes (for locals) in terms of cleanliness and income." Another woman in her forties commented: "When tourists come to visit Haa it encourages us to keep the forest and surroundings clean so that we make good impression on them to visit Haa again."

In the past, *Haaps* were involved with yak herding and agriculture, so sanitation in their homes was poor. With the introduction of ecotourism, household and personal hygiene improved drastically, besides the generation of additional income. The introduction of ecotourism as an economic activity has encouraged locals to revere the deities. Now locals have guests visiting their valley constantly so they have to clean the house every now and then. Due to the shift in the working sectors the locals no longer have to borrow money from each other because they have their own homestays to earn extra income.

Conclusion

This chapter discussed and illustrated the inter-generational ontological conflict that is currently emerging in the Haa Valley. It also highlighted the history and religious practices in Haa and showcased how these played a major role in shaping a particular ideology of the environment. The traditional eco-spiritual perception helps express the relationship between traditional values and the natural environment, whereas the economic perception of the environment emphasises the existing economic potential of the mountains in Haa Valley. In the article "Thinking-feeling with the Earth: Territorial Struggle and the Ontological Dimension of the Epistemologies of the South" Escobar talks about social transformation due to a change in ontological occupation and, based on this idea, I illustrated a similar case of change in the ontological occupation in Bhutan, arising due to the different world view of the generations.

There is a generational difference in people's recognition of the forces that occupy the forest. The older generation of *Haaps* has a strong ontological relationship with the spiritual landscape, but the younger generations increasingly seem to have only an economic perception of the environment. The older generation views the ontology of Haa in terms of spiritual presences governed by history and cultural belief system. Although the mountains have been worshipped until today, the younger generation has lost touch with their histories and thus questions the belief system of their elders due to the influence of economic development. Culture and religious beliefs previously limited the utilisation of natural resources, yet there are recent cases where people have exploited the mountains. The strong, ontologically bonded relationship between the deities, humans and the environment is shifting as the younger generations are exposed to nontraditional, modern economic worldviews and are finding financial potentials in their native mountain environment.

Acknowledgements

I would like to express my gratitude to the host Aum Chador and Dr. Sherub for supporting the interviews conducted in Haa. I would like to earnestly thank Dr. Jesse Montes for giving me the opportunity to be part of the project and Dr. Jelle J. P. Wouters for the valuable guidance and feedback which has helped me in completing this chapter.

Notes

1 Guru Rinpoche is the Buddhist master Padmasambhava, who brought Buddhism from India.
2 This ethnographic research is a part of a larger research project titled "Ecotourism and Neoliberal Logic." The ethnographic fieldwork was conducted through semi-structured interviews, participation observation, the taking of field notes, and focus-group discussions with a total of five members who spent 2 weeks in Haa Valley.
3 The locals used to sacrifice animals such as yaks, oxen, pigs and bulls.
4 Before Vajrayana Buddhism was introduced, people followed the Bon religion, a group of common pre-Buddhist religious practices that were animistic in nature.
5 Maitreya Buddha is known as Jowo Jampa, and is the Buddha of the future.

References

Allison, Elizabeth. 2004. "Spiritually Motivated Natural Resource Protection in Eastern Bhutan." In *The Spider and the Piglet: Proceedings of the First International Seminar on Bhutan Studies*, edited by Ura, Karma, Kinga, Sonam, 529–563. Thimphu: Centre for Bhutan Studies.

Allison, Elizabeth. 2015. "Religion Inscribed in the Landscape: Sacred Sites, Local Deities and Natural Resource Use in the Himalayas." In *The Changing World Religion Map*, 439–459. Heidelberg: Springer.

Bernbaum, Edwin. 2010. "Sacred Mountains and Global Changes: Impacts and Responses." In *Sacred Natural Sites: Conserving Nature and Culture*, edited by Verschuuren, Bas, Wild, Robert, Mcneely, Jeffery, & Oviedo, Gonzalo, 33–41. Switzerland: IUCN.

de Castro, Eduardo Viveiro. 2020. "Cosmological Deixis and Amerindian Perspectivism." In *The Journal of the Royal Anthropological Institution*, 4 (3): 469–488.

De la Cadena, Marisol, and Mario Blaser. 2018. *A World of Many Worlds*. Durham: Duke University Press.

Denkar, Tshering. 2018. Must Hike Meri Puensum Nature Trail in Haa Denkars Getaway. https://denkarsgetaway.com/meri-puensum-nature-trail-haa/, accessed April 20, 2019.

Escobar, Arturo. 2016. "Thinking-Feeling with the Earth: Territorial Struggles and the Ontological Dimension of the Epistemologies of the South." *AIBR, Revista De Anthropoligia Iberoamericana*, 11(1), 11–32. doi:10.1156/aibr.110102e

Harvey, David. 2005. *A Brief History of Neoliberalism*. Oxford: Oxford University Press.

Holbeche, Soozi. 1998. *Dreams*. London: Thorsons.

Karst, Heidi. 2016. "This Is a Holy Place of Ama Jomo: Buen Vivir, Indigenous Voices and Ecotourism Development." *A Protected Area of Bhutan. Journal of Sustainable Tourism*, 25 (6): 746–762.

Montes, Jesse, and Bhuwan Kafley. 2019. "Ecotourism Discourses in Bhutan: Contested Perceptions and Values" In *Environmental Governance in Bhutan: Ecotourism Environmental and Cosmological Subjectives*, 77–96. Netherlands: Wageningen University.

Montes, Jesse, Bhuwan Kafley, Dilisha Subba, Thinley Dema, Tashi Dendup, and Pema Selden. 2019. "Ecotourism and Social Cohesion: Contrasting Phobjikha and Laya Experiences". *Rig-Tshoel*, 2 (1): 23–33. Bhutan: Rig-Tshoel.

Pommaret, Francoise. 2004. "Yul and Yul Lha: The Territory and its Deity in Bhutan" In *Bulletin of Tibetology*. 40: 39–67. Sikkim: Namgyal Institute of Tibetology.

Pommaret, Francoise. 2012. "On Local and Mountain Deities in Bhutan." HAL. Österreichische Akademie der Wissenschaften. https://halshs.archives-ouvertes.fr/halshs-00717941, accessed May 18, 2020.

Rose, Deborah Bird. 1996. *Nourishing Terrains: Australian Aboriginal Views of Landscape and Wilderness*. Canberra: Australian Heritage Commission.
Royal Government of Bhutan. 1998a. *The Middle Path: National Environmental Strategy for Bhutan*. Thimphu: National Environment Commission.
Royal Government of Bhutan. 1998b. *Biodiversity Action Plan for Bhutan*. Thimphu: Ministry of Agriculture.
Schaaf, Thomas. 2003. "UNESCO's Experience with the Protection of Sacred Natural Sites for Biodiversity Conservation", In *The Importance of Sacred Natural Sites for Biodiversity Conservation*, edited by Lee, Cathy, and Schaaf, Thomas, 13–20. Paris: United Nations Educational, Scientific and Cultural Organization
Spoon, Jeremy. 2010. "Sacred Natural Sites: Mutual Learning, Analysis, Planning and Management." In *Sacred Natural Sites: Conserving Nature and Culture*, edited by Verschuuren, Bas, Wild, Robert, Mcneely, Jeffery, & Oviedo, Gonzalo, 87–97. Switzerland: IUCN.
Ura, Karma. 2001. "*Deities and Environment: A Four-Part Series*." Thimphu: Centre for Bhutan Studies.
Verschuuren, Bas. 2010. "Arguments for Developing Biocultural Conservation Approaches for Sacred Natural Sites." In *Sacred Natural Sites: Conserving Nature and Culture*, edited by Verschuuren, Bas, Wild, Robert, Mcneely, Jeffery, & Oviedo, Gonzalo, 62–72. Switzerland: IUCN.

Part II

Aqueous earth

4 Narratives from a fluvial world

Poetics of charland dwelling in neo-colonial Assam

Bhargabi Das

Introduction

In June 2019, when I started my fieldwork on Ramapara Pam, a type of seasonal riverine island called a char, in Baghbar Circle, Barpeta district of Assam, India, it was the beginning of monsoon and hence the time for the annual floods. During this time of year, the Brahmaputra River and its numerous tributaries that flow through Assam inundate the floodplains. Many chars become completely submerged. In Baghbar Circle alone, there are around 86 chars, out of which almost all are under human habitation. Chars, or charlands, are river islands created by activities of tropical rivers that are formed when a river's velocity decreases once it reaches the plains as a result of the massive amount of sediment that increases the bed load. The continuous sediment deposits create small islands and give the river a braided appearance. Reeds and tall grasses soon begin to grow on the fertile chars. The Brahmaputra has the fourth largest average discharge of any river in the world and the third largest sediment yield (Chakraborty 2018, 180). This, along with the fact that Assam falls under a highly active seismic zone (Zone V, the highest under the Bureau of Indian Standards) has led to the creation of several chars in the region. In fact, Ramapara Pam (my field site) was created after the 1890 and 1950 earthquakes in Assam that led to the change of the course of the Brahmaputra and the rise of the riverbed causing more braiding, more devastating floods and char formation. To be noted, the word "Pam" means a space for seasonal cultivation. The inhabitants of this char previously lived in Ramapara Gaon but during the 2014 floods, the entire char submerged and they shifted to nearby Ramapara Pam.

As chardwellers face the continuous erosion of chars caused by river currents in addition to the seasonal floods, they shift to different chars when they lose their houses and lands to erosion. Most of the people living in Ramapara Pam have moved at least ten times. This continuous relocation caused by the river has not only made chardwellers economically pauperised but also raised questions for me regarding their ideas of home and belonging. What is their idea of home? Do they face conflicts regarding belonging, be it to a community or the nation? These questions were largely what I discussed with my respondents during my first month of fieldwork. On one rainy day, when I was waiting for the next ferry on the *ghat* (boat dockyard) after I had finished a group discussion, Ashraful (name changed), my translator of the Miya language spoken by inhabitants of the char, quipped in "Baideo [madam], have you noticed that the river is slowly eating away at the char. The last time you came, the *ghat* was further ahead but now it is almost at the gates of my hut." I nodded and asked him if he would shift his hut. He said that he has to. In his lifetime, he had already moved 17 times and this would be the 18th. So, when I asked if he then has no affection for the river, he looked up and broke into a Bengali folk song:

> Oh my Brahmaputra River,
> You have no mercy towards people
> You tear apart nations and villages
> And also our sandy chars…

"The river," he said "is like our lover. It frustrates us, feeds us, destroys us, and nurtures us. It is giving and arrogant but how do you really hate your lover?" The way chardwellers like Ashraful talk about the river endows it with an active and agentive characteristic ("eating away," "a lover," etc.). Hence, I take up cultural narratives like folksongs, myths and festivals prevalent in chars in Barpeta district of Assam to argue that, traditionally, chardwellers of Assam shared a relational understanding of their environment and saw their lifeworld as interconnected with other lifeworlds, in what Arturo Escobar calls a "pluriverse" (2016, 20) or a world of many worlds (Blaser and de la Cadena 2018, 4). And yet from the responses of my participants in the charlands, one could sense a strange anxiety of not feeling able to belong to a community or the nation. This, I argue is one of the many consequences of colonial and post-independence neo-colonial state policies, and is evidence of their impact on chardwellers' traditional relationship with the environment. Finally, this chapter examines how the chardwellers' agentive responses to state policies have catalysed the Miya poetry movement that endeavours to bring back collective memories of chardwellers' lived experiences with land and water, to break the ascribed, hegemonic notions of their cultural identity, and to challenge the mainstream Assamese society's and the state's cultural and political representation of the chars and their people.

The chapter is written as a hybrid text of cultural anthropology, environmental humanities and political ecology given the fact that human–nature relations among the chardwellers are both ontologically affective on a personal basis and politically contentious as public issues. Anthropologically, I see myself as a native Assamese anthropologist who trained in Europe; however, I do not see my simultaneously being a native and an anthropologist as a dichotomy or a professionally sanctioned taboo as if "going native" meant "losing all sense of distance or objectivity" (O'Reilly 2009, 12). Perhaps, a nonnative anthropologist might find it easy to practice this artificial separation of native and nonnative; however, for a native anthropologist, this warning sign sounds more like shutting down the voice of one's native lived experience and worldview. I have no intention to throw myself into this artificial dichotomy; instead, I see my practice of writing culture amalgamating both my native experience and my disciplinary training, along the lines of what Kirin Narayan calls "the enactment of hybridity" signifying "writing that depicts authors as minimally bicultural in terms of belonging simultaneously to the world of engaged scholarship and the world of everyday life" (Narayan 1993, 672). I then situate my ethnographic narratives in environmental humanities. In particular, I find ontological and intellectual comfort in the texts of Deborah Bird Rose regarding the Australian Aboriginal notion of Country as "nourishing terrains" and "a living entity," which "gives and receives life" and as "a matrix of communicative interaction, a system of connectivities and benefits, a home on earth" (Rose 1996, 2 and 7; 2014, 139). Re-sounding Rose's narratives, chars in Assam are nourishing riverine terrains that are both creative and destructive as the forces of nature move through them and into the psyches of their human dwellers. The forces of nature are not things in themselves but take shapes in all lives. Ecology then is not merely a matter of nature but is exceedingly political when humans divide, harness and consume nature's nourishment as sustenance and nature's omnipresent energy in all its varieties as power of all sorts in the human world. In the last section of the chapter, my environmental humanism shows its political nature in ecological terms. In the conclusion, I will share with readers what environmental humanities means to me for this particular research project (Figures 4.1 and 4.2).

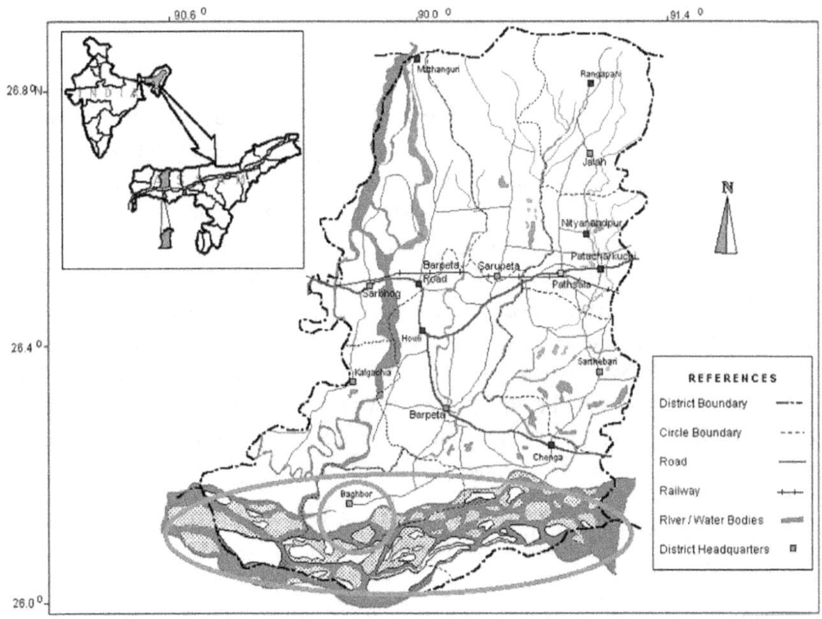

Figure 4.1 Barpeta district map, Assam, India (Talukdar and Kalita 2008, 213). The bigger circled area indicates all the char areas in the district and the smaller circled area shows the location of my field.

Figure 4.2 Ramapara Pam char (my field).
Courtesy: Author.

Chardwellers' cultural narratives

Until the nineteenth century, the chars in Assam had been mostly used by the Assamese Hindu population who lived along the riverbanks in close proximity to the islands for temporary cultivation during the winter when the water would recede. Some chars in Assam were also inhabited by tribal population like the Mishing who lived in houses on stilts called *chang ghar* to keep themselves above the floodwaters. People also used to grow crops like mustard, poppy, sugar cane and rice in limited quantity. There were seasonal visitations by Nepali pastoralists who brought their cattle during winter to feed on the charland grasses. However, facilitated by the colonial state the charlands saw the gradual migration of Bengali Muslim peasants from East Bengal (present Bangladesh) who lived in riverine areas and brought their own agricultural and cultural narratives along with them. The chars in Barpeta, including my own fieldwork site, are now populated by descendants of such peasants from East Bengal and the folksongs, myths and festivals reflect that ancestral past. But these cultural narratives also reflect a close dependence and cohabitation with the environment and, more often than not, factors like land and water are celebrated. This can be seen in the *bhatialigaan*. "While *Bhati* means low-lying land, *ali* means a low mound to prevent flowing of water" (Nayak 2012, 101). These are folksongs that originated in the low-lying lands of East Bengal and are sung mostly by boatmen when they are in their boats out on the river for extended periods. These songs tell of pain, fear and longing, particularly for lovers left behind, while also expressing nostalgia for and fear of the river:

> The river is pregnant with huge waves
> And here I am,
> Sailing this lonely boat
> With no one beside me…

The boat, in fact, is an important gateway to understand the chardwellers fluvial, hybrid existence that will be discussed in detail. For the chardwellers who depend on the river for their livelihood and communication, boats allow them to spend hours in the river and to understand the lifeworld of the river better. During my long ferry trips to and from the chars, I would find chardwellers, mostly men of all ages, rowing various kinds of boats. Most of them are used to either carry saplings or for fishing. Called *haat nao*, these boats are made of rough pieces of wood and are mostly flat at the bottom. Passenger boats are called motor boats or *bhootbhoti*. They are less traditional, with their gasoline-fuelled motors that give them speed and the ability to carry heavy loads. *Bhootbhoti* can also carry four to five motorbikes along with people and livestock. There are different boats for boat races, locally called *naokhel*, too. These are competitive, friendly races participated in mostly by men and the accompanying songs celebrate not just the river but also the boats and the art of using a boat:

> Row the boat fast,
> Shall kill the fatty goat
> And go to the market at Mandiya
> And buy tooth powder…

Unlike the Assamese Hindu people, for whom boats in the pre-British and British days were used by only lower castes whose traditional occupation was fishing and trading, Bengali Muslim chardwellers have traditionally depended on and celebrated boats. In my fieldwork, most people owned a country boat while the well-off owned more than one, with some being larger boats. The prolonged exposure to the river through boat use endowed chardwellers with better knowledge about rivers. On my numerous ferry trips, other passengers would immediately know where a river had the strongest current, where one could find what variety of fishes, how unruly the river could get at a particular time of the day, and how much land can the river erode and within what time. Boats are also used as metaphors in various folksongs to teach valuable life-lessons to chardwellers. Take, for example, this *biyageet* or songs sung during a marriage:

> Oh my lovely daughter,
> When sailing in mid-river the rope of the boat snaps,
> Use the torn loin-cloth,
> Use it as a rope,
> And continue sailing,
> Oh my lovely daughter!

The metaphor of the boat and sailing is used to advise a newlywed bride and groom who might face issues in their marital life (the snapping of ropes) but who should not give up and should make use of whatever is available to continue sailing (living).

The chardwellers' understanding and imagination of land is shaped by agrarian activities. But this does not indicate that they experience land simply economically. Agriculture, for them, is beyond an economic activity alone. It is, in fact, a collective cultural event. Hence, land, which is tied up with performing agrarian activities, is also experienced culturally. There is a deep sense of affection attached to land, for it also brings in ideas of home and belongingness.

The proverb *Jar pushena hale, tar pushena kuno kale* is roughly translated as "Those who are not self-sufficient in agriculture, can never be self-sufficient in life." This shows the chardwellers' understanding of agriculture as more of a way of life, one that emphasises the intricacy of living a particular kind of life. Here can also talk about *dhuageet* or *jarigeet* which are songs sung during harvesting that celebrate people's ties with the land through farming. The farmer who leads these songs is called *boyati* and the other farmers, generally tenant farmers, who join the chorus are called *magon kamla*. At the end of such farming activity, the landowners do not pay them money but instead hold a feast that brings the entire village together. During an interview with Shalim Hussain (2019), a student rights activist, Hafiz Ahmed, a Miya poet and a scholar, fondly recalls how when a landless tenant farmer does not have enough money to hire labourers, the entire village organises feasts to pull together sharecroppers who can lend help to his field. In one harvest song following this theme, the *boyati* sings that his body (or his labour) is his land and thanks Allah for giving the opportunity to use his body to grow the crops that will feed him and his family for the entire year. Songs like this shows that ownership of land is unequal but relationship with land is critically relational beyond the legal definition of ownership. For the landless, the body is perceived as "land" that allows for farmers' work and is an inherent part of the maternal principle of the earth (Berry 2009, 116); whereas for the large landowners, land simply represents social status and economic gain.

Colonial and neo-colonial politics

Arupjyoti Saikia (2020, 247) notes that the pre-colonial chars witnessed a wandering population who cleared new patches to cultivate crops and who temporarily took residence there. However, when flooding occurred, this population relinquished earlier patches and moved on to fresh ones. Throughout the nineteenth century, cultivation in chars did not extend beyond 3 years. This method of temporary cultivation along with high rates of land relinquishment frustrated the British who saw the extremely fertile chars being "wasted" by the "lazy natives." The British were losing out on tremendous amount of revenue. Gorky Chakraborty (2012, 3–5) mentions how wasteland as a category was a colonial construct and how transforming wastelands to cultivable land was considered transformation from a state of nature to a state of civilisation. In fact, the British justified colonialism using an ideology such as this wherein valueless land was transformed to land of value thus bringing more and more territory into the "light of civilisation."

Thus, the British encouraged the migration of peasants from what was then East Bengal of an undivided India. These peasants, largely Bengali Muslims, were settled in the charlands in Assam. Many of them often rented the lands from upper-caste Hindu landlords or purchased the lands from them. Others often just came and cleared the charlands for farming. Fleeing an oppressive land revenue system in East Bengal, the new migrants soon started turning the temporarily cultivated charlands to permanent cultivation. At the same time, they brought in new changes, as Saikia points out: "Between 1910 and 1928, three riverine revenue *mauzas* (revenue circle) of Kamrup—Barpeta, Ramdia, and Chamaria—showed 699, 123, and 150 percentage of population increase respectively which was much higher than the 64 percentage average increase in other *mauzas* with higher lands" (Saikia 2020, 251). They introduced new crops and practiced double or multiple cropping which increased the agricultural productivity of the charlands but also allegedly led to soil erosion, about which I will offer my assessment shortly. With permanent human habitation and rapid cropping, tall grasses that were not only food for animals like rhinoceroses and elephants but also kept erosion in check, disappeared. In turn, wild animals disappeared as they lost their habitats. Nepali pastoralists also stopped their seasonal migration to chars with their herds. (Chakraborty 2018, 186; Saikia 2020, 252 and 271–272).

The increased growth of weekly *haat* (markets) in chars with Marwari (those who migrated to Assam from Marwar region of Rajasthan, India, and subsequently settled down here; largely merchants by profession, they dominated various trade in Assam) and a few Assamese and Bengali merchants frequenting them helped in the growth of the cash economy. The local communities' loss of access to the chars as seasonally productive lands led to political conflicts with the "settlers" from East Bengal, the repercussions of which are still felt today in Assam. The conflict over charland use soon became an ethno-religious conflict between Bengali Muslim "settlers" and Assamese, mostly upper caste, Hindus. British policies tapped into such conflicts and aggravated them. It triggered xenophobic–Islamophobic politics in Assam. The British highlighted the differences between Bengali and Muslim chardwellers and, in the 1940s, encouraged the then Minister Muhammad Sadullah to bring in more "settlers" in large numbers under the "Grow More Food" campaign. This British-encouraged ethno-religious conflict culminated in the anti-foreigner Assam Movement in the 1980s. Upper-caste Hindu Assamese male leadership of the movement alleged that the chars, away from state eyes, are havens of Muslim foreigners. This Islamophobic public perception even led to the bloody massacre of several chardwellers in Chaulkhowa Chapori in 1983.

But did the arrival of the Bengali Muslim chardwellers disturb the ecology as Saikia and Chakraborty have suggested? I would argue that they have not. They brought changes but did not practice the strict separation of land from water for the sake of increased productivity as did the colonial and the neo-colonial state (Lahiri-Dutt 2014, 6–7; Saikia 2020, 32 and 88). Until today, Bengali Muslim chardwellers depend on the arrival of the annual floods to inundate the floodplains because they know that floods not only bring the fish but also help deposit the fertile silt, critical for agriculture. During the months of May and June, when the fresh wave of floods rush in, men, women and children go for collective fishing which is a happy and rewarding event. The songs sung during these events are called *machhmaragaan* (songs of collective fishing).

Cultural narratives such as this also indicate that floods are beyond the paradigm of "disaster." The understanding of river and water as "destructive" or as that which needs to be controlled shows very little knowledge of chardwellers' lived experience that is inseparable from water. An official from the Disaster Management Authority of India recounted how during the 2014 floods, many chardwellers refused to go to evacuation camps saying they could spend some nights on their boats. They were even able to predict within how many days the water would recede. When the inspection staff members from the centre noticed char villagers "stranded" in the flood, they expressed anger at the local rescue team for not rescuing them. The team had to return to the village and plead with the villagers to be "rescued."

As Saikia (2020, 58 and 96) notes, even during colonial times when the entire area would be flooded for almost 4 months, char villagers showed an amphibious way of living and happily came in and out of water, using boats and engaging in trade. Hence, the boats, as mentioned earlier, allow the chardwellers to have a fluvial, hybrid existence that reveals their ecological fluency on both land and water. Scholars like Kuntala Lahiri-Dutt and Gopa Samanta (2013, ix–x) have argued that chars are ecologically hybrid, a place wherein one cannot clearly argue if they are part of the water (as the riverbed) or part of the land (as riverbanks). Chars erode, their soil becoming part of the river; they get submerged during floods and yet they can re-emerge post flood, and just as easily, completely new chars are created. Hence, it can really confuse the strict modernist categorisation of land and water. I argue that the chardwellers in Assam are not just living in hybrid ecologies but they themselves are hybrid in terms of their cultural and political identities.

The cultural expressions of chardwellers, like their folksongs and festivals, often celebrate nature by displaying a beautiful syncretism between their ancestral Bengali Muslim heritage of East Bengal and the Assamese Hindu community of "mainland" Assam. One can see this in the following *dhuageet*:

> Sitting by the river,
> Hasan and Hussain have stopped to drink water.
> Oh look! What happened at Janakpur,
> Lakshman's brother Lord Ram
> Arrived at Janakpur, broke the divine bow,
> And married Princess Sita, the daughter of King Janak.

In this song, the invocation of the religious Muslim men Hasan and Hussain, and the Hindu deities Ram, Lakshman and Sita finds equal place. This syncretism can also be seen when Ahmed, in the interview mentioned above, mentioned how women in chars also make *alpanas*, or patterns on floor using rice powder and water, which is a largely Bengali tradition to decorate during the festival of Ghassi. The chardwellers syncretic culture marks their hybrid cultural identities and encourages them to develop an "ethics of care" (Bawaka Country et al. 2013, 189) wherein they understand that humans exist in relation with other humans and nonhumans and are continuously in the process of co-becoming. Human and land co-becoming is amply shown in chardwellers' folksongs or festivals inspired by their living traditions. They recognise the world as a pluriverse made of not just humans but also rivers, fishes, boats, land, crops, livestock, plants and insects.

Colonial politics, on the other hand, was based on separation, not just of land and water in service of increasing revenue, but also separation of a syncretic cultural identity and a projecting of a monolithic identity of the chardwellers' in the political arena that then contributed to conflicts between Bengali Muslims and Assamese Hindus as well as with the native tribal peoples like the Boros and Karbis in Assam. This one-sided perspective of dividing environment into distinct categories, and the inability to acknowledge hybridity led to a series of political practices by the colonial state that suppressed the chardwellers' syncretic cultural identities and underlined their distinct, monolithic identities as Bengali Muslims. The colonial Line System helped in concretising the chardwellers' identity as Bengali-Muslim and also fomented feelings of "otherness" in the dominant Assamese Hindu population. This colonial legacy continues on in the current Indian state that is neo-colonial in the sense that it has inherited the colonial politics of separation in both ecological and human terms. The neo-colonial state that is dominated by upper-caste, middle class, Assamese Hindu men continues to see the chardwellers as Bengali-Muslims and hence as an "other" thereby aggravating populist sentiments that represent Bengali Muslims simply as foreigners eating away at the "native Assamese" people's land rights and economic opportunities. The tendency to fit them into distinct, tight-fit categories is borrowed from their colonial masters. This politics of monolithic separation has not only caused violence but fuelled the indigenous-versus-settler debate wherein the Bengali Muslim chardwellers who trace their ancestry from East Bengal are never considered to be indigenous and are always relegated to the status of settlers and hence are seen as insufficiently Assamese by the neo-colonial state and society alike.

If environmentality is understood as the governmentality of the environment, the current Indian state has made many of the chardwellers its "environmental subjects" (Agrawal 2005, 166). I argue that the colonial and neo-colonial state politics of land/water and Muslim/Hindu separations led to chardwellers' self-perception as not so much syncretic but monolithic identities. In many cases, they gave up their traditional syncretic cultures and identities in order to become more like local Assamese. Chardwellers are slowly giving up their Miya language and taking up Assamese while festivals like Ghassi and Pushura are hardly celebrated. Chardwellers' anxiety resulting from their desperate attempts to belong to the Assamese community, to be seen as Assamese and Indian and not foreigners is induced by colonial and neo-colonial state politics. Both colonial and neo-colonial states favour static senses of belonging and identity, defined in terms of the nation and encompassing only the human community, such that the relationally understood idea of home held by a migratory population becomes an aberration.

The Indian state has established departments like the Water Resources Department and the Brahmaputra Board which are funded to locate new sites to build embankments and enact flood control measures. In the course of my fieldwork, the Bahari–Baghbar embankment has been destroyed in every big flood and needs repairing. Officers in the Brahmaputra Board confided in me, saying that though the department knows that embankments are not any solutions to flood, there is a big nexus of contractors, officers and politicians who benefit from embankment building projects and who happily lobby for embankments. Floods become money-minting annual events. This separation of land from water can be again seen in how, just like the colonial government, the neo-colonial state also places greater emphasis on land over water. However, the state-sanctioned changes to the chars are now being challenged by a poetry movement geared towards the revitalisation of traditional relationships and the celebration of hybrid ecologies and politico-cultural identities.

The Miya poetry movement

The Bengali Muslim people living in the charlands have faced continued marginalisation and xenophobia by the larger Assamese society and state. The term Miya is a derogatory Assamese slur that connotes Bengali Muslims as illegal Bangladeshi migrants. It is important to note that state policies and knowledge production have led to the creation of a hurtfully homogenous understanding of the chardwellers wherein the diversity and complexity of their cultural narratives are never brought into the larger public discussion and consciousness. The Miya poetry movement is challenging such politics and understanding of the chardwellers as a community. In this popular movement, people who are descendants of East-Bengal-origin Muslims but belong to the charlands in Assam, voice out their angst, fear and pain at the marginalisation, xenophobia and stereotyping that they continue to face from the state and Assamese society alike. Poetry is the medium for them to re-embrace their chardwelling memories and to have deeper understandings of their diverse, hybrid cultural origins found in traditional festivals, folksongs and everyday objects. Poems depicting the Muslim Charuas' (chardwellers) deplorable plight and the exhaustive plea to respect their East Bengal origin occurred as early as the late 1930s and found resurgence in 1985 when Khabir Ahmed wrote his "I Beg To State That." Verses like "I Am a Settler, a Hated Miya" resounded through the communities of chardwellers. However, the poems written in either the 1930s or the 1980s had no expressions celebrating the identity of the Muslim chardwellers as "Miya" as a way to highlight their East Bengal origins. The tone of the poems was of pleas from the Muslim chardwellers to be noticed or to be seen simply as Assamese. The actual Miya poetry movement occurred in recent years after the violent Bodo-Muslim clashes in Assam and the current right-wing government's decision to update the National Register of Citizens (NRC) that has inflamed narrow-minded Assamese chauvinism. The landmark poem was Hafiz Ahmed's "Write Down I am a Miya" (Ahmed 2016) with its overt tone of reasserting the Miya identity, and that widely spread through Facebook. A few days later, Shalim Hussain, a research student from the chars studying in a leading public university of India, replied back to Ahmed's poem on Facebook with his own poem titled "Nana I Have Written":

> Nana I have written attested countersigned
> And been verified by a public notary
> That I am a Miyah
> Now see me rise

From flood waters
Float over landslides
March through sand and marsh and snakes

This encouraged many other young Muslims of East Bengal Origin from the chars (a few of whom also currently reside in the chars) to start writing poems to each other and share them on social media. These poems have led to the creation of the Facebook page "Itamugur," a social media venue for the Miya poets. As opposed to earlier poems, one can see these poets asserting their identities as Miya. In fact, from the few lines quoted from Hussain's poem above, one can sense the growing celebration of a liminal identity, of a hybrid identity wherein they acknowledge and fiercely guard their distinctiveness of being Miya while nonetheless not denying their identity as Assamese, either. Hussain in a conversation with Nabina Das, an Assamese poet and writer, says:

> Let us look at our history of loss. We are a community descended from impoverished low-caste Hindus who converted to Islam. Our ancestors moved to Assam with little else but just clothes on their back. Even before the borders were drawn, we began losing touch with Bengal. We are not a diasporic community with a nostalgic attachment to our place of origin. We have no standing monuments, no venerable ancestors, nothing. This might sound bleak but I consider it quite liberating because minus the baggage of chauvinism and jingoism, we have the freedom to be as hybrid as we want to (Das 2016)

This generation of youths is reclaiming its hybrid identity through poetry, which is both and yet neither. This liminality endows them with a sense of freedom that the socio-political milieu has for so long denied them. It is reclaiming the histories of marginalisation and questioning the erasure of the narratives of the lived experiences of violence, discrimination and state control over resources and economic exploitation by both the colonial and neo-colonial states (Peluso 1993, 202, 205–207, 210).

Most of the poems revisit collective memories and refer to chardwellers' traditional relationship with their riverine dwelling place. One can see this in Hussain's poem wherein he invokes the narrative of surviving floodwaters, landslides and marshes, and continuing with life. Miya poems bring back symbols of the river, forests and land that are critical in understanding the chardwellers' ecologically niched cultural life. Kazi Neel, another Miya poet, aptly writes:

flame of the forest does not bloom any more
the koel does not sing anymore
karim majhi's river song
floats on the river
like an unclaimed dead body
what epic can I write about
this shattered country
blood flows from my pen…

In fact, the group's name "Itamugur" is itself a symbolic nod to a traditional but slowly disappearing agricultural tool used in the chars to break clods of soil to dust before preparing the fields for sowing. Hence, Miya poets are invoking their traditional experiences with nature and how those experiences have been crushed by an oppressive state and society. One can see from another of Kazi Neel's poems—"Digging a Grave"—how the chardwellers' intricate relationship with nature is remembered and instrumentalised to fight socio-cultural stereotyping and political othering:

> I find the smell of wet soil inside my bosom,
> the broken remnants of a plough in my fist.
> Digging the grave I take out my sunless past.
> I see everyone has a history of journey…
> Digging my grave I myself carry my corpse to the graveyard.
> Whether they declare me a martyr or not,
> before this land is sold out, before this air is exhausted,
> before these rivers get poisoned,
> I wish to be devastated at least once in a tumultuous battle.

Miya Poetry as a movement is helping chardwellers celebrate their dependency on nature, particularly land and water. Shalim Hussain underlines this sentiment when he says that the strength of this movement is born from the very strong connection young people of the chars have with this earth. "I Beg to State That" by Khabir Ahmed captures this youthful strength well:

> …And yet you talk of the river
> The river is Assam's mother, you say
> You talk of trees
> Assam is the land of blue hills, you say
> My spine is tough, steadfast as the trees
> The shade of the trees my address…
> You talk of farmers, workers
> Assam is the land of rice and labor, you say
> I bow before paddy, I bow before sweat
> For I am a farmer's boy…
> I beg to state that I am a
> Settler, a hated Miyah
> Whatever be the case, my name
> Is Khabir Ahmed or Mijanur Miyah
> Subject—I am an Assamese Asomiya.
> Sometime in the last century, I lost
> My address in the storms of the Padma
> A merchant's boat found me drifting and dropped me here
> Since then I have held close to my heart this land, this earth
> And began a new journey of discovery….

Another noteworthy facet of the poetry movement is the use of the Miya language. The poems show a growing linguistic syncretism in the Miya language, particularly between various dialects of East Bengal, Kamrup and Goalpara. Miya as a language is itself very syncretic borrowing from Assamese and various dialects of Bengali language. A lot of hybridisation happened in Assam with the arrival of Arab traders after Muslim invasion in 1532 and the arrival of Sufi saints. Some, like Shah Milan and Azan Fakir, even learned Assamese and composed folk and devotional songs using words from both Assamese and Bengali languages. Many such songs, called *jikir*, are popularly sung in chars as *jarigaan* (devotional songs). Over time, local influences have also reshaped certain vocabularies of already syncretic folksongs. The Miya language symbolises a syncretic history and the assimilation efforts of a community. The use of Miya by these poets is also a celebration of such culture and syncretism.

In the face of the loss of myths and proverbs in Miya language, the poetry movement is reviving and keeping alive the chardwellers' ways of seeing the world and celebrating their lived experiences. Hence, the movement is restoring the culturally hybridised narratives of chardwellers. These narratives are bringing back an empathetic understanding of nature and society alike, challenging hegemonic, populist and homogenously distanced understandings of the chardwellers and their riverine living environments. As the act of the writing among the Miya poets is revitalising the traditionally hybrid festivals and folklores rooted in the riverine char landscape and is declaring that one can be both Bengali and Assamese, both Hindu and Muslim, the public is questioning the colonial and neo-colonial politics of the separation of land from water, of Bengalis from Assamese and of Muslims from Hindus. James Scott remarks that when the weak resist the symbols of oppressive authority, they start from ideas and beliefs (Scott 1985, 322). Similarly, Arturo Escobar points out that the hurtful narratives of modernity and capitalism can be countered through the creation of new hybrid narratives (Escobar 1996, 65). The Miya poetry movement is a vanguard of the new narratives taking shape in a communal and publicly engaged poetry writing process.

Countering their existing subjectivity that was nurtured by colonial and neo-colonial politics, the poetically expressed new subjectivity of the chardwellers is calling for re-embracing their ancestral past with reverence and ending the exploitation and stigmatisation against them from the state and mainstream Assamese society. This new collective subjectivity takes pride in the chardwellers' traditional ethics of care, not just for humans but also for nonhumans who share the same dwelling space. Richly showing human affection for the riverine land, Miya poems are embodiments of collective memories and the ecological awareness of the earth not simply as the source of resources but, more critically, as the total livingness of the environment, humans, animals and plants. Remembering and recounting is a powerful political act of assertion and of cultural empowerment. The young chardwellers' poetic revitalisation of ancestral experiences and traditional cultural symbols are evoking the collective longing to belong. The Miya poetry movement brings forth a collective action of asking the public to recognise their pangs of systemic uprootedness, their already-rooted belonging to Assam, to India and the charland as their homeland that they have, for centuries, tilled, celebrated and considered their own dwelling place.

Postscript

The lifeworld of the chardwellers speaks of a deep and complex understanding of their ecosystem. One can sense that dependence upon nature is a profound lived experience. As shown in my ethnographic narratives, nature in this case is the riverine earth embodied in the human body and personified as the forces of the hybridised divinity of Islam, Hinduism and indigenous animist practices. Existentially, the riverine nature is a physical manifestation of the earth's maternal principle in terms of its provision to human sustenance and livelihoods. Ecologically, shifting its shape—in the form of monsoons, floods, braided rivers, chars and seasonable farming lands and pastoral grazing grounds—the force of nature has its own agency, agenda and course of actions that are both destructively creative and creatively destructive. It humbles the chardwellers to live with it accordingly with their willingness to adapt to both the nourishing and depriving power of the earth. Culturally, nature in its muddy-watery form affords both personal and communal celebratory moments expressed in songs, poems and festivals eulogising chardwellers' rootedness in the charlands. Politically, the riverine nature is highly humanised in the languages of the colonial/neo-colonial economic practices and of the chardwellers' social movements for recognition, de-stigmatisation, rights, justice and equality. The former perceives nature as the source of extractive resources, while the latter lives in it as home.

Similar to the chapters in this book by Charisma Lepcha, Thinley Dema, Zainab Khalid, and Rongnyoo Lepcha and Mongfing Lepcha, my practice of environmental humanities is also centred upon Rose and her colleagues' call for "lively ethologies of meaning and value, entangled within rich patterns of cultural and historical diversity that shape who we are and the ways in which we are able to 'become with' others" (Rose et al. 2012, 2). Furthermore, my work in the charlands is a combined form of collaborative ethnography and lively ethography. The former emphasises the ethnographer's relationality with his or her interlocutors (Lassiter 2005, 84), which, in my case, is a set of my native personal and social relations with the chardwellers as well as an anthropologically enabled rapport with them as my professional practice of writing culture. The latter is inspired by Van Dooren and Rose's idea of lively ethography, a narrative genre for depicting a given ecologically conditioned, more than human lifeworld with an emphasis on its ethos—its character or "an overarching set of values and practices—a style" (Van Dooren and Rose 2016, 80). I resonate with Mongfing and Rongnyoo's practice of lively ethography in their chapter in that ethos "reflects our social spirit in the face of the modern transformations and challenges that the Lepcha-land has undergone over the last two centuries." Ethos in the case of the chardwellers is not an isolated, essentialised social spirit of the charlands; instead, it is deeply and widely entangled with the riverine earth, colonial history, state policy implementation, Assamese xenophobic nationalism and chardwellers' affective political approach to public awareness of their rootedness in the charlands.

Finally, unlike Mongfing and Rongnyoo's focus on the eco-spiritual presence of Lepcha ancestors embodied in the contemporary landscape of Sikkim, environmental humanities in my case study, to a large extent, is a political ecology of "violent environments," in which violence is understood as "a site-specific phenomenon rooted in local histories and social relations yet connected to larger processes of material transformation and power relations" centred upon power struggles between the state and society over resource use and control (Peluso and Watts 2001, 5; Peluso 1993, 201). Specific to the charlands, violence not merely takes place as human-to-human events but also comes in the nonhuman form of the annually predictable monsoons and the human-induced environmental changes such as the artificial separation of land from water and climate change. Violence in this sense is both ecological and political. The Miya poetry movement is a response to chardwellers' violated environment, and it rejects state-led assumptions about their community and ecology. Though still developing, with more and more young people from the chars joining the movement by writing poems, it is a strong political voice revitalising the chardwellers' traditional ecological vision of the world and advocating the use of Miya language to write about their life stories through poetry. The poetic narratives capture the heartbeat of a community's diverse, complex and deep relationship with its habitat. They celebrate human–earth relations as an interconnected whole and deepens the traditional sense of belonging to this riverine earth as the foundation of their home and their migratory, fluid identities.

References

Agrawal, Arun. 2005. "Environmentality: Community, Intimate Government, and the Making of Environmental Subjects in Kumaon, India." *Current Anthropology* 46 (2): 161–190.

Ahmed, Hafiz. 2016. "Write Down I am a Miya." *Facebook* website. Accessed September 20, 2018. https://www.facebook.com/itamugur/posts/895166793960541

Berry, Thomas. 2009. *The Sacred Universe: Earth, Spirituality, and Religion in the Twenty-first Century*. New York: Columbia University Press.

Blaser, Mario and Marisol de la Cadena. 2018. "Introduction: Pluriverse – Proposals for a World of Many Worlds." In *A World of Many Worlds*, edited by Marisol de la Cadena and Mario Blaser, 1–23. Durham: Duke University Press.

Chakraborty, Gorky. 2012. "*Roots and Ramifications of a Colonial 'Construct': The Wastelands in Assam*." Institute of Development Studies, Kolkata. Retrieved from http://idsk.edu.in/wp-content/uploads/2015/07/OP-39.pdf

Chakraborty, Gorky. 2018. "The char dwellers of Assam: flowing river, floating people." In *Water Conflicts in Northeast India*, edited by K. J. Joy, Partha J. Das, Gorky Chakraborty, Chandan Mahanta, Suhas Paranjape and Shruti Vispute, 178–192. London and New York: Routledge.

Country, Bawaka, Sarah Wright, Sandie Suchet-Pearson, Kate Lloyd, Jill Sweeney, Laklak Burarrwanga. 2013. "Caring as Country: Towards an ontology of co-becoming in natural resource management." *Asia Pacific Viewpoint* 54 (2): 185–197. DOI:10.1111/apv.12018.

Das, Nabina. 2016. "For Better or Verse: Miyah Poetry is now a symbol of empowerment for Muslims in Assam." *Firstpost* website. Accessed March 18, 2020. https://www.firstpost.com/living/for-better-or-verse-miyah-poetry-is-now-a-symbol-of-empowerment-for-muslims-in-assam-3007746.html.

Escobar, Arturo. 1996. *Encountering Development: The Making and Unmaking of the Third World*. Princeton: Princeton University Press

Escobar, Arturo. 2016. "Thinking-feeling with the Earth: Territorial Struggles and the Ontological Dimension of the Epistemologies of the South." *Revista de Antropologia Iberoamericana* 11 (1): 11–32. DOI:10.11156/aibr.110102e

Hussain, Shalim. 2019. "In Conversation with Dr. Hafiz Ahmed: 'The Char Chaporis cut their roots to fit into "Greater" Assamese Culture. " *Sahapedia* website. Accessed March 16, 2020. https://www.sahapedia.org/conversation-dr-hafiz-ahmed-char-chaporis-cut-their-roots-fit-greater-assamese-culture.

Lahiri-Dutt, K. and Gopa Samanta. 2013. *Dancing with the River: People and Lives in the Chars of South-Asia*. London and New Haven: Yale University Press.

Lahiri-Dutt, K. 2014. "CHARS Islands that float within rivers." *Shima: The International Journal of Research into Island Cultures* 8 (2): 22–38.

Lassiter, Luke Eric. 2005. "Collaborative Ethnography and Public Anthropology." *Current Anthropology* 46(1): 83–106. DOI: 10.1086/425658

Narayan, Kirin. 1993. "How Native Is a 'Native' Anthropologist?" *American Anthropologist New Series* 95(3): 671–686. Accessed September 11, 2018. http://www.jstor.org/stable/679656

Nayak, Amar Kumar. 2012. "Folklore of the Chars of Nalbari District: A Study for understanding the socio-cultural and philosophical identity of the people." *Shodhganga* website. Accessed July 2, 2020. http://hdl.handle.net/10603/28597

O'Reilly, Karen. 2009. *Key Concepts in Ethnography*. Los Angeles: SAGE.

Peluso, Nancy Lee. 1993. "Coercing conservation?: The politics of state resource control." *Global Environmental Change* 3(2): 199–217. DOI: 10.1016/0959-3780(93)90006-7.

Peluso, Nancy and Michael Watts. 2001. "*Violent Environments*." In Violent Environments, edited by Nancy Peluso and Michael Watts, 3–38. Ithaca: Cornell University Press.

Rose, Deborah Bird. 1996. *Nourishing Terrains: Australian Aboriginal Views of Landscape and wilderness*. Canberra: Australian Heritage Commission.

Rose, Deborah Bird, Thom Van Dooren, Matthew Chrulew, Stuart Cooke, Matthew Kearnes and Emily O'Gorman. 2012. "Thinking Through the Environment, Unsettling the Humanities." *Environmental Humanities* 1 (1): 1–5.

Rose, Deborah Bird. 2014. "Death and Grief in a World of Kin." In *The Handbook of Contemporary Animism*, edited by Graham Harvey, 137–147. New York: Routledge.

Saikia, Arupjyoti. 2020. *The Unquiet River: A Biography of the Brahmaputra*. Oxford University Press.

Scott, James C. 1985. *Weapons of the weak: Everyday forms of peasant resistance*. New Haven: Yale University Press.

Talukdar, Nilohit K. and Sarbeswar Kalita. 2008. "Flood and Its Impact on Socioeconomic Life in Barpeta district, Assam." [Published as part of National Seminar on "Earthquake hazards and disaster management of North Eastern States of India, 18–19 October 2008 at N.I.T Silchar, Assam] ResearchGate website. Accessed March 8, 2020. https://www.researchgate.net/publication/319980137_Flood_and_Its_Impact_on_Socioeconomic_Life_in_Barpeta_district_Assam.

Talukdar, Nilohit K. and Sarbeswar Kalita. 2005. "Flood and Its Impact on Socioeconomic Life in Barpeta district, Assam." *Disaster Management* 1: 19-27.

Van Dooren, Thom and Deborah Bird Rose. 2016. "Lively Ethography: Storying Animist Worlds." *Environmental Humanities* 8 (1): 77–94.

5 Painting the genesis of the Lepcha

A world emerging from the water spirits

Rongnyoo Lepcha and Mongfing Lepcha

Introduction

We are two friends, colleagues, research project team members, and, most of all, two Lepchas, who work together on our indigenous understanding of human–earth relations. For many months, we have trailed streams and ravines and searched for the actual landscapes across Dzongu, North Sikkim, where the stories of our ancestors and their companion spirits took place. In addition to organising our field notes, we have also drawn illustrations of the mythical but actual landscapes ready to deliver our research outcome; we nonetheless have moments of self-doubt. It has never been easy for us to be indigenous scholars, as we have to carry the burden of biases in the academic world. It is rather challenging to effectively put forward collected folk stories because of the fact that the world has turned so scientific and technologically driven and that most people would rather believe in the scientific narrative rather than acknowledging the essence of indigenous stories, an essence that has lived throughout generations in a given community. In our observation, scholars and scientists are very much aware of the contributions of indigenous people in safeguarding the pristine lands and environment and yet not many of them would pay interest in their stories that reflect the bond between the people and their land. Indigenous oral traditions thus tend to stagnate in the academic world as they are often treated only as ethnic identity markers without considering their ecological significance and environmental value. They thus become passive subject matters of the social sciences and humanities that are only selectively explored and recounted.

In spite of the unfavourable condition of our scholarly work, as indigenous scholar-artists, we are endeavouring to narrate our folklore through our artwork and scholarship in an effort to reach out to a larger audience, especially to those who are exploring indigenous environmental wisdom for a sustainable world. Our attempt then is to show the important link the Lepcha have with the environment, how we relate to everything present in and around our dwelling place, and how the earth and humans give meaning to each other as a journey of co-becoming. It turns out that the stories from our community hold much more under-expressed knowledge than what one can see from outside. Our Lepcha world is not and has not been understood adequately, other than as a historical colonial subject, a current territory of postcolonial India, and a research subject of social sciences and humanities, for nearly 200 years. In her essay on environmental art in the Anthropocene, Laura Hall, a Haudennosaunee (Iroquois) environmental humanist, writes, "Indigenous aesthetics are rooted in the original ecologies and life ways of traditional territory, and of the responsibilities that come from Creation" (Hall 2015, 283). Where words fail, we connect through art.

In their introduction to *Art in the Anthropocene*, Heather Davis and Etienne Turpin write, "It is not from some desire to add another conjunctive term to the growing literature on the Anthropocene that we turn to art; art, as the vehicle of *aesthesis*, is central to thinking with and feeling through the Anthropocene. And we believe the inherent relation between the two occurs at a number of strata and across various scales" (Davis and Turpin 2015, 3). We resonate with their emphasis on *aesthesis* connoting sensing and feeling the sentient world we live in. The unique Himalayan homeland of the Lepcha shows us the art of the earth, which is a living world integral to all living beings in it. Its geology and ecology envelop and nurture us with its material vitality. It is a world animated with spiritual forces. We humans are an inherent part of this living earth-art that never stops nourishing our *aesthesis* and reminding us of the sentience of the earth expressing through snow, ice, rivers, forests and mountains. We find our cultural belonging and ancestral roots in this awe-inspiring part of the earth. It asks more than our ears to listen and our minds to analyse the stories of its Deep Time encompassing both prehuman and human times. It takes our heart and soul to sense and feel the presence of its ancient past incarnated in the Lepcha homeland in this lifetime. Whether this ancient presence has been deformed or has differently evolved under anthropogenic pressures, its essence has taken deep root in our cultural consciousness and our affective behaviours toward our ancestral land.

This is where we find an opportunity to craft this chapter as an entwinement of our environmental artwork and sociological research situated in the environmental humanities. This hybrid form of our scholarship is an experiment of what Deborah Rose and Thom van Dooren call "lively ethography" understood as "a mode of knowing, engaging, and storytelling that recognizes the meaningful lives of others and that, in so doing, enlivens our capacity to respond to them by singing up their character or ethos" (Van Dooren and Rose 2016, 77). In our case, it is the ethos of the Lepcha lifeworld, not of others, to which we dedicate our art-scholarship. Ethos in our ethographing endeavour not only means "an overarching set of values and practices" of our cultural customs as Van Dooren and Rose delineate (2016, 80), but it also reflects our social spirit in the face of the modern transformations and challenges that the Lepcha homeland has undergone over the last two centuries. Inspired by our ancestors' cave paintings that afford us a visual commemoration of our ancient cultural ethos, our keystrokes and paintbrushes likewise put together this chapter as a tapestry for our readers to see how the Lepcha lifeworld is woven together with animist, spiritual, geological and ecological forces in a geomythological language. The primary message of our ethography is—our ancestors haven't stopped speaking to us in the capacity of the spirits, supernatural beings, environmental flows and animals who have been part of Lepcha land since the dawn of humankind in the Himalayas.

Storying the water spirits of the Lepcha homeland

Our history is written in our land and our oral tradition. It is embodied in the mountains, rivers and forests that have nourished us since time immemorial as well as in our abundant folk stories. Storytelling is synonymous to history-writing in our cultural routines except that what is being recounted is scribed and re-scribed in the hearts and minds of the young and the old. The historical figures in our stories are not only our human ancestors but also the ancestors of the mountains, rivers and animals, who are inherent members of our homeland. These nonhuman ancestors are spirits and supernatural beings in the appearance of different landforms, environmental flows, plants and animals. They are not things or beings of the past but are our companions who surround us and embody us with their daily provisions and moral responses to our worldly affairs. Thus, our history is always being recounted as the entwined presence of human ancestors, the physical earth, and both natural and supernatural nonhuman beings. Before we invite you, our readers, into our storied lifeworld, we would like to inform you that the names of rivers and mountains are also the names of the spirits and earth-deities who are responsible for creating our lifeworld and nourishing us. Many of them are our human ancestors in spiritual forms. Therefore, when we say the name of a river, for instance, we would like you also to see the river not just as a body of moving water but also as a spirit whose life-giving power flows in our land like a bloodline and wherever it reaches, life shows its animated, abundant diversity. As our storytelling progresses, we hope you will quickly learn to shape shift a river between an environmental flow and a spiritual being just like how Tibetans are able to mentally shape shift a sacred mountain between a landform and an earth-deity (Smyer Yü 2015, 167).

Now, we begin our story of Rongnyoo and Rungnyit, a female water spirit and a male water spirit. It is a love story; however, as it unfolds itself, it is also a story of the earth, taking place in our Himalayan homeland. The two water spirits find their presence in many aspects of our social life. In particular, they are religiously, spiritually, environmentally and affectionately celebrated. The story depicts a love as well as a tragedy that occurred as great deluge upon the land. The story lets us time-shift to the formative time of our homeland when spirits and supernatural creatures were working together to create a world with high mountains and abundant waters. Rongnyoo and Rungnyit, as narrated in this story, continue to flow in our land and beyond as the Teesta and Rangeet Rivers. They are revered as an epitome of love and sacred waters.

Once upon a time, when Rongnyoo and Rungnyit had yet to flow from their sources—Naho Lake in the northeast and Nahar Lake in the southwest at the base of Mt. Kanchanjunga—it is said that an earthquake shook the region and cracked the lakes, causing the two rivers to originate from their respective sources. So, the two pristine rivers Rongnyoo (female) and Rungnyit (male) emerged after the earthquake. The two rivers spirits were much in love with each other. Their bond was envied by other celestial beings of the region. One day they decided to flow down to the plains to start a new life together. They decided to meet at Pozok, now known as Peshok, and continue their venture from there as one. To make matters interesting they playfully challenged each other to race to Pozok, the place of their planned rendezvous, and see who would reach there first. Since it was new for both river spirits to venture out of the mountains for the first time, they opted for guides to help them navigate their way. Rungnyit chose *Tut Fo* (the Himalayan monal, a brilliantly coloured pheasant) since he wanted to be the first one to reach the spot and he thought that the beautiful pheasant would be swift and fast. At the same time, Rongnyoo chose *Paril Bu* (the python) as her guide.

As they started their journey on the agreed-upon date, the *Tut Fo* who was guiding Rungnyit was as swift as expected, but became easily distracted by colourful flowers and insects. It made too many stops and turns in its search for food. This is why the river Rungnyit flows in a sharp zigzag pattern. Meanwhile, Rongnyoo guided by the *Paril Bu,* who avoided every hurdle, reached the destination first. As Rungnyit was getting closer to Pozok, he was filled with excitement. He could see Rongnyoo at the far end waiting for him. Rungnyit became anxious and started picking up more speed to meet his lover, dragging boulders and rocks along the way. As he reached Rongnyoo, he shouted "*Thee sutha?*" which means "When did you arrive?" A little anxious, a little furious with himself for not reaching the destination first, he was disappointed. Filled with frustration on losing the race, he started flowing backward. Rongnyoo pursued Rungnyit, hoping to console him. Finally, she caught up with Rungnyit and persuaded him to come back with her so that they could fulfil their dream together. Rongnyoo, in order to delight Rungnyit, asked him to flow above her and she would flow from beneath him. From that day onwards, the confluence of these two rivers came to be known as Thee Sutha, now pronounced as Teesta, and even today, one can see the clear waters of Rungnyit flowing on top of the smudged waters of Rongnyoo.

In unison, the backward flowing of the two rivers started drowning the entire region, engulfing whatever was in their way. Animals, birds and insects started fled for their lives. People climbed up Tungronglho (now Tendong Mountain) peak as the waters rose higher and higher. They started offering prayers to *It-Bu-De-Bu-Rum-Daar*, the creator as well as the destroyer, for help (see Figure 5.4). Finally, after days of praying, a Kohomfo (Himalayan partridge) came to their rescue, offering them *mong-chebep* (fermented-millet alcohol). After intense prayer, finally their prayers were answered. The earth shook again, making the water flow southward and thus saving the Lepchas from this great deluge.

In order to commemorate the event where the Lepcha were saved by *It-bu-de-bu-Rum-Daar* (the creator) from the wrath of the river spirits, for which we are eternally thankful. The event of the rivers' joining together in a flood, Tendong Lho Rumfaat, is celebrated each year to remember this day of peril and to celebrate life flourishing once again. As an auspicious occasion, it is also celebrated to thank *Tendong-lho* (Tendong Mountain) for saving the Lepcha from the flood. Every year, it is celebrated on August 8and is now a state holiday in Sikkim. From the offering of prayers at the *lho* (hill) itself by *mun* and *boongthing* (traditional Lepcha spiritual and faith healers) to get-togethers in different areas in Sikkim, Lepchas have always been consistent in celebrating the event with full enthusiasm and interest.

Painting the genesis of the Lepcha

Rongnyoo Lepcha

I had heard the story of Rongnyoo and Rungnyit many times from my grandparents. At first, I was captivated by their love story. As I grew up, it was constantly at the back of my conscience, unknowingly creating a bond between the rivers and me, since I share my name with one of them. The more I observed the turquoise waters of Rungnyit and the muddy yellow waters of Rongnyoo mingle, while crossing National Highway 31A, I realised Rongnyoo had more rage in her. She was hiding much more than the stories lead us to believe. Maybe it was this rage that helped her stop Rungnyit from turning back. Yet her nature is described as being the "humble one," a characteristic that Lepcha women are expected to have. Imagine if it was Rongnyoo who had decided to flood the lands in her rage instead of Rongnyit, there would have been a final deluge without escape. These thoughts have led me to create the artwork in Figure 5.3. On the left is Rongnyoo, led by *Paril Bu*, strong and determined. And on the right is Rungnyit, a hot headed lover led by *Tut Fo*. The two river spirits in human form had strong determination for what they believed in.

Mongfing Lepcha

We are colleagues in a same department at Sikkim University; however, our art journey began in different phases and places of our life-paths, which are shown in our enclosed art pieces. Regardless of our aesthetic differences, our common ground is that we are an integral part of our homeland and our art. The former enters the frame of the latter and the latter encapsulates the former with the stories of our ancestors and earth-spirits in different landforms, environmental flows and animal bodies. These stories are deeply inscribed in our shared cultural–ecological consciousness and oral tradition that recounts them from generation to generation. From our fieldwork, we collected many stories of the sacred land but also witnessed devastating environmental changes there. Both the stories and the artificial changes connect me more with our sacred land. The illustrations in Figures 5.1 to 5.4 not only show how I connect myself with our living history of the spirits, humans, animals and trees of the Himalayan earth, but also afford me an opportunity to recreate the time of our genesis. Although I have been accustomed to these folk stories all my life, painting them and writing about them brings me ever closer to our land and the lived experiences of our people since ancient times.

Aesthetics of our storied world

Our artistic illustrations are visual narratives of Rongnyoo and Rungnyit, two anthropomorphised water spirits incarnated as two merging rivers in real life. Both rivers are our critical sources of nourishment as well as the spine of our cultural heritage deeply rooted in the forests and mountains of our homeland. Each of our illustrations was drawn at an actual location along the rivers. Our purpose is to show that Lepcha myths are not merely bedtime stories for children, but are

Figure 5.1 An illustration of Rungnyit, the male water spirit, being guided by Tut Fo. By Mongfing Lepcha.

Figure 5.2 An illustration of Rongnyoo, the female water spirit, being guided by Paril Bu. By Mongfing Lepcha.

Figure 5.3 An illustration of Rongnyoo, the female water spirit (left), being guided by Paril Bu and Rungnyit, the male water spirit (right), being guided by Tut Fo.

By Rongnyoo Lepcha.

Figure 5.4 An illustration of the ritual offerings by the Lepchas at Mt. Tendong during the deluge and It-bu-de-bu-Rum-daar (Creator and Destroyer) emerging as (Himalayan partridge).

By Mongfing Lepcha.

histories embodied in the earth. We don't need to stretch our imaginations too far to recreate the world of the remote past. As a matter of fact, we live in it and we are an inherent part of our storied world. The aesthetics of our drawings come from our sensing and feeling the physical landscapes Rongnyoo and Rungnyit created and continue to create. It also comes from our cultural narratives of how this particular part of the Himalayas made us who we are. Our art is thus meant to convey our environmental realism of the Lepcha land, as well as our affective understanding of it, to our readers. The former is our indigenous and scholarly conviction that humankind is an integral part of the earth's complex biophysical system (Cockerill et al. 2017, 3). The latter reaffirms to us that humanity, in an affective sense, is also embodied in our living environment. Art in this sense is a medium that transports our lived environmental history and our cultural emotions to viewers. Our artistic illustrations of the Lepcha riverine landscape thus aim to transport sentiments that entwine the environment and the community, revealing their integrated knowledge system, and the resulting beliefs and understanding of the earth. Such environmental artwork can conjure up a thousand emotions through which people can see and relate to their community, memories and histories. Thus, art in our context is an expression of our co-becoming with the environment as a habitat, an embodiment of spiritual forces and a shared story of humans, animals, plants and spiritual beings. We live in the environment, receive nourishment from it and are empowered by our guardian spirits embodied in it. Our environmental art is thus meant to celebrate and share the ethos of a Himalayan natural–cultural heritage with our fellow Lepcha and people around the world.

To further elaborate our environmental aesthetics, the story of Rongnyoo and Rungnyit did not cease in a mythological time when they found their perfect union as lovers; instead, it has evolved, particularly during the Anthropocene, our current human-induced geological era. Before we give a fuller account of the current changes, we would like to tell our readers that for the Lepcha, like many indigenous peoples around the world, the earth is sacred and sentient; it is a "focal point around which life activities revolve" (Ramakrishnan et al. 2011, 1). Our sense of the living earth resonates with what Rose writes about the Australian Aboriginals' bond with their land, "Country in Aboriginal English is not only a common noun but also a proper noun. People talk about country in the same way that they would talk about a person: they speak to country, sing to country, visit country, worry about country, feel sorry for country and long for country" (Rose 1996, 7). For the Lepcha, the land is a living entity and everything in it has a spirit, a protector, called *dokh.* These *dokh* look after the mountains, valleys and water bodies they are bound to. They protect everything that falls under their stewardship. Harming any living creatures or any nonliving things under the protection of a *dokh* leads to a person's illness or misfortune as a form of retribution from the *dokh*. Nature in its diverse manifestations has always been the spiritual and moral gravity of Lepcha culture.

Our original and current lifeworld all start from the mountains and the rivers. The snow-capped mountains hold a special place in our hearts. Our oral history tells us that *It-bu-de-bu-Rum-Daar*, the creator, created the first Lepcha man and woman from the fresh snow of *Kinchumzongbu chyu* (Mt. Kanchenjunga). We affectionately call this majestic mountain *Anum timbu* or elder brother. We also call it *Pum Zoo*, or the place of our ancestors, as we believe that our souls will eventually go through the mountain and rest in the lap of our or ancestors. We will become ancestors of our future generations. Our constant longing for and belonging to the majestic mountain giants in our land incessantly flows like Rongnyoo and Rungnyit in our cultural psyche. We would never really be able to tell when our bond with the mountains and the rivers began, but we are certain to tell our readers that we are one with this unique part of the earth we live in.

Living in the foothills of the great mountains since ancient times, Lepchas have preserved a Himalayan ethnic history as well as a local eco-geological history in our folk stories. These stories have been passed down through word of mouth from generation to generation. They are rooted in the physical earth and in our cultural consciousness. The environmental knowledge in these aesthetically rich stories depicts the kindred relationship between humans and nature, and between humans and nonhumans. Our artistic infusion of the mythological imageries of Rongnyoo and Rungnyit with the actual riverine landscape of the two spirit-rivers is meant to highlight nature's complex texture and reveal the inner essence of nature (Bergmann 2018, 28–30). The inner essence captured in our artwork conveys both a geological history of the Himalayas that predates the birth of humankind and a multifaceted spiritual force that breathes life into the physical environment of a Himalayan habitat. Like the artwork of our Australian Aboriginal counterparts, our environmental art is also "concerned with ideas and processes more than with appearances" (Bergmann 2018, 30) in which the land is a creation of both geological forces and spiritual beings. Our drawings time-shift, providing an invitation to our readers to experience the dawn of humankind in the Himalayas.

The history of the Himalayas, as nature's monument related to a sense of Deep Time (Rudwick 2005, 315) is well preserved in our folk stories. In earth sciences, "nature's monument" or "nature's archive" refers to fossils, geological layers and geomorphological shapes as the physical evidence of the evolutionary course of the Earth. Although our human-centred stories are unable to retrieve the earth's geological memories in a deeper time, they at least show us a glimpse of the time when the genesis of humankind was just starting to unfold from an already-ancient planet. It was a time frequented by geological and environmental events like earthquakes and floods, and one in which humans were proportionally insignificant and fragile. The devastating earthquakes, the enormous floods and the supernatural power of spiritual beings in the forms of the Himalayan python, the pheasant and the snowcock humbled our ancestors to revere the earth as a living entity. Even to this day, the Himalaya continues to be geologically active. As we speak, we live in Deep Time in the present tense and will grow ancient, deep into the future. Like how environmental humanists cherish Australian Aboriginal and Native American mythological narratives of the earth, we, too, take our environmental reading of our origin stories very seriously in the Anthropocene. They are in fact ancient lively ethographies of our ancestors orally recounted for generations. We are still toddling to learn how to articulate our ethographical accounts before we tell our stories in the lap of our ancestors at *Pum Zoo*, the place of our reunion with them.

Art of ethographing our past and present

Van Dooren and Rose introduced their "lively ethography" 4years ago as an environmental humanist "mode of knowing, engaging, and storytelling that recognizes the meaningful lives of others and that, in so doing, enlivens our capacity to respond to them by singing up their character or ethos" (Van Dooren and Rose 2016, 77). Conceived "via our own worldly encounters" and "through the lenses of Western science and philosophy" (Van Dooren and Rose 2016, 81), this new approach to environmental studies apparently draws a distinction between a self and others by explicitly bracketing out "indigenous understandings" (Van Dooren and Rose 2016, 81). This may sound self-defeating to their attentiveness to "a common feature of all life" and "a kind of openness toward the world" (Van Dooren and Rose 2016, 81). We are nevertheless inspired by their passionate call for ethographic storytelling as "the arts of becoming-witness" (Van Dooren and Rose 2016, 89). In our case, the practice of writing/drawing our ethography is a combined process of our listening to the voice of the earth speaking through our ancestral stories and our witnessing of the current changes the Lepcha lifeworld is undergoing. For us, what is critically challenging is that the distinction between the West and the Himalayas is increasingly becoming difficult to draw in the arenas of our social scientific trainings, modern economic development ideologies, environmental discourses and methods of environmental social activism. We have to look through all the refractive lenses that are available; otherwise, we do not see any future for environmental humanities in our scholarship as it has been inspired by indigenous knowledge but currently remains a European and Neo-European interdisciplinary field as Dan Smyer Yü remarks in the introduction chapter of this volume. Thus, if environmental humanities is truly being advocated as an interdisciplinary and global endeavour to explore an "earth language" of the multispecies environment by its progenitors (Rose 2013, 107), we see our contemporary Himalayan homeland as a contact zone of the local and the global, imbued with thicker meanings of humanity, indigenous as well as exogenous.

In principle, the continuation of a tradition is entrusted from one generation to the other. In practice, when a new generation encounters and adapts to new, overwhelming changes, tradition also undergoes modifications and transformations both voluntarily and involuntarily. We resonate with what Bergmann and Clingerman say: "being alive means to be exposed to a continuous flow of change" (Bergmann and Clingerman 2018, 5). We have to take into account the complexity of the always-evolving layers of our traditional knowledge in adaptive, contesting and changing modes. We can't help but conceptualise and reconceptualise the space of our indigenous memory and its social dissemination in the face of erasure and overwriting. In this regard, ethos is not necessarily a positive or a neutral word capturing our indigenous social character but rather a meeting ground of our indigeneity and externally introduced changes, enriching or depriving our wellbeing. Everything is in an entangled web and contradictions are all in the same enclosure. For instance, while we are accustomed to maize, a New World crop, and cultivate it regularly, we are also disturbed by how much water modern hydraulic technology can impound from Rongnyoo (the Teesta River) and how much environmental alteration it can do to the river and its surrounding habitats of humans and nonhumans. The current social ethos of the Lepcha is overwhelmingly an environmental ethos of degradation, extinction, dilemma, contestation and hope.

As practicing anthropologist-ethographers, it is dreadful for us to be standing at a point where we have to witness our communities hanging between the ideas of progress and preservation. Progress, in real time, turns out to be a betrayal of promised human flourishing. The forests are rapidly being depleted. The dams have the once mighty rivers in captivity. The fishes can't swim upstream back to the mountains for spawning. The natural resources are being overused and animals are being excessively hunted. The Mt. Kanchenjunga area, which is believed to host *Mayel Lyang*—our hidden earthly paradise of abundance and happiness—is becoming a site of ecological stress and a site of our cultural struggles for survival and continuity. Modern human-induced climate change and environmental transformation are bringing our oral traditions to a dwindling existence crying out for attentiveness and restoration from our fellow Lepcha men and women, scientists, scholars and policymakers. This is an ethographical scene we are witnessing outside our environmental artwork that features our stories of human–earth relations.

In our experimenting with environmental humanities in the manner of lively ethography, we have encountered two recurring challenges concerning how indigenous origin stories and their environmental wisdoms are retrieved beyond the indigenous intent of their cultural restorations. These challenges are namely the romanticisation of the indigenous past and the political instrumentalisation of the indigenous sense of the sacredness of the earth. Similar to the nonnative appropriation of Native American cultures, the romanticisation of the Lepcha cultural tradition is a common occurrence for nontribal people and nonnative researchers in the postcolonial context. When it happens, the Lepcha, especially those of us living in Dzongu Reserve, are looked upon as a native people "living simple uncomplicated lives in close proximity to the spirits of their ancestors" (Gergan 2014, 74). When we are romanticised merely as indigenous environmentalists, our "local roots of wider political processes, and the varying livelihood strategies and the self-interests of these communities" are ignored (Arora 2006, 56). This style of romanticisation appears to be a continuation of the colonial stigmatisation of the Lepcha as "primordial environmentalists" (Arora 2006) as mentioned by Charisma Lepcha in this book. The question is, does this mean that in fear of falling into the romanticising trap, we should not recount our ancestral stories of our ethnic origin, our traditional environmental knowledge and engage with our protective spirits and the supernatural beings embodied in the mountains and rivers? On the contrary, we would like to reclaim our ancestral lived experience and knowledge of the past. When we do, we are compelled to have the awareness that the colonial stereotype of the Lepcha and the postcolonial deconstruction of the colonial stereotype are both distancing us from our ancestral history. One racialised us with stigmas and another robs us of our ecologically niched cultural history in the name of postcolonial destigmatisation.

In response to this colonial–postcolonial dilemma, our conviction is to stand our (ancestral) ground. Like many other indigenous peoples around the world, the Lepcha have relied on oral tradition as a mode of passing down life experiences and knowledge from generation to generation. Our oral narratives are one of the pillars that hold up our cultural identity. They help us sustain a "relational awareness" (Van Dooren and Rose 2016, 81) of our origin in terms of how we have rooted deeply in the Himalayan mountains, rivers and forests. Although our population is small and our orally transmitted cultural knowledge is on the verge of extinction, we are still in an advantageous position to restore what has been lost and carry on what is available to our future generations. Just as our critics say about us, as a complex human group with desires, agencies and objectives for self-preservation (Arora 2006, 57–59), we do ask, take and consume resources from nature; however, we do not wish to consume them in the same environmentally insensitive way as the majority of the world's population is doing. Instead, our ancestors had a precious body of knowledge about sustainable living with the earth in ecological, spiritual and moral senses. This knowledge needs to be preserved as both a local tradition and a world heritage. In this regard, our myths are the ancient ethographies woven together by our ancestors and their companion spirits sharing the same habitat—our common home!

Origin myths, in our case as oral ethographies, are a system of communication (Barthes 1957), which is meant to sustain our cultural coherence and cohesion. On the one hand, they help us communicate with our ancestors and their supernatural protectors. The latter have been alive since the time of our ancestors and are the time bridges connecting us with our ancient past. On the other hand, many of the oral ethographies we have collected are the basis of current Lepcha efforts at the political representation of the sacredness of our land expressed in public debates over and protests against the dam building projects in Dzongu Reserve. Our ancestral ethographies in this context are not merely the stories of our forefathers and foremothers but are being politically instrumentalised to voice our opposition to the dam constructions because of their obvious disruptions and degradations to our livelihoods and environmental wellbeing. While we are witnessing the effective political union of the Lepcha and the Bhutia peoples' senses of the sacred earth from their respective indigenous and Tibetan Buddhist traditions, we feel called to differentiate this political instrumentalisation from the ecological understanding of the Lepcha indigenous knowledge of the land. For the former, we are thrilled to see the Lepcha, the Bhutia and a concerned general public come together to preserve indigenous heritage sites like Dzongu Reserve. For the latter, we would like to emphasise our sense of Lepcha indigeneity as enmeshed with the Tibetan Buddhist culture, the Hindu influence, the Christian cosmovision and the secular politics of environmental conservation. Many of our fellow Lepcha men and women take Buddhist or Christian first names but retain our common last name "Lepcha." Many of us find livelihoods in the greater Indian society as NGO workers, scholars and development specialists outside our indigenous lifeworld. Since the seventeenth century, our indigenous history has evolved into a history of indigenisation of exogenous peoples and beliefs.

The ancestors of the Bhutia from Tibet brought Tibetan Buddhism, the British introduced Christianity, and British India brought in Hindu labourers from Nepal. Many of our forested hills and valleys were transformed into monocropped fields and tea plantations. Inter-ethnic marriages became common. Cultural ideas and ethos were also mixing and hybridising.

Undoubtedly, our current sense of the sacredness of the earth is mostly a hybrid outcome of the intertwinement of the Lepcha animist and the Bhutia Tibetan Buddhist traditions. The political representation of this hybrid sacredness in the arena of hydraulic politics is gaining attention from the public and in the development-policy realm. From our environmental humanist perspective, we would like to caution ourselves not to let the success of our political instrumentality of the sacred push aside the ecological cosmovision of our ancestors, which has dynamic values for forging new environmental ethics for a sustainable Himalayas. This is one of the reasons why we dedicate our landscape drawings to highlighting the animist habitat of our Lepcha ancestors, which we continue to live in. Our intention is to reclaim our pre-Buddhist history and the ecological wisdom embedded in it. This does not mean to exclude the current hybrid reality of our land but rather to emphasise the potential contribution of Lepcha animist cosmovisions to worldwide explorations for new environmental visions, ethics and responsible modes of living. What we can say is that when the land is sensed, felt and understood in animist terms, it becomes alive and compels us to consider its moral considerability (Goodpaster 1978, 310; Callicott 1987, 217) when we take natural resources and animal lives from it as the source of our sustenance. Noting the case studies featuring Amazonian indigenous animists, we would like to express our resonance with Eduardo Viveiros de Castro's reaffirmation of the indigenous animist understanding of the earth and all life forms as both "a spiritual unity and a corporeal diversity" (Viveiros de 1998, 470). Nature and culture are entwined as one rather than artificially hyper-separated opposites (Plumwood 2012, 18). This means that nature is embodied in culture and culture embodies its humanity in nature. This is why our ethographic narratives emphasise the livingness of the earth in anthropomorphised and geomorphised spirits, humans and animals in order to sense the sacredness of the earth in animist expressions.

Conclusion

For too long, indigenous peoples have been viewed politically as a deprived, underprivileged, and neglected strata of the society and as "nature worshipers," "animists," "primitives," "savages," "hunter gatherers," and more. The latter set of ecologically and religiously grounded nomenclatures are often the racialised cultural identities constructed externally and backed by the "social sciences" of the past. In the postcolonial world today, many nonindigenous scholars are endeavouring to right the past wrongs. Their fight reveals their own struggle against the dominant political and economic paradigm of "extractive globalisation" (Escobar 2018, xi) in their own societies in the forms of colonialism and development. We are secondary in their fight as our social and political conditions apparently serve merely as empirical evidence of both colonial extraction and postcolonial modernised extraction of natural resources couched as development. The labels above are infrequently peeled off to reveal new understandings when indigenous lands are researched simply as historical sites of colonialism and/or as a subject matter of native resource use and management. We are truly victims with little voice. Our protective deities and spirits can no longer protect us from the colossal deprivation and fragility we are facing. We then appear to be waiting for the saving power of the societies and nations that were the extractors of our abundance to start with. We are thus inextricably caught in this "paradox of global environmentalism" (Guha 2000 and Smyer Yü in this book)—those who ruined us are also now appearing to save us. We are painfully growing to be big-hearted enough to work with well-intended scholars, scientists and development policymakers regardless of their ethnic and national origins. We want them to recognise our initiative and agency, not just our victimhood. Saving and preserving our comparatively small homeland has to be done on a global scale rather than as an isolated Himalayan event, as the whole earth is suffering under human-induced climate change and ecological deterioration.

This chapter is our first experiment of lively ethography written in words and drawn as paintings. The full authorship of our ethography does not come from us alone but also from our ancestors, our spirits, our community members, our mountains, rivers and animals who are our flesh and blood. All of us are the inherent authors and narrators of our lively ethography. What words fail to express is expressed by our art and vice versa. Like other environmental humanist works of art, our drawings "highlight the perceptual and affective dimensions of our knowledge of the world" and are expected to be an "important contemporary bridge between culture and nature, human and more than human world" (Bergmann and Clingerman 2018, 1). Each story we recount and draw leads to our profound re-embracing and re-understanding of our own land and people. Our keystrokes and brushes are immersed in the brilliant colours of the spirited earth and with our growing planetary understanding of local environmental crisis and desired sustainability. What we have learned from our ethographic experiment finds resonance in what Rose writes: "Aboriginal people were not conservationists in this contemporary sense of the term because they did not have to be. They had managed continents in such a way that they did not have to face a massive loss of life support systems" (Rose 1996, 4).

We are not alone in this world as victims of colonialism and extractive globalisation and as environmental humanists with a long history of practicing sustainable living. The environmental wisdom of our ancestors is on the verge of extinction in our cultural consciousness; however, it continues to be alive in our land as it speaks through the mothering expressions of the earth as well as through the raging destruction of floods, earthquakes and landslides often provoked by modern extractive development projects. Rose profoundly notes, "This earth has an Aboriginal culture inside" (Rose 1996, 85). The aboriginal world inside our Himalayan homeland is a more-than-human world not just in the multispecies sense but also in the spiritual sense. In this chapter, the ethographic narratives of our ancestors, spirits and protective deities endeavour to revive the cosmic map of our aboriginal earth and to recover "a poetic key to reality" (Rose 2000, 44) in which the earth is not merely an eco-geological system but is also spiritual and sentient in its own right. Our ethographic findings reconfirm our ancestors' vision that the living earth is nourishing as well as destructive. As the earth has its own agency and agenda for its own survival and sustainable flourishing, we humans have to learn to live with the environmental order of the earth. Looking at the dwindling of the Lepcha population since the British came to India, it is not difficult to see that the pattern of our near extinction as a unique ethno-linguistic group and an ecological community has paralleled colonial-led deforestation for cash crops and modernisation projects with their wide ranging environmental disorders. The core of our aboriginal culture currently lives in an externally prescribed reserve area but is left defencelessly wide open for artificial environmental alterations in the name of development and progress. Therein, we are not interested in playing a blame game pointing fingers at external destructive forces of change, because we are also internalising them into our changing lifestyle sustained by consumer goods produced elsewhere in the world with raw materials extracted from other indigenous lands. What we learn from our ethographic experiment is that our ecological conscience is compelled to find resonances with those of multiple other counterparts in the world, in indigenous communities as well as affluent nations. From this experiment, what we can share with our peer environmental humanists around the world is that our conviction that there can be a sustainable future that comes from our ancestors' wisdom of recognising the earth as a living being who animates and breathes life into mountains, rivers, forests and humans and nonhuman beings. There is an ancient foundation for our future!

References

Arora, Vibha. 2006. "The Forest of Symbols Embodied in the Tholung Sacred Landscape of North Sikkim, India." *Conservation and Society*4 (1): 55–83.

Arora, Vibha. 2007. *Assertive Identities Indigeneity and the Politics of Recognition as a Tribe: The Bhutias, the Lepchas, and the Limbus of Sikkim*. New Delhi: Indian Sociological Society.

Barthes, Roland. 1957. *Mythologies*. Paris: Les Lettresnouvelles.

Bergmann, Sigurd and Forrest Clingerman. 2018. "Introduction: Exploring Nature's Texture." In *Arts, Religion, and the Environment: Exploring Nature's Texture*, edited by S. Bergmann and F. Clingerman, 1–13. Boston: Brill Rodopi.

Bergmann, Sigurd. 2018. "With-In: Towards an Aesth/Ethics of Prepositions." In *Arts, Religion, and the Environment: Exploring Nature's Texture*, edited by S. Bergmann and F. Cligerman, 17–42.Boston: Brill Rodopi.

Callicott, J. Baird. 1987. "The Conceptual Foundations of the Land Ethic." In *Companion to A Sand County Almanac*, edited by J. Baird Callicott, 186–220.Madison: University of Wisconsin Press.

Cockerill, Kristan, Melanie Armstrong, Jennifer Richter, and Jordan G. Okie.2017. *Environmental Realism: Challenging Solutions*. Cham, Switzerland: Palgrave/Springer.

Davis, Heather and Etienne Turpin. 2015. "Art & Death: Lives Between the fifth Assessment & the Sixth Extinction." In *Art in the Anthropocene Encounters Among Aesthetics, Politics, Environments and Epistemologies*, edited by D. Heather and E. Turpin, 3–29. London: Open Humanities Press.

Escobar, Arturo. 2018. *Designs for the Pluriverse: Radical Interdependence, Autonomy, and the Making of Worlds*. Durham: Duke University Press.

Gergan, Mabel D.2014. “Precarity and Possibility: On Being Young and Indigenous in Sikkim, India.” *Himalaya* 34 (2): 67–80.

Goodpaster, Kenneth. 1978. “On Being Morally Considerable.” *Journal of Philosophy* 22: 308–325.

Guha, Ramachandra. 2000. “The Paradox of Global Environmentalism.” *Current History* 99 (640): 367–370.

Hall, Laura. 2015. “My Mother’s Garden: Aesthetics, Indigenous Renewal, and Creativity.” In *Art in the Anthropocene Encounters Among Aesthetics, Politics, Environments and Epistemologie*, edited by Heather Davis and Etienne Turpin,283–292. London: Open Humanities Press.

Plumwood, Val. 2012. *The Eye of the Crocodile*. Canberra: ANU E Press.

Ramakrishnan, P.S., K.G. Saxena, K.S. Rao, and G. Sharma.2011. *Cultural Landscapes: The Basis for Linking Biodiversity Conservation with the Sustainable Development*. New Delhi: National Institute of Ecology.

Rose, Deborah Bird. 2013. “Val Plumwood’s Philosophical Animism: Attentive Interactions in the Sentient World.” *Environmental Humanities*3: 93–109.

Rose, Deborah Bird. 2000. *Dingo Makes Us Human: Life and Land in an Australian Aboriginal Culture*. Cambridge: Cambridge University Press.

Rose, Deborah Bird. 1996.*Nourishing Terrains: Australian Aboriginal Views of Landscape and Wilderness*. Australian Heritage Commission.

Rudwick, Martin J.S.2005. *Bursting the Limits of Time: The Reconstruction of Geohistory in the Age of Revolution*. Chicago: University of Chicago Press.

Smyer Yü, Dan 2015*Mindscaping the Landscape of Tibet: Place, Memorability, Ecoaesthetics*. Boston: De Gruyter.

Van Dooren, Thom and Deborah Bird Rose. 2016. “Lively Ethography: Storying Animist Worlds.” *Environmental Humanities* 8 (1):77–94.

Viveiros de Castro, Eduardo. 1998. “Cosmological Deixis and Amerindian Perspectivism.” *The Journal of the Royal Anthropological Institute* 4 (3): 469–488.

6 Muddying the waters

The invention and enclosure of Tibet's wetlands

Ruth Gamble

If you were to ask most people the first image they had when they thought of Tibet, it is unlikely it would be wetlands. Travel brochures and pictorial spreads of the region more usually include pictures of the world's highest mountains, including Chomolungma (Mt. Everest) and the glaciers that make up the world's "third pole." Its unique wetlands and the critical role they play in the world's most extensive hydrological system—the Great Himalayan Watershed that stretches from Afghanistan through South, Southeast and East Asia—are less well known. The Plateau's wetlands include vast peatlands, permafrost fields, swamps, marshes, lakesides and riversides. Despite having no coastline—the most usual wetlands site—the Plateau houses between 30% and 40% of China's total wetland area (Wei et al. 2019). Like all wetlands, they play an essential role in earth systems. They purify water and help to stabilise hydrological and climatic cycles. They minimise flooding, prevent droughts, maintaining groundwater supplies and stop erosion (Gujja 2005, 76–80; Page and Baird 2016, 35–57). As the Tibetan Plateau houses the headwaters of the Great Himalayan Watershed's eight major rivers, its wetlands' sustainable functioning is of particular importance to earth systems and the 45% of humanity that depend on these eight river systems. Furthermore, wetlands are sites of intense biodiversity, and given the region's distinct topography and geographical position, Tibet's wetlands play a particularly important role in the globe's ecological variety. They are the habitat of a diverse range of species, particularly migratory birds and fish. Fifty-four per cent of Asia's endangered species of ducks and geese rely on Tibet's wetlands as migration sites (An et al. 2007).

Tibet's wetlands are also globally crucial because of their relative health. Worldwide, at least 80% of wetlands are degraded (Zhao et al. 2015), and many more have been destroyed. Wetland destruction is caused by either change to their structure—when they are drained, filled in with soil or dredged—or when their water sources dry out. As Tibet's wetlands are generally healthy, their importance to the Greater Himalayan Watershed has made their survival a priority for conservation biologists and state-backed water management programs.

Their health is a result of geographic luck and pre-1950s local water management practices. They were lucky because they were positioned in sometimes-frozen environments that did not sustain high-density populations and high agricultural or urbanisation levels. All of these human activities have proved detrimental to wetlands in other regions. But the region's inhabitants should also be commended for enacting less damage on Tibet's wetlands than their lowland counterparts. As this chapter will show, the limits that Tibetan culture and society placed on wetland destruction were both socio-cultural and survivalist. The Tibetans related to the Plateau's many wetlands in a variety of different ways. Like other agriculturalists, they drained wetlands and manipulated the flow of water through their land. But their socio-cultural system did place some restrictions on how they approached bodies of water and sites of intense biodiversity. Some of these restrictions on behaviour were enacted for metaphysical reasons, and others to ensure future ecological systems services.

The limited protection they provided for wetlands was not encouraged by any specific relationship with the ecological construct we call "wetlands." Until recently, there was no generic word for "wetland" among the Plateau's languages. The region is dotted with "saturated or inundated ecosystems," which is the much-cited definition of a wetland,[1] but these systems were not understood as representatives of a singular phenomenon. Instead, the wetlands that remain were preserved through various localised responses that ranged from respect—when these sites were taken to be the home of water spirits, water sources, or cherished birds—to benign neglect operating within a more extensive, relatively sustainable eco-social system.

The chapter will compare these diverse localised responses to wetlands across the Tibetan Plateau with the existential threat to the wetlands that unfolded after incorporation into the People's Republic of China (PRC) in the 1950s. From this point until the 1990s, the PRC government enacted a series of Marxist-inspired, high-modernist, socially controlling and engineering-intensive schemes that severely damaged the Plateau's wetlands. This threat to the wetlands arose, ironically, during the same period that scientists operating on the other side of the Cold War divide within techno-capitalist systems were uncovering wetlands' vital role in earth systems. The international focus on wetlands led eventually to a reversal in PRC policy in the early 1990s, after the Cold War, and the 1992 Rio Earth Summit. But as this chapter explains, the reversal in policy towards the wetland was not accompanied by a government practice change. The PRC government sought to implement its conservation practices through statist, centralised methods that created a clear divide between areas to be protected and those that were not to be protected, a process known as "fortress conservation" that encloses places within fences and state-made boundaries (Büscher and Fletcher 2015). These methods separated the socio-cultural human systems from nonhuman systems, instead of recognising that both sustained the wetlands. They did not recognise the intersections between the sites' fixed land and the fluid water that moved through it. They worked to increase the quantity rather than the quality of the wetlands, and—perhaps even more damagingly—were presented as a replacement for community conservation practices. This unfortunate governance practice has led to community tensions and environmental degradation that continue today. Now, the Plateau's governors and inhabitants must fix these legacies and adapt to the environmental stresses caused by rising temperatures.

Wetland diversity, diversity in wetlands

Wetlands occur across the Plateau because it houses many salt and freshwater lakes and the upper reaches of eight major river basins: the Senge Tsangpo (Indus River), Ganges, Yarlung Tsangpo (Brahmaputra), Gyelmo Ngulchu (Salween), Dzachu (Mekong), Drichu (Yangzi), Machu (Yellow) and—along with the Tian Shan Mountains—the Tarim Rivers.

Since they began to study the Plateau's wetlands in depth in the 1990s, natural scientists have come up with at least five different schemes to classify the Tibetan Plateau's wetlands (Zhao et al. 2015, 880). The wetlands' many forms are created through the inundation by river, spring and geothermal-steam of various shrubs and tundra. These combinations of water and plants create swamps, marshes, pools and the most common form of Plateau wetlands, peatlands. Peatlands develop in broad valleys and play a particularly important role in both the globe's hydrogen and carbon cycles.

Peatlands (otherwise known as mires) are wetlands that have been created where stores of decomposing organic matter have accumulated. They contain large amounts of water and carbon—more carbon than any other terrestrial substance—because they are produced when plant matter is suspended in water and ice and stops decomposing. In Tibet, most of the organic matter that accumulates in these depressions are old layers of grass. They rely on either regular water inflows or freezing for the organic matter to become waterlogged, sitting on top of mineral soil. The old grass layer sinks into the water, and a new layer grows on top of it. This process repeats and repeats until a peatland arises. In the peatlands found on the Tibetan Plateau, these layers can be as thin as 30 centimetres and as thick as 10 metres (Page and Baird 2016, 35–57). Peatlands store carbon when they are continually wetted, but when they dry out or catch fire, they release atmospheric carbon dioxide (Cao et al. 2018). Peatlands can also begin to emit another greenhouse gas called nitrous oxide if they have been disturbed by chemical imbalances from fertilised fields. This process is called eutrophication (Hao et al. 2017, 815–829).

The depth of a peatland can affect the amount of carbon it processes. Decomposition of organic matter and, therefore, carbon release occurs near a peatland surface, while its lower levels remain stable. The grassy "skins," so to speak, of the peatlands protect them from the degradation and slow the release of carbon dioxide and other gasses (Iqbal and Shang 2020).

Peatlands have a complementary relationship with permafrost. In summer, if the water within them is liquid rather than frozen, peatlands insulate underground permafrost, making it less suspectable to ground-level temperature variations. In winter, peatlands are capable conductors of cold, allowing it to penetrate the soil and create permafrost where it otherwise would not exist (UNEP 2018). Permafrost, for its part, can stabilise peatlands, providing a solid base on which it can form (UNEP 2018, 40).

Peatlands need constant watering to survive and, therefore, only tend to accumulate in two parts of the globe: in places where frozen water in permafrost stabilises peat production (i.e. the Arctic and the Tibetan Plateau) or underneath tropical rainforests. The Tibetan Plateau's peat is arguably more susceptible to climate change than the peat in these other sources. Climate change is occurring at twice global averages on the Plateau, and the region's high concentration of permafrost, snow-and-ice-fed rivers, and its altitudinal diversity all exaggerate the impact a changing climate will have on its peatlands (Bai et al. 2013). This vulnerability is also true for the Plateau's other wetlands. What is more, the Plateau's wetlands help stabilise local precipitation as well as aiding it. Studies of the eastern Tibetan wetlands at Zoigé have, for example, shown that this large conglobation has had a cooling effect locally, sustaining higher precipitation rates. Scientists established this cycle by observing the reverse; wetland degradation has led to precipitation reduction and rising air and soil temperatures. These changes, in turn, have led to more wetland degradation (Bai et al. 2013).

How did we get here?

Humans, generally, as a species, have a long history of wetland degradation. Our antipathy to wetlands began with our long struggle to develop agriculture. For agriculturalists, wetlands presented an obstacle to arable land that needed to be dried (Scott 2017). Their murky waters and high biodiversity have often been viewed as sites of disease and threat; they are the proverbial swamps that need to be drained. Some wetlands, particularly peatlands, have, furthermore, been used as sources of fuel in areas with limited forests (Tcvetkov 2017, 1–12.). As cities developed from agricultural societies, they tended to be built on drained land next to rivers, lakes, or seas. Indeed, many of the world's largest cities were constructed on drained soil (Giblett 2016), including many cities within the Greater Himalaya Watershed (Smith 2006; Elvin 2008, 105–115).

Although the Tibetan Plateau's wetlands are less degraded than many others worldwide, and there were no large cities built on it until recently, this does not mean that they have not been affected at all by the Plateau's human occupation. Like most other agricultural societies worldwide, the Tibetans drained wetlands and redirected their water to create arable fields. The Plateau's history is replete with heroic drainage and irrigation stories. For example, on one reading, the story of the building of the Jokhang Temple, Tibet's most sacred temple in Lhasa, is a drainage story. According to the most commonly held version of this story, the site of the temple was chosen—through geomancy or some other form of divination—in the middle of a lake, that was then filled to create land for the temple to be built (Dorje 2010, 70). As the Lhasa plain is flat and marshy, this lake was likely a marsh.

But the limited amount of arable land on the Plateau concentrated the region's environmental transformation in specific areas. They only engaged in agriculture in the river valleys. Most of the river-valley agriculture depended on complex irrigation systems that were preceded by wetland drainage. These processes stretch back at least 3,000 years, and the Tibetan river valleys may even be the site of one type of barley domestication (Zeng et al. 2018).

These drained and irrigated river valley agricultural settlements were also the site of the first large-scale Tibetan polity, the Tibetan Empire (seventh–ninth centuries CE). After starting in these valleys, the Tibetan Empire spread beyond them and eventually controlled all of the Plateau and its surrounding regions (van Schaik 2011, 21–40). It was followed by a series of smaller fragmented polities across the Plateau before the Plateau was incorporated into the Mongol Empire (1240–1354). Mongol groups and then the Qing Empire (1644–1912) continued to exercise influence in Tibet until the early 1900s, but they had little to do with local governance, including water governance.

Within the records of legal disputes in Tibetan communities, one of the best records we have to describe their pre-1950s social lives, many conflicts were not just over land ownership but "subjects, water, wood, grass" (Kensaku 2016, 200). Ownership within this context was understood to refer to land and some elements that were managed communally and did not stay within the bounds of a person's field, such as people, water, trees and the grass pastures on nearby hillsides. Each valley had a water controller (*chu dpon*), and power within this irrigated society was implicitly associated with water's control. The lushness of fields, the growth of trees and grass, all depended on water access, so water was power. Many of the words associated with power and control in the Tibetic languages refer to the control and redirection of water. One of the most common names for politics itself, for example, is *chab srid*, "becoming water," and the word for government, *gzhung*, has the alternate meaning of "central stream" (Sörenson 2003).

The term *gzhung* was often used to describe the more-central authorities that headed loose confederations of these local rulers. Local rulers in the upper Senge Tsangpo, Yarlung Tsangpo and Gyelmo Ngulchu River Valleys answered to a series of regimes, including the Dalai Lamas' Lhasa-based Ganden Podrang (1642–1959) government. Those within the Dzachu (Mekong), Drichu (Yangzi) and Machu (Yellow) maintained religious affiliations with Lhasa-based religious networks. Still, they were ruled by either independent kingdoms or, in the late nineteenth and early twentieth centuries, Chinese and Hui warlords.

According to the data from pollen samples taken from wetland cores across central Tibet, many pre-historical marshes within the Yarlung Tsangpo River Basin have since been drained. These samples record a long history of wetlands that grew and shrunk with climate changes until humans arrived in the area, and the wetlands started to shrink significantly (Hudson et al. 2016). Even after human occupation, the Basin's government still encountered distinctly swampy problems. The residents of the Lhasa Plain, for example, had to deal with twice-yearly inundations and built up complicated urban drainage and levee systems to resist them (Sörenson 2003, 85–134). Furthermore, a similar process of drainage and water redirection was carried out in the Yarlung Valley south of the Yarlung Tsangpo River in Central Tibet (Sörensen et al. 2007). The traditional irrigation systems of Shigatse and Gyantse, which have been repeatedly drawn and photographed and survive in fragments today, were also drainage systems.

Despite the critical role that drainage and irrigation played in Tibetan society and communities, several factors mitigated the more pervasive transformation of the Plateau's land and large-scale destruction of its wetlands. Primary among these was the Tibetans combining of agriculturalism with pastoralism. Pastoralism was necessary for survival on the Plateau; most of the region, particularly those parts of it that support permafrost, are not suited to agriculture, but they are suited to animal grazing. Pastoralists, who moved with herds across pastures engaged in several practices that helped protect the wetlands. Fundamentally, their prioritisation of movement and their communal use of pastures (Tan 2018, 87, and 12) meant that they did not have the same necessity or freedom to transform the environment in which they lived. Their communal use of pastures included access to both grass and water; the two were not separated.

Herds tend to stay away from wetlands and, by coincidence, herd-grazed pastures contributed to the production of the region's peatlands. Peat is a—very slowly—renewable resource that collects over time. Limited numbers of yaks grazing on peatlands top layer at the right time of year increases its production (Lu et al. 2015, 1195–1205).

The region's previously mentioned sparse population distribution also aided the wetlands' preservation, as did cultural proscriptions against the degradation of some of them. Wetlands that were marked for protection included the abodes of mythological beings or particular bird species. The type of spirits most commonly associated with wetlands and other small water bodies are *lu* (Tib. *klu*), who are often associated with the serpent-like *nāga*s found in South Asian mythology. *Lu* live in and around water bodies across the Tibetan Plateau and are said to react negatively to insult, pollution, and disturbance. They show their displeasure by supplying too much or too little water or producing epidemics (*Gtsang ma klu' bum chen mo* 1977; Tucci and Samuel 1988, 222).

Along with *lu*, wetlands were sometimes considered special if they attracted black-necked cranes (*Grus nigricollis*; Tib. *khrung khrung dkar mo*) to nest in them. Black-necked cranes represent long-life across the Himalaya and are often incorporated into auspicious drawings and symbols. They are long-lived birds who travel large distances and work with a mate to raise their young. Wetlands in which they make their nests can still be found among the agricultural fields of the Yarlung Tsangpo and Lhasa River Valley systems in Central Tibet.

Along with cultural proscriptions about the destruction of specific wetlands, Tibetans maintained a generalised embargo on mining that slowed the use of peat and other turf forms found in wetlands as fuel sources. Within the Tibetan cultural sphere, mining and ground disturbance are often said to remove "the *bcud* (nutrition or essence) of the earth, of which minerals are the primary form" (Nyima and Yeh 2016, 168). If peat and turf were collected from wetlands, it was taken from the top of the wetland's stores in limited amounts. This turf skimming was, for example, the practice for the removal of peat from the Lhalu Marsh behind the Potala Palace in Lhasa. There it was called *la ma* (Yeh 2009, 111). In other parts of the Plateau, this substance is called "pieces of turf" (*spang bog*) or "pieces of wet meadow" (*na rtog*). This proscription against mining and the limited, surface use of turf was, in turn, enabled by the availability of another fuel source, yak dung.

These factors worked together to preserve wetlands despite the lack of a concept of "wetlands" within the Tibetan cultural sphere. There is no linguistic, written or archaeological evidence that the Plateau's inhabitants had a specific interest in preserving wetlands generally or any of their subgroups. The Tibetans' different usages of and relationships with wetlands are reflected in the variety of names that they gave them in various geographical sites and across time. They called them simply mud (*'dam*), "mud lands" (*'dam sa*), "mud centres" (*dam gzhung*), "wet meadows" (*na, na gor, na chu*), or a combination of both, "muddy, wet meadows" (*na'dam*). The wettest of peatlands were and are sometimes just called "lakes" (*mtsho*) or "water" (*chu*), and the driest of them are generally known just as meadows (*spang*) or grasslands (*rtswa thang*). The neologism used to designate "wetlands" in official signage and discourse, *rlon sa*, pronounced "lönsa," is a relatively new word that people are more likely to see on signs than here in conversation.

The differentiated relationships with individual wetlands are found throughout the records of them in pre-PRC literature. Some of Tibet's most extensive and wettest peatlands were noted and described by locals and visitors to the Plateau. In Emily Yeh's history of the Lhalu Marsh, she refers to conversations with elderly Lhasa residents who remembered it as being much wetter before the 1950s. These elderly residents and other sources confirm that the influential Lhalu family managed the wetland and treated it as both a garden, a retreat and a source of la ma fuel (Yeh 2009).

Pre-modern Tibetan, Chinese and English language texts also contain descriptions of the world's most extensive alpine wetlands in Zoigé, on the eastern edge of the Plateau, across the present-day borders of Sichuan and Gansu. The Zoigé Marsh sits in a high, flat valley just to the east of Amnye Machen Mountain, between the headwaters of the Yellow and Yangtze Rivers. It consists of four distinct but interconnected regions: the Zoigé region area itself, Hongyuan, Ngawa and Songpan. The peat in these interconnected wetlands is usually between 2 and 3 metres thick but can be as deep as 7 metres. The region is exceptionally biodiverse, housing at least thirty-seven bird species, including the much-loved black-neck crane.

Local histories record that the Zoigé Marsh was visited by members of the Tibetan emperor's entourage in the eighth century. Episodes within the epic of Gesar of Ling, a Tibetan King and folk hero, are also set there (NA 1999, Raine 2014). The marsh itself was understood to be a mix of land types, including pastures, lakes and mud. The nomads used the peatlands when they froze in winter as pastures for yaks, but otherwise, they were left mainly untouched (Hayes 2010). Water levels in the region's lakes and swamps fluctuated somewhat, depending on rain patterns and seasons, but there is no history of them drying up or being drained as they were in the twentieth century. On all available evidence, it seems as if the Zoigé Marsh was a constant in the lives of the people of that region for millennia (Chen et al. 2014).

In the north of the Plateau, wetlands were noted by nomads and travellers around Mitikha (Tib. *smi ti kha*) in Nagchu, and around sacred Lake Mapam Tso (Tib. Ma pham tsho known as Lake Manasarovar in India) in Western Tibet. Other, smaller wetlands are scattered across the Plateau among grass pastures and on the margins of farmland. There are also large peat stores in the permafrost areas of Tibet, which are primarily located in the Plateau's very sparsely occupied northwestern region, known as the Changthang Plain.

The downside of high modernism

The relationship between peat and people on the Plateau changed dramatically when most of the Tibetan Plateau was incorporated into PRC during the 1950s. The PRC government's fervour for high-modernist, utopian schemes and the famines, disease outbreaks and dam collapses these schemes caused have been well-documented (Shapiro 2001; Economy 2010), as has its inability to adjust centralised planning decisions to diverse local environs like the Tibetan Plateau (Yeh 2009). But even among the early PRC-government's surfeit of environmentally destructive activities, the draining of the Plateau's wetlands stands out as an example of a poorly engineered program with long-lasting consequences.

In the 1950s and 1960s, the PRC's priority was developing land for agriculture and securing food sources for its large and impoverished population. The Chinese state did not approach Tibet's peatlands as vital ecosystems or hydrological wonders. Instead, it saw them as fallow ground, and it had the funds, workers and ability to build the levees, drains, roads, dams and other infrastructure that would transform these wetlands in a short time (An et al. 2007).

The drainage occurred across the Plateau. The Lhalu Marsh was drained, mined for its peat and turf, polluted and filled with sand (Yeh 2009, 103–137). The Mapam Tso Lake wetlands shrunk (Phan and Menenti 2012). But the most considerable degradation occurred in Zoigé. Zoigé's peat-rich wetlands, known in Chinese as the Ruoergai Marsh (若尔盖沼泽 Ruo'ergai zhaoze), were infamous among the PRC elite as the site of many Red Army deaths during the Long March of 1934–1935 (Hayes 2013). Twenty years later, after eastern Tibet came under PRC control, they marked them for drainage. During the fifties and sixties, the government sent whole work units to Zoigé to "tame the swamps." These work units built nearly 1,000 artificial drainage channels in Zoigé, which ran for 2,864 kilometres and drained approximately 41% of the peatlands (Mei et al. 2012). The idea was to turn this "wasteland" into productive agricultural fields that could produce wheat and even rice. As with most large, state-centered environmental adaptation projects, however, the plan failed and left behind what biologists call an "extinction debt"—long-term ecological impacts that affect plants and animals and can take centuries to manifest (Kuussaari et al. 2009)—that is yet to be paid. Some of the project's consequences became evident straight away, and others have taken longer to manifest.

The issues that became immediately manifest occurred during the peatlands' drainage. To drain and change the peatlands, the government needed to send work units into the area. They needed food, fuel and transport, which were all in short supply. Therefore, the draining of the peatlands was accompanied by an increase in unsustainable road-building and infrastructure in the area. It even led to the creation of a large peat mine in the area for the first time. The peat mine in Hongyuan started as an unofficial mine in the 1960s, as workers moved into the area. In the late 1960s and early 1970s, it was semiprofessionalised with help from the Irish Mire Association (local communication). It became the primary source of fuel for the region's largest industry, a milk powder factory (Bjork 1993).

Reform and environmental retreat

The drainage of the wetlands, infrastructure development, population influx and other environmental stresses had a devastating effect on Zoigé's environment and the other wetlands across the Plateau. By the late 1960s, as the Cultural Revolution was at its peak, Zoigé's whole ecosystem faced collapse; desertification was intensifying, nothing would grow, and even the peat stores available for mining were decreasing (Li et al. 2019, 271). Luckily, by the early 1970s, the approach to wetlands on the other side of the Cold War divide began to change. In the wake of environmental movements in the United States and Western Europe in the 1960s, international pressure to conserve essential ecosystems grew. The Ramsar Convention on Wetlands of 1971 was the first, and still the most influential, international treaty on the protection of wetlands. Named for the city in Iran, Ramsar, where it was signed, the convention works similarly to UNESCO's World Heritage system. Proposals for wetland recognition are made to the body, and once wetlands are recognised as significant, they receive funds and assistance from Ramsar. The health and protection of the wetlands are then reviewed at tri-annual conventions.

The PRC government was an observer of the first meeting of the Ramsar convention, but it did not contract itself to it until 1992. Indeed, the most intense period of wetland degradation across the Plateau occurred between the first Ramsar convention and China's entry into it in the early 1990s (Bai et al. 2013). During this period, not only did channelisation continue to dry the Plateau's wetlands, but in the newly dried areas, local pastoralists and would-be agriculturalists were encouraged to dig wells and canals that further lowered their water table by 5–15 centimetres (Cao et al. 2018). The lowering of their water tables strained both the wetlands and nearby hydrological systems, changed rain patterns, stopped subterraneous water flows, hastened the drying of connected wetlands, and continued the process of desertification.

Desertification and water shortages became evident around the Lhalu Marsh and Lake Mapam Tso during this period (Niu et al. 2012). But the region most affected by this drying was Zoigé, which by that stage, the Ramsar convention was monitoring (Ramsar 2015). Desertification in Zoigé increased rapidly from 1975 to 1990 (Hu et al. 2015). This increase fed into and was exacerbated by the rise in local temperatures, the overgrazing of available land and an increase in the numbers of small marsupials, particularly marmots and pika. Marmot and pika numbers grew in line with peatland degradation, and their burrowing further damaged wetlands and grasslands (Hu et al. 2015).

The pace of degradation only began to slow in the early 1990s. Following the end of the Cold War and the Rio Conference on the environment in 1992, the PRC adopted the "sustainable development" model. This new model included the setting aside of areas of particular environmental benefit as reserves (Hayes 2010, 340) and the signing of the Ramsar convention. This signing allowed Ramsar authorities to actively monitor and recognise wetland sites. The language that the authorities used to describe these systems also changed; instead of "marshes and swamps" (*zhaoze*沼泽), they were now described by the Chinese word *shidi* (湿地), which is a direct translation of the English term "wetland."

Ramsar soon recognised shidi or wetlands across the Plateau. Zoigé was one of the first wetlands to be recognised and protected within the PRC. In 1993, the government inaugurated conservation reserves within it, and it was renamed the "Ruoergai Wetland" (Ruoergai shidi 若尔盖湿地) in Chinese and the "Zoigé Wetland" (Mdzod dge'i rlon sa) in Tibetan (Zhang et al. 2016). Neither the recognised Ramsar site nor the government's national reserve covered all Zoigé's wetlands, but this recognition did facilitate a change in approach to the region's wetlands. Further drainage was halted and rehabilitation programs were instated. Other Ramsar sites were later recognised within the larger Zoigé wetland area across the Gansu border in Luqu County (2011) and along the Yellow River (2020).

Elsewhere on the Plateau, wetlands around several lakes have been recognised over the past two decades: Tso Ngönpo (Tib. Mtsho sngon po, Ch. 青海 Qinghai 1992), Mapam Tso (Tib. Ma pham mtsho, Ch. 玛旁雍错 Mapamyong Cuo, 2004), Tso Kyaring (Tib. Mtsho skya ring, Ch. 扎陵湖 Zhaling Hu, 2004), Tso Ngoring (Tib. Mtsho sngo ring, Ch. 鄂陵湖 Eling Hu 2004), Serling Tso (Tib. Ser gling mtsho, Ch. 色林错 Selin Cuo 2018), Trari Namtso (Tib. Bkra ri gnam mtsho, Ch. 扎日南木错 Zha ri nan mu cuo 2020). Other marsh sites have also been recognised at Pota Tso (Tib. Po ta mtsho, Ch. 普达措 Pudacuo 2004) and Nakpak Tso (Tib. Nags phags mtsho Ch. 纳帕海 Napahai 2008) near Gyaltang in northern Yunnan, and Mitikha (Tib. Smi ti ka, Ch. 错查村 Cuo chacun 2012) in Nagchu Province north of Lhasa. The largest site to be recognised outside of Zoigé is Tromza Göma (Tib. Khrom gza' gong ma, Ch. 长沙贡马乡 Chang sha gong ma) in Sershul County (2018) in northeastern Sichuan. This site is less-well-known to scientists and less-well documented in Tibetan-language texts

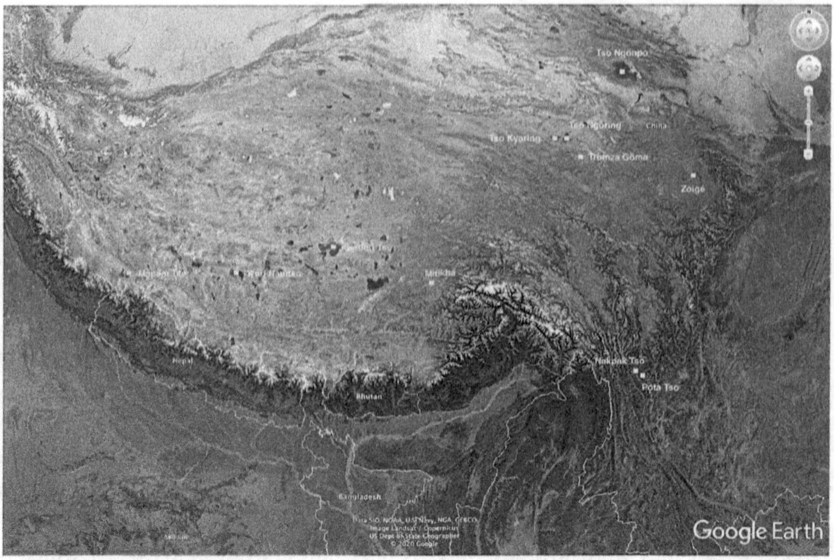

Figure 6.1. Ramsar sites in Tibet.

Adapted from Google Earth.

than the region's other wetlands, but it is of immense importance to the upper Yangtze catchment area (Figure 6.1).[2]

While a Ramsar designation has been offered to some of the region's wetlands, others have been protected through national or provisional conservation provisions. The Lhalu Wetlands were declared a protected area in 1995 and their status has been upgraded several times (Gujja et al. 2003). Other non-Ramsar but important peatlands were discovered by scientists within the Source of the Three Rivers National Nature Reserve (三江源国家级自然保护 Sanjiang yuan guojia ji ziran baohu qu, which houses the headwaters of the Yellow, Yangzi and Mekong Rivers) where they designated eight wetland sites and the permafrost peatlands of the Changthang Plain.

The recognition of these sites as places of particular biodiversity, ecological and hydrological import was the first step in the slow regeneration of these areas. Both the central government and international organisations paid for the scientific research of Tibet's wetlands, concentrating most of their efforts on the Zoigé Marsh. The marsh was dubbed "the Plateau's kidney" or *gaoyuan zhi shen* (高原之肾) for its role in water purification. Governmental and nongovernmental organisations worked to restore wetland areas with practical measures such as the filling of drains, the closing of mines (Bjork 1993), the replanting of desiccated hillsides and the enclosure of mainly degraded peatland areas (Hu et al. 2015). Lhalu Marsh near Lhasa was also studied and somewhat rehabilitated (Gujja 2005). The plateau's other peatlands received less attention but arguably required less restoration.

The limits of reform

Although active wetland destruction on the Plateau has been curtailed since the 1990s, and the wetlands have partially recovered, they are still slowly degrading. The reasons for this continued degradation are multiple and difficult to forestall.

The most impactful influence on them has been the PRC's persistent statist responses to conservation. The PRC state's adoption of sustainable frameworks has curved its worse excesses. However, their new managerial paradigm is still grounded in resource utilitarianism (Gremaud 2014) that cannot conceptualism or allow for the interconnected socio-ecologies that previously sustained these wetlands. The gap between this resource utilitarianism and socio-ecologies is evident across the PRC. Still, it is particularly apparent in regions that the state designates as the home of "minority ethnicities" (shaoshu minzu 少數民族), such as the Plateau. In these regions, resource-utilitarianism-influenced conservation strategies have been ineffective. Wetland conservation is another example of this. According to Xu et al. (2019), the government's conservation efforts between 2000 and 2015 have recovered 6,765 square kilometres of wetlands. But in the same period, human activity has degraded 16,032 square kilometres of wetland. Half of the region's monitored wetland area has been affected by the increase and intensification of agriculture (Xu et al. 2019, 3066). And the concrete and pavement used in urbanisation projects has altered water flows and encroached further on wetland environments (Xu et al. 2019, 3067).

What is more, the quality and positioning of the wetlands that were officially recovered are problematic. As Xu et al. (2019) show, the "recovery" has been built on the new wetting of land by large reservoirs created to service state-owned hydropower constructions. As repeated scientific studies have shown, reconstructed wetlands, especially those associated with hydropower projects, do not reach the same levels of system health and ecological function as those that developed organically. They depend on monocultures and take decades to approximate the carbon-storage capacity and water filtration effects of older wetlands. Xu et al. (2019) register equal concern for the geographic concentration of new wetting endeavors and other governance-led wetland rehabilitation projects on the Plateau. Most of them are within the Three Parallel Rivers region (三江并流 Sanjiang Bingliu) in northern Yunnan, which is also home to the greatest concentration of hydropower projects on the planet.

Government records suggest that both this region and the Source of the Three Rivers region further north on the Plateau's north-eastern edge, which is home to the extensive Zoigé network of wetlands, have been targeted by the PRC government for intensive wetland conservation (State Forestry Administration 2000). What is more, much of the 25% of wetland gains that the government claims has occurred since 2000 consists of repurposed or relabelled agricultural and grasslands (Xu et al. 2019) called "wetlands" and enclosed within fences and legally bounded conservation regions. Conservation through enclosure is usually referred to as "fortress conservation" (Escobar 1998; Lezak 2019). It is controversial within the international conservation movement as it tends to entail the removal of land from Indigenous peoples. Therefore, conservation through enclosure on the Plateau continues the dispossession of Tibetan people and the undermining of their multifaceted but generally sustaining relationships with wetlands. All that has changed is the focus: where before the state was interfering to transform "unproductive land" into productive land, at this stage, it was interfering to seal off what it considers environmentally and hydrologically important land from a resource utilitarian perspective.

The many analyses conducted on this form of conservation suggest that it does not allow either side of the enclosure divide to thrive (Duffy 2016). There is no understanding of the benefits of movement in and out of the zones designated as "special," or indeed the broader hydrological and ecological landscapes in which they are positioned. There is, furthermore, a tendency towards categorisation that makes the preservation of "wetlands," which, as Kuntalala Lahiri-Dut (2014) has pointed out, do not fit easily within the environmental categories of water and land, difficult. The designation of special wetland zones, in other words, create a fortress around the "land" but does not protect the water that flows through it.

The enclosure of wetlands fits within a broader pattern the PRC government enacted during the 1990s of land enclosure and privatisation across the Plateau's pastures. These enclosures have been studied in depth elsewhere (Hayes 2010; Bauer 2015; Gaerrang 2017). Privatisation was presented as a remedy to earlier collectivisation programs, but they represented a further step away from pre-1950s communal land-usage practices on the Plateau. Instead of restoring social and ecological order, they encouraged out-migration, fence building, the use of chemical fertiliser, urbanisation, industrialisation and tourism (Gaerrang 2017).

The degradation wrought on the Source of Three Rivers area through privatisation policies was particularly dramatic. In this region, out-migration and industries' development led to the concentration of stock in larger herds managed by fewer people. These larger herds were then concentrated in smaller sections of land, where overuse degraded the pastures and wetlands and led to the proliferation of rodents (Gaerrang 2017). It also disabled some movement between winter and summer pastures. The regions' wetlands coalesce in valleys, and when they freeze, they are suitable for winter pastures. Herders used to bring their animals down to these regions in winter. After fencing, they had to use them in summer too, and the yaks' hoofmarks cut away at the peatlands' upper level, it "skin." Yaks also use the remnants of the 1950s–1990s channels as water holes and wear away at their edges. These holes and other forms of degradation were then exacerbated by tourist infrastructure and the runoff from agriculture, which changed the peatlands' chemical balance (Huo et al. 2013).

During the 2000s, the government began to realise the environmental degradation privatisation had incurred and shifted their focus again away from privatisation to an "eco-migration policy." At its base, the eco-migration policies blamed the region's ecological issues on local pastoralists rather than the five decades of disastrous environmental policies that had shaped their behaviour. It operated on the premise that if the pastoralists and their livestock were removed from the grasslands, the grasslands' health and the region's hydrological system would be restored (Bauer 2015). This approach ignored the palaeo-ecological evidence that showed the grasslands and their associated wetlands were, in fact, cultural environments that had been created by herders (Kaiser et al. 2007). Traditional land-use regimes in which smaller herds of yak were grazed on peatlands in the winter and moved from region to region aided grass-based peatlands and did not lead to grassland degradation (Sun et al. 2018; Cao et al. 2019). And some of the peatlands that have been closed off to yaks and other herbivores are not recovering as well as expected (Lu et al. 2015).

Lack of local engagement also meant that the initiatives would not be sustainable. Internationally, best-practice in wetland preservation is the development of community-based conservation regimes that balance local livelihoods with conservation and employ traditional environmental knowledge (Sayer et al. 2018). Outside of the sealed zones, the pursuit of resource utilitarianism has led to the construction of roads, dams and train networks through wetlands, and they all direct runoff that damages wetlands (Shang et al. 2014).

All of these stresses on the region are, furthermore, being exaggerated by the changing climate. Rising temperatures are encouraging drying and intensifying precipitation patterns. Previously, the monsoon would creep up through the river valleys carrying rain, and westerly winds would bring winter snows. Snowmelt and spring rain would fall relatively gently. Now there are more rain, hail and snow storms (Li et al. 2019) that flush through the wetlands, stripping them of nutrients (Xu et al. 2019).

Conclusions

The changes occurring across the Plateau's wetlands suggest that the development of a whole of Plateau wetland conservation scheme, and the scientific, diplomatic and statist processes that could produce this scheme may have been necessary to help the wetlands survive their current predicament. If it worked, a plateau-wide approach would protect all the wetlands in their diverse forms and allow them to continue providing their ecological systems services to the Plateau's and the region's human and nonhuman inhabitants. But the history of the attempts to manage Tibet's wetlands suggests that the localised, community-led management schemes and sparse populations helped preserve the wetlands better than any of the disconnected, state-led programs that have been imposed from afar. It is unlikely that the same statist centralised institutions that led to the destruction of the wetland will ensure their survival. To date, these institutions have sought to replace diverse local discourses and socio-cultural practices with state-managed exclusory zones that disconnect locals from the wetlands. The enclosure and exclusion of wetlands may have saved them from the destructive potential of the state's policies, but it will not be enough to preserve them as the climate changes. The reformation of a bespoke approach to wetlands, one that approaches each of them through pluralistic, pre-statist and community-led schemes, is the best chance of ensuring their conservation in a warming world. But it means subverting the dominant statist approach to their care.

Notes

1 This is the Ramsar definition of a wetland (Ramsar 2015). Ramsar is the name of the city in which the first treaty on wetland conservation was signed, and the name of the international organisation that monitors wetlands around the world. It will be discussed shortly.

2 All information on the Ramsar sites is taken from the Ramsar official website, https://www.ramsar.org/.

References

An, Shuqing, Harbin Li, Baohua Guan, Changfang Zhou, Zhongsheng Wang, Zifa Deng, Yingbiao Zhi. 2007. "China's natural wetlands: past problems, current status, and future challenges." *AMBIO: A Journal of the Human Environment* 36 (4): 335–343.

Bai, Junhong, Qiong-qiong Lu, Junjing Wang, Qingqing Zhao, Hua Ouyang, Wei Deng, and Ai-nong Li. 2013. "Landscape pattern evolution processes of alpine wetlands and their driving factors in the Zoige plateau of China." *Journal of Mountain Science* 10 (1): 54–67.

Bai, Junhong, Qiongqiong Lu, Qingqing Zhao, Junjing Wang, and Hua Ouyang. 2013. "Effects of alpine wetland landscapes on regional climate on the Zoige Plateau of China." *Advances in Meteorology* 2013, Article ID 972430, https://doi.org/10.1155/2013/972430.

Bauer, Kenneth. 2015. "New homes, new lives-the social and economic effects of resettlement on Tibetan nomads (Yushu prefecture, Qinghai province, PRC)." *Nomadic Peoples* 19 (2): 209–220.

Bjork, Sven. 1993. "The Hongyuan Wetland Research Project, Tibet." *Vesanse*.

Büscher, Bram, and Robert Fletcher. "Accumulation by Conservation." *New Political Economy* 20 (2) (2015): 273–298.

Cao, Jianjun, Jan F. Adamowski, Ravinesh C. Deo, Xueyun Xu, Yifan Gong, and Qi Feng. 2019. "Grassland degradation on the Qinghai-Tibetan Plateau: re-evaluation of causative factors." *Rangeland Ecology & Management* 72(6): 988–995.

Cao, R., Y. Chen, X. Wu, Q. Zhou, and S. Sun. 2018. "The effect of drainage on CO2, CH4 and N2O emissions in the Zoige peatland: a 40-month in situ study." *Mires and Peat* 21 (10): 1–15.

Chen, Huai, Gang Yang, Changhui Peng, Yao Zhang, Dan Zhu, Qiuan Zhu, Ji Hu. 2014. "The carbon stock of alpine peatlands on the Qinghai–Tibetan Plateau during the Holocene and their future fate." *Quaternary Science Reviews* 95: 151–158.

Dorje, Gyurme. 2010. *Jokhang: Tibet's Most Sacred Buddhist Temple*. Vienna: Edition Hansjorg Mayer.

Duffy, Rosaleen. 2016. "War, by conservation." *Geoforum* 69: 238–248.

Economy, Elizabeth C. 2010. *The River Runs Black: the environmental challenge to China's future*. Ithaca: Cornell University Press.

Elvin, Mark. 2008. *The retreat of the elephants: an environmental history of China*. New Haven: Yale University Press.

Escobar, A. 1998. "Whose Knowledge, Whose nature? Biodiversity, Conservation, and the Political Ecology of Social Movements." *Journal of Political Ecology* 5: 53–82.

Gaerrang, Kabzung. 2017. "Tibetan Buddhism, Wetland Transformation, and Environmentalism in Tibetan Pastoral Areas of Western China." *Conservation and Society* 15 (1): 14–23.

Giblett, R. 2016. *Cities and Wetlands: The Return of the Repressed in Nature and Culture (Environmental Cultures)*. London: Bloomsbury Academic.

Gremaud, Ann-Sofie N. 2014. "Power and Purity: Nature as Resource in a Troubled Society." *Environmental Humanities* 5(1): 77–100.

Gujja, Biksham. 2005. "WWF International's regional approach to conserving high-altitude wetlands and lakes in the Himalaya," *Mountain Research and Development* 25 (1): 76–80.

Gujja, Biksham, Archana Chatterjee, Parikshit Gautam, and Pankaj Chandan. 2003. "Wetlands and lakes at the top of the world." *Mountain Research and Development* 23 (3): 219–221.

Hao, Wang, Lingfei Yu, Zhenhua Zhang, Wei Liu, Litong Chen, Guangmin Cao, and Haowei Yue. 2017. "Molecular mechanisms of water table lowering and nitrogen deposition in affecting greenhouse gas emissions from a Tibetan alpine wetland." *Global Change Biology* 23 (2): 815–829.

Hayes, Jack Patrick. 2010. "Modernisation with Local Characteristics: Development Efforts and the Environment on the Zoige Grass and Wetlands, 1949–2005." *Environment and History* 16 (3): 323–347.

Hayes, Jack Patrick. 2013. "From Great Green Walls to deadly mires: wetlands as military environments and ecosystems in Chinese history." *Water History* 5 (1): 7–26.

Hu, Guangyin, Zhibao Dong, Junfeng Lu, and Changzhen Yan. 2015. "The developmental trend and influencing factors of aeolian desertification in the Zoige Basin, eastern Qinghai–Tibet Plateau." *Aeolian Research* 19: 275–281.

Hudson, Adam M., John W. Olsen, Jay Quade, Guoliang Lei, Tyler E. Huth, and Hucai Zhang. 2016. "A regional record of expanded Holocene wetlands and prehistoric human occupation from paleo-wetland deposits of the western Yarlung Tsangpo valley, southern Tibetan Plateau." *Quaternary Research* 86 (1): 13–33.

Huo, Lili, Zhike Chen, Yuanchun Zou, Xianguo Lu, Jiawei Guo, and Xuguang Tang. 2013. "Effect of Zoige alpine wetland degradation on the density and fractions of soil organic carbon." *Ecological Engineering* (51): 287–295.

Iqbal, Awais, and Zhanhuan Shang. 2020. "Wetlands as a Carbon Sink: Insight into the Himalayan Region." In *Carbon Management for Promoting Local Livelihood in the Hindu Kush Himalayan (HKH) Region*, 125–144. Berlin: Springer.

Kaiser, Knut, Werner H. Schoch, and Georg Miehe. 2007. "Holocene paleosols and colluvial sediments in Northeast Tibet (Qinghai Province, China): properties, dating and paleoenvironmental implications." *Catena* 69 (2): 91–102.

Kensaku, Okawa, 2016. "A Study of gTan tshigs: A Genre of Land Tenure Document and its Implication in Tibetan Social History." In *Social Regulation: Case Studies from Tibetan History*, Bischoff, J., and S. Mullard eds., 197–205. Leiden: Brill.

Kuussaari, Mikko, Riccardo Bommarco, Risto K. Heikkinen, Aveliina Helm, Jochen Krauss, Regina Lindborg, Erik Öckinger et al. 2009. "Extinction debt: a challenge for biodiversity conservation." *Trends in Ecology & Evolution* 24 (10): 564–571.

Lahiri-Dut, Kuntala. "Beyond the Water-Land Binary in Geography: Water/Lands of Bengal Re-Visioning Hybridity Kuntala Lahiri-Dut." *ACME: An International Journal for Critical Geographies* 13, no. 3 (2014): 505–529.

Lezak, Stephen. 2019. "Re-Placing the Desert in the Conservation Landscape: Charisma and Absence in the Gobi Desert." *Land* 8 (1).

Li, Zhiwei, Peng Gao, Xuyue Hu, Yujun Yi, Baozhu Pan, and Yuchi You. 2019. "Coupled impact of decadal precipitation and evapotranspiration on peatland degradation in the Zoige basin, China." *Physical Geography* 41(2):145–168.

Lu, X., Y. Yan, J. Sun, X. Zhang, Y. Chen, X. Wang, and G. Cheng. 2015. "Short-term grazing exclusion has no impact on soil properties and nutrients of degraded alpine grassland in Tibet, China." *Solid Earth* 6 (4): 1195–1205.

Mei, Li, Li Lianxia, Liao Huasheng, Huang Zhengwen, and Huang Wendian. 2012. "The influence of drainage on wetland degradation in Zoige plateau." *Disaster Advances* 5 (4): 659–666.

Niu, Zhenguo, Haiying Zhang, Xianwei Wang, Wenbo Yao, Demin Zhou, Kuiyi Zhao, Hui Zhao, Nana Li, Huabing Huang, Congcong Li, Jun Yang, Caixia Liu, Shuang Liu, Lin Wang, Zhan Li, Zhenzhong Yang, Fei Qiao, Yaomin Zheng, Yanlei Chen, Yongwei Sheng, Xiaohong Gao, Weihong Zhu, Wenqing Wang, Hong Wang, Yongling Weng, Dafang Zhuang, Jiyuan Liu, Zhicai Luo, Xiao Cheng, Ziqi Guo, Peng Gong. 2012. "Mapping wetland changes in China between 1978 and 2008," *Chinese Science Bulletin* 57 (22): 2813–2823.

No stated author. 1977. Gtsang ma klu 'bum chen mo. TBRC W2CZ7968. 1: 10–45. Dolanji, h.p.: Tibetan Bonpo Monastic Centre.

No stated author. 1999. Mdzod dge'i dgon sde'i lo rgyus. Mdzod dge' rdzong: srid gros mdzod dge rdzong yon lhan khang.

Nyima, Yonten, and Emily Yeh. 2016 "Environmental Issues and Conflict in Tibet." In *Ethnic Conflict and Protest in Tibet and Xinjiang: Unrest in China's West*, edited by Tuttle, Grey and Hillman, Ben. New York: Columbia University Press.

Page, S.E., and A.J. Baird. 2016. "Peatlands and global change: response and resilience." *Annual Review of Environment and Resources* 41: 35–57.

Phan, Vu Hien Roderik Lindenbergh, and Massimo Menenti. 2012. "ICESat derived elevation changes of Tibetan lakes between 2003 and 2009." *International Journal of Applied Earth Observation and Geoinformation* 17: 12–22.

Raine, Roberta. 2014. "Translation and Appropriation of the World's Longest Epic: Tibet's Gesar of Ling." *FORUM. Revue Internationale d'interprétation et de Traduction/ International Journal of Interpretation and Translation* 12 (2): 65–85.

Ramsar. 2015. *National Report on the Implementation of the Ramsar Convention on Wetlands: to be submitted to the 12th meeting of the conference*. Uruguay: Ramsar.

Sayer, Carl, Helen Bennion, Angela Gurnell, Emma Goodyer, Donovan Kotze, and Richard Lindsay. 2018. "Restoration of Freshwaters: Principles and Practice." *Freshwater Ecology and Conservation: Approaches and Techniques*. Oxford: Oxford University Press.

van Schaik, Sam. 2011. *Tibet: A History*. London: Yale University Press.

Scott, James C. 2017. *Against the grain: a deep history of the earliest states*. New Haven: Yale University Press.

Shang, Z.H., M.J. Gibb, F. Leiber, M. Ismail, L.M. Ding, X.S. Guo, and R.J. Long. 2014. “The sustainable development of grassland-livestock systems on the Tibetan plateau: problems, strategies and prospects.” *The Rangeland Journal* 36 (3): 267–296.

Shapiro, Judith. 2001. *Mao's war against nature: Politics and the environment in revolutionary China*. Cambridge: Cambridge University Press.

Smith, Monica L. 2006. “The archaeology of South Asian cities,” *Journal of Archaeological Research* 14 (2): 97–142.

Sörensen, Per K. 2003. “Lhasa Diluvium: Sacred Environment at Stake.” *Lungta* 16: 85–134.

Sörensen, Per K., Guntram Hazod, and Tshal-pa Kun-dga. 2007. *Rulers on the celestial plain: ecclesiastic and secular hegemony in medieval Tibet: a study of Tshal Gung-thang*. Weiss: Verlag der Österreichischen Akademie der Wissenschaften.

State Forestry Administration. 2000. *China's National Wetland Conservation Action Plan*. Beijing, China Forestry Publishing House.

Sun, Jian, Baibing Ma, and Xuyang Lu. 2018. “Grazing enhances soil nutrient effects: Trade-offs between aboveground and belowground biomass in alpine grasslands of the Tibetan Plateau.” *Land Degradation & Development* 29 (2): 337–348.

Tan, G.G. 2018. *Pastures of Change: Contemporary Adaptations and Transformations among Nomadic Pastoralists of Eastern Tibet*. Berlin: Springer International Publishing

Tcvetkov, P.S., 2017. “The history, present status and future prospects of the Russian fuel peat industry.” *Mires and Peat*, 19(14): 1–12.

Tucci, G., and G. Samuel. 1988. *The Religions of Tibet*. Berkely: University of California Press.

United Nations Environment Programme (UNEP). 2018. *Frontiers 2018/2019: Emerging Issues of Environmental Concern*. UNEP Science Division: Nairobi: 38–39.

Wei, Da, Hui Zhao, Huang Lin, Yahui Qi, Ziaodan Wang. 2019. “Feedbacks of Alpine Wetlands on the Tibetan Plateau to the Atmosphere,” *Wetlands*. Online: https://doi.org/10.1007/s13157-019-01220-4.

Xu, Weihua, Xinyue Fan, Jungai Ma, Stuart L. Pimm, Lingqiao Kong, Yuan Zeng, Xiaosong Li, Yi Xiao, Hua Zheng, Jianguo Liu, Bingfang Wu, Li An, Zhang Lu, Xiaoke Wang, and Zhiyun Ouyang. 2019. “Hidden loss of wetlands in China,” *Current Biology* 29 (18): 3065–3071.

Yeh, Emily T. 2009. “From wasteland to wetland? Nature and nation in China's Tibet.” *Environmental History* 14 (1): 103–137.

Zeng, Xingquan, Yu Guo, Qijun Xu, Martin Mascher, Ganggang Guo, Shuaicheng Li, Likai Mao, Qingfeng Liu, Zhanfeng Xia, and Juhong Zhou, “Origin and Evolution of Qingke Barley in Tibet,” *Nature Communications* 9/1, 2018, 1–11.

Zhang, Y., Hu, Z., Qi, W., et al. 2016. “Assessment of effectiveness of nature reserves on the Tibetan Plateau based on net primary production and the large sample comparison method.” *Journal of Geographical Science* 26, 27–44.

Zhao, Zhilong, Zhang Yili, Liu Linshan, Liu Genggui, Zhang Haifeng. 2015. “Recent changes in wetlands on the Tibetan Plateau: A review.” *Journal of Geographical Sciences* 27 (7): 879–896.

Part III

Evolving cosmovisions, climate change and community resilience

7 Aloof but not abandoned

Relationality and the exploitation of the environment in the Garo Hills of India

Erik de Maaker

Toppled boulders

It was late June, and after months of drought the rains started to gain in intensity. Dry spells were becoming rare, as hours of steady drizzle alternated with shorter periods of heavy downpour. It was not a good time to travel in Garo Hills, and certainly not a good time to try to go around on slippery and muddy paths. Yet on such a rainy afternoon, we were walking through Emangre, trying to get a sense of its layout. Emangre is a somewhat large village territory, located about 20 kilometres to the north of the India–Bangladesh border. Spread out across the territory, it has several wards. The people of Emangre use large parts of their land to grow crops, while much of the remainder is covered by forest. I had come to Emangre with some friends and colleagues, one of whom was a Garo PhD student from the Folklore Department of the North Eastern Hill University in Shillong[1]. At the time, I had been in Garo Hills for several weeks, and much of my ability to speak Garo, acquired during my 2 years of stay in the early 2000s, had resurfaced. The colleague was a folklorist with a fascination with Garo traditions, and was familiar with the topics of conversation that I intended to engage in.

Most of all, I was eager to learn more about the fate of (and faith in) Emangre's planted boulders (*ro'ong gia*). To my knowledge, almost every Garo village has at a central location a cluster of such boulders. These derive from ancient times—in what is referred to as the days of the grandfathers and grandmothers (*atchu-ambini somoi*). The boulders are typically tall and slim. Standing straight up they can be up to a metre tall, and about 30 centimetres across. They are always huge and heavy, and their planting will have required the concerted effort of many people. The boulders are firmly associated with the traditional animist Garo community religion, which is nowadays practiced by only few people. The practitioners of this community religion are referred to as Songsarek. I use "community religion" rather than "religion" or "Garo religion," in order to avoid implying that these practices and ideas have been codified in a way comparable to established religions that are organised hierarchically, base their teachings on sacred texts and employ professional religious specialists. The traditional Garo community religion lacks any such characteristics, but rather depends on a body of ideas and practices that entirely depend on enactment and orality. The vast majority of the Garo have, over the last century and a half, converted from the community religion to one or another form of Christianity (mostly Baptist and Catholic). Emangre is located alongside the Rongdik River, which has historically allowed for easy contact with the plains of Bengal in the south, even at a time when roads did not yet exist. Already by 1914, Baptist missionaries founded a primary school in Emangre. Today, most of its inhabitants are Baptists, a smaller number are Catholics and the smallest denomination are Seventh Day Adventists.

It was not difficult to learn about the "planted boulders" of Emangre. There, the last conversions had taken place only decades ago, and ideas, knowledge and skills that derive from the realm of the traditional Garo community religion continue to have a presence. Garo Christians, consequently, have an ambiguous relationship to these boulders. According to the tenets of their religion, they are not expected to acknowledge these, but cultural readings and experiences influence them to not let themselves be directed rigidly. A relative of a friend introduced us to Ales Momin, an elderly man, who was a reputed herbal healer. We were advised to talk to him, since he knew about the Songsarek past. Initially, he told us rather bluntly that there were no boulders anymore. While we sat at his house, drinking the very sweet tea he offered, he told us that the boulders had been destroyed in a thunderstorm that occurred back in 1978. This sounded unlikely to me, since how can a storm "destroy" boulders? After some persuasion, he agreed to take us in the vicinity of the place where the boulders had been located. Unfortunately, we did not get to see them. From the roadside, we peered at a densely wooded slope, a plantation of sal trees. The plantation was inaccessible: in the rainy season the dense vegetation is known to hide snakes, poisonous millipedes, scorpions, leeches and so on, turning it into a no-go area (Figures 7.1 and 7.2).

For Ales, who had remained a Songsarek until fairly recently, being in close vicinity of the boulders brought back memories from the past. He explained that

Figure 7.1 Ales Momin, pointing out the toppled boulders.
Photo by Erik de Maaker, 2019.

Figure 7.2 The sal plantation, with its undergrowth.
Photo by Erik de Maaker, 2019.

the boulders had at one moment toppled over (perhaps due to incessant rains, during the storm). This had apparently deprived them of their ritual powers, allowing someone to grow the hardwood trees on the land surrounding them. But he also told us that a long time prior to that, before the British came, planting the boulders had required the sacrifice of a person. More recently, rather than a human, a monkey had been sacrificed at such an occasion. This aligns with historical ethnographic records that state that up to the late nineteenth century Garo people were involved in headhunting, which—among others—served a religious purpose (Mackenzie 1995 [1884], 261–265). The British used the need to end what they called "raids" from the hills onto the plains as an important justification for their conquest of Garo Hills. In reverse, it seems likely that Garo-led raids were at least partly a response to extractive policies enacted towards the hill dwellers by landlords from the surrounding plains, who had been empowered by the colonial authorities (Barman 1994, 79). Yet, it seems likely that the practice of human sacrifice has its origins in the need to relate to the entities associated with the boulders.

I was familiar with the religious significance of the boulders from the fieldwork that I had previously done in the Garo Hills. The largest of these boulders is the *a'song-kusi* or *a'song-chiga ku'si*, which translates as: "for the land and flowing water"[2]). At the time, people told me that this boulder hosts a spirit,[3] while the other boulders are merely meant to support it. This spirit identifies as a deified ancestor, a valiant warrior who, upon death, transformed into a spirit. According to legend, the boulders were initially arranged by the founders of a village. The head of a village derived his authority from his ability to patronise the spirit of the boulder, as it empowered him. The spirit would inflict blindness, or render lame, anyone who touched the boulder with the exception of the village head as well as men closely associated to him. For the inhabitants of Emangre all this related to a pre-Christian past, but did that mean that nowadays no one cared about the boulders any longer? Even if they were no longer an object of worship, did they somehow continue to be imbued with their earlier powers? I was struck by the fact that in Emangre, where people had turned to Christianity early in the twentieth century, the location of the boulders was still known, and—with a bit of probing—their earlier agency seemed to resurface. It confirms what I have argued elsewhere, which is that there is much more continuity between the traditional Garo community religion and Garo Christianity than might be acceptable from a doctrinal perspective (de Maaker 2013).

The spirit of the *a'song-kusi*-boulder, which is of human origin, has a human-like disposition. In return for being provided with blood, it can express a certain degree of loyalty. Other spirits, originating prior to the creation of the earth, omnipresent and innumerable, show this kind of responsiveness as well, although they are much less predictable in their reactions, revealing their nonhuman origin. The ability of the spirits to communicate with humans, and even inflict deadly illnesses, suggests that both spirits and humans share, in certain respects, a comparable "interiority" (Descola 2016, 39). Including the boulders in narratives that involve engagement and interaction, and attributing active agency to them, creates a relational perspective. Such narratives derive from the past, but are more or less equally acknowledged by Garo Christians. The boulders found in Garo villages continue to "demonstrate intentionality and agency" (Harvey 2005, xvii).

These sort of continuities are at the core of the "new animism" studies that have emerged over the last two decades by Graham Harvey (2005), Philippe Descola (2013), Eduardo Viveiros de Castro (1998) and Leslie Sponsel (2012), among others. Such studies aim "to open interpretive possibilities for understanding an integrated environmental vision that transmits spiritual states of knowing and moral ways of being in the world" (Grim 2001, xxxvii). "New animism" entails a departure from earlier evolutionist usages of the term animism, as it does not consider the ontologies it refers to as subordinate to "Western" thought. Rather, it wants to challenge the dominance of what Escobar, after Law (2015), calls the "One-World World" (OWW), "the dominant form of Euro-modernity," which presents itself as the only "right" world, to which all other attempts at world-making are subservient (Escobar 2016, 15). Central to this OWW is a distinction between nature and culture, in which humans are the active agents who operate in a nonagentive physical environment. The predominance of this ontology is problematic, since it renders all other experiences of the world invalid. From this perspective, the agency attributed to the *kusi*-boulders can at best be rendered a belief that is in essence mistaken. In other words, OWW implies claims to ontological dominance over alternative ontologies.

In an attempt to challenge the hegemony of the OWW perspective, de la Cadena and Blaser (2018, 4) have argued for the recognition of ontological heterogeneity. As an alternative for the "universe" that exists according to the OWW ontology they argue for a "pluriverse," which acknowledges alternative ontologies on equal footing. Each of these ontologies would in itself be valid, and deserves recognition as an act of world-making. According to Escobar, this is of considerable political relevance. Where the OWW ontology renders feasible and legitimises destructive "extractivist projects," such as mining, the logging of forests, and the sourcing of hydropower, "pluriversal" ontological alternatives would resist these. The latter, which Escobar refers to as "epistemologies of the South," would not proceed from the disparities that characterise the OWW. Rather, these would centre on the relationships that create a specific environment. Such relationships then encompass humans as well as other agentive entities (Escobar 2016, 13–14).

Escobar thus juxtaposes the "extractivism" enabled by the OWW, to the relationality of the epistemologies of the South. Proceeding from the Songsarek perspective on the environment, which locates humans in a world created, inhabited and maintained by omnipresent spirits, Garo ontology is, in principle, relational, but the nature of these relationships has, over the last couple of decades, been redefined. Contrary to what Escobar argues, this has created room for "extractivism" to exist next to and in combination with a relational perspective. How these two ontological perspectives coexist, is the question that is central to this chapter. Below, I discuss the agency of the spirits, notably in relation to forests. Next, I focus on how ontological adaptations facilitate but also generate changes in land use. Last, I return to the *kusi*-boulders and explore how these continue to be attributed agency, even in contexts where they are primarily regarded as objects relating to the past.

A multispecies environment

According to the traditional Garo community religion, Garo live in an environment in which humans are one among many different "agentive beings" (Ogden, Hall, and Tanita 2013, 6). This notion is helpful here, since it refers to "biophysical entities" (i.e. organisms), as well as "objects that magically animate life itself." Boulders such as the *a'song-kusi*, but also hilltops, rocks and deep river pools are included in this last category, reflecting an experience of the environment that resonates in many parts of the Himalayas, where rivers and mountains can be sacred (this volume: see chapters by Thinley Dema and Kinley Choki on Bhutan, by Mongfing Lepcha and Rongnyoo Lepcha, and by Charisma Lepcha on Sikkim). From such a multispecies perspective, spirits are an important presence. Most spirits existed prior to humanity, but this differs from the case of the spirit of the *kusi*-boulder, which as mentioned is a deified ancestor. This ancestor, whose name is Guira (or sometimes Guira-Kalkame), is attributed the ability to more or less keep the spirits of the forest at bay (Bhuyan and Sangma 2019, 215). Living in the vicinity of the "seat" of Guira provided people with a certain degree of protection against these spirits of the forest. Guira's presence constituted a broad zone around the boulders as village land, and by planting the boulders people thus also advanced a claim to land vis-à-vis other people. In the early twentieth century, the colonial administration recorded these land claims, as they mapped and demarcated the boundaries of the distinct villages. These boundary maps are now maintained by the Garo Hills Autonomous District Council, the government body that is currently responsible for the maintenance of land rights, and they continue to serve as land title deeds (Karlsson 2011, 158).

The spirits that are central to the traditional Garo community religion are innumerable, omnipresent and ubiquitous. What people know about the spirits derives from experience and from stories carried over from the past. These often relate to unfortunate events such as attacks by wild animals, traffic accidents or illnesses, and can serve to explain contemporary misfortune as well. Knowledge is also condensed in rituals and chants that have proven effective over time in facilitating and influencing the behaviour of the spirits. Yet rather than being confident about this imparted knowledge, people are frank about their unfamiliarity with the character, names and whereabouts of the vast majority of the spirits. As a result, people are continuously in dialogue with an "inspirited" or agentive environment, interpreting the signs of the entities that are included in and thus part of and subjected to the relationships the land encompasses (Bird-David 1999, S78). While distinct entities have different "affordances," they are at the same time in certain respects subjected to a single "we-ness" (Bird-David, S73). Spirits matter to people, while people can influence the state of spirits. And albeit having a rather distinct constitution, both people and spirits share the same spaces, the same land. Rather than suggesting that people exist on their own, they are defined by the relationships they maintain with other entities acknowledged to be present.

The spirits, in their ubiquity, impose taboos that restrict people in their ability to harvest crops ahead of time, hunt certain animals and otherwise randomly subject the environment to (over) consumptive activities. The spirits can enforce these taboos through—often deadly—sanctions, typically meted out by venomous or ferocious animals such as scorpions, bird spiders, snakes, wild cats or elephants. Some of these animals are strongly associated with spirits, such as the giant electric eel, monitor lizards or tigers. Electric eels live in deep river pools and are said to bite their victim's forehead or navel to suck the victim's blood. Monitor lizards can be up to three feet tall at the shoulder. They pierce their tail through their victim's head, to then bite its forehead to finish the kill. A tiger can also be associated with a spirit, and may kill people unless they sacrifice to it.

Most spirits have the ability to harm people. This manifests in the "biting" (*chika*) of a spirit, and in the victim being subsequently depleted of blood. People cannot see the spirits, and neither can they see their bites, since spirits are invisible in the everyday, though they do show themselves occasionally in dreams. Blood in this context equals life fluid, and being bitten can translate into a variety of ailments, most of which involve a severe loss of energy and strength. More specifically, being bitten can translate into conditions such as stomach pain, toothache, earache, headache and so on, each of which is caused by a dedicated spirit. For instance, the spirit known as Risi, at once feared and propitiated, can cause a condition known as "hot–cold" (*sinha-dinga*), which is characterised by high fevers (de Maaker 2013). The biting can be so severe that it results in death.

People's relationship to the spirits is ambiguous: while spirits kill, people also depend on them for the sustenance of life. The spirits have been present from the beginning of time, and the creator spirit (Dakkara-Rabokka) made the earth as we know it. He created humans, animals and plants, and continues to do so. Songsarek ritually attribute each birth to him. According to the traditional Garo community religion, people also depended on the spirits for their crops to grow. Much of Garo Hills consists of hills with steep slopes. Historically, Garo practiced shifting cultivation on these slopes, clearing stretches of forest. Clearing forest could only be done with the agreement of the spirits, as the forest's primeval occupants who also had to be provided with the first fruits of many of the various crops being cultivated before people could harvest these.

The conversion of many Garo to Christianity over the last couple of decades has had implications for the ways in which Garo perceive their environment. Christianity, with its rituals of baptism as spiritual rebirth, emphasises rupture and distancing from animist perceptions (Robbins 2004). According to its tenets, as voiced by pastors and others who claim religious authority, Christian divinity is superior to the spirits, and conversion demands that people withdraw from the frequent and costly ritual sacrifices for the spirits that are part and parcel of the traditional community religion. This has resulted in curbing the influence attributed to the spirits and the degree to which people acknowledge them. A man once explained to me that many of the spirits that used to make their presence felt in the recent past had now "gone silent" (*jrip donga*) (de Maaker 2013).

The traditional Garo community religion emphasises the omnipresence of the spirits. It is a religion of "immanence," in which "there is no clear or sharp distinction between natural and supernatural, knowable and unknowable, matter and spirit" (Lambek 2016, 15). Garo Christianity does not explicitly locate the divine in the material world. This certainly holds for Baptists, but even in Catholicism, materialisation of the divine tends to be confined to images and objects located in churches and shrines. Christianity, in emphasising the transcendence of the divine, creates the conceptual space for a secularised understanding of nature, in which nature is "a resource" that harbours "'objects' to be had, destroyed or extracted" (Escobar 2016, 18). Christianity, as an inspiration for "modernity and development" (van der Veer 1996) can thus contribute to the emergence of a dichotomy between nature and culture. How does this translate in a redefinition of people's engagement with land and forest?

Anthropogenic forests

Garo Hills is green. In summer, the canopies of its many trees provide shade against the heat. Throughout the rainy season, which can last from April to September, all barren patches of land become covered by weeds, shrubs and trees. "This land is so fertile, anything you plant will flourish here," a friend once remarked. Indeed, I have been amazed at the speed by which plants can grow in the humid heat. In the rainy season, bamboo easily grows up to three feet in just a couple of days' time. The lush vegetation is due to an annual rainfall ranging between 520 and 930 mm, fertile soil, and tropical temperatures throughout much of the year. According to official statistics, 70–85% of the land in Garo Hills has forest cover (Meghalaya Basin Development Authority 2017, 5). This does not mean that all of this is prime forest. Government statistics are based on satellite reconnaissance, which cannot differentiate between prime forest and orchards, that is, trees of one single species planted to grow a certain marketable produce (Kurien, Lele, and Nagendra 2019). Even much of the forest is anthropogenic, in the sense that it is either felled at intermittent periods to allow shifting cultivation, or has been excluded from such because it is forest reserve (Figures 7.3).[4]

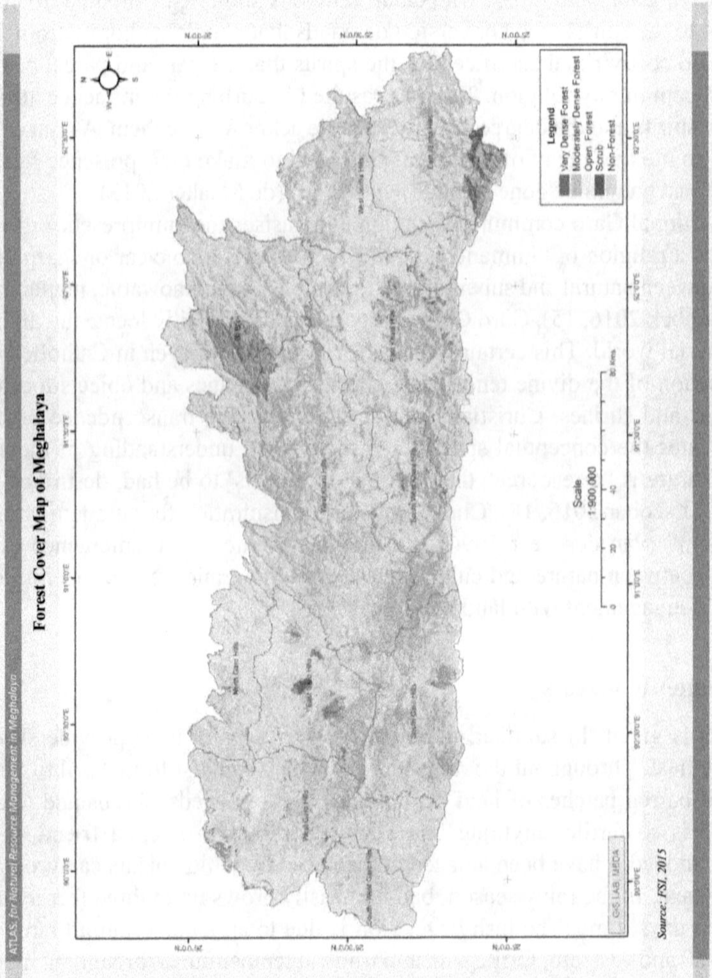

Figure 7.3 Forest-cover map of Meghayala State.

Source: Meghalaya Basin Development Authority 2017, 6.

The inhabitants of a village domain jointly own and manage their forest, which allows them to use it to log the wood required for house building, to hunt, and to collect medicinal plants' bark, leaves, roots and resins. Historically, the forests of Garo Hills facilitated shifting cultivation. For shifting cultivation, stretches of forest are cleared at regular intervals then cultivated for anywhere from 1 to 3 years in a row. Fields are then abandoned, allowing the forest to recuperate for anything between 5 and 20 years, dependent on the amount of land available. Ideally, people leave land fallow for as long as possible. The age of shrub or forest cover not withstanding, the location of new fields also depends on the quality of the soil, the degree to which these can retain rainwater, and the altitude at which these are located. Fields are sown and planted with a mix of crops that come to fruition in different months of the agricultural season. For sowing and planting, people use seeds and tubers that derive from the previous harvest. Many of the varieties of hill rice, vegetables and roots used are unique to specific areas or even villages, and through cultivation are carried forward from generation to generation. Sowing mixed crops greatly reduces the risk of pests, and in the absence of pesticides certainly traditionally shifting cultivation was duly organic.[5] As mentioned, according to the traditional Garo community religion, the crops are "given" by the spirits. These are therefore consulted and propitiated in a variety of rather extensive annual ceremonies that allow people to connect to the spirits, while at the same time they reflect the social and political hierarchies prevalent within a given village domain. Typically, important ceremonies were held following the harvest of (dry) hill rice, after the burning of the new fields, to ensure the hill rice yield, and when abandoning the fields (Marak 2018). Most of these ceremonies centre on hill rice, which, in this context, is perceived as a living entity. Hence, before being sown the rice seeds need to be "woken up," the rice plant with its budding rice grains needs to be cured when it falls ill due to the straying of its soul, while its harvest is ritually framed as nothing less than murder (de Maaker 2006). Christians do not offer first fruits to the spirits, but they do annually provide a tithe to their church, which is in a way also an expression of gratefulness for the crops that Christian divinity has provided to them. This tithe is either given in cash or in kind, the latter most often being a bag of rice.

A field belongs to the individual household that cultivates it. After cultivation, households who abandon their fields revoke their claim to the specific plot of land. Once abandoned, a field reverts to forest. From the 1960s and 70s onward, shifting cultivation has gradually lost its earlier prominence.[6] Instead, the cultivation of tree crops, notably of the areca palm, which produces areca nuts, has gained popularity. Areca nuts are in demand, as a much appreciated ingredient in the stimulating chew known locally as *paan*. Government agencies have certainly encouraged the cultivation of these tree crops, but the changeover was first and foremost driven by people's desire to get more stable cash incomes. The production and sale of tree crops such as areca, as well as cashew, integrated people with the market much more emphatically than before. Rather than focusing on the production of their own food, as subsistence cultivators, farmers came to use the money they obtained from the cash crops to buy rice and other necessities. As a consequence, they became entangled to a much greater degree than before in "capitalist relations," in which both labour, land, as well as the crops they produced and the food they bought became commodities (Li Murray 2014, 7).

Throughout much of Garo Hills, areca nut trees have come to occupy large tracts of land that at one time were intermittently used for shifting cultivation. Areca nut orchards, once created, require relatively little labour to maintain. Harvesting the nuts from the tall trees, though dangerous, does not require many people. Households have therefore taken on much more land than they did in the days when they depended solely on shifting cultivation. As a result, the forest is becoming increasingly scarce. As mentioned, according to satellite imagery areca nut orchards qualify as forest, but the reality on the ground is rather different (Kurien, Lele, and Nagendra 2019). Mature areca nut trees have a slim, very straight trunk, without any branches, and can stand up to 30 feet tall. The orchards are monocultures and yield best if there is no undergrowth. Consequently, orchards provide no shelter to wild animals, which thus disappear from the area. People also complain that with the orchards replacing biologically diverse forest, it is becoming more and more difficult to find the plant matter needed for herbal medicines.

Areca nut palm trees cannot withstand cold winter nights, and until recently cultivation of the trees was not possible on land located at an altitude of approximately one thousand feet and above. In recent years, due to climate change, average temperatures have gradually been increasing, which has also meant that at the higher altitudes winter nights are less cold. This has allowed for further extension of the area on which areca nuts can grow. In addition, climate change is rendering rainfall increasingly erratic. While this is not detrimental for the areca nut palms, it does have a negative impact on what shifting cultivation yields. Thus, in multiple ways, climate change encourages the cultivation of the orchard crops, resulting in further reduction of Garo Hills' forests.

Even so, with increasingly less forest available, shifting cultivation is continued, but at a smaller scale than before. The swiddens provide people with vegetables and other crops that are otherwise hard to obtain, and the activity continues to be culturally desirable (de Maaker 2018). Due to the gradual extension of the area being used for orchards, the number of years that forest is allowed to lie fallow between cycles of cultivation has become shorter. With less time to recuperate, this results in lower soil fertility and thus lower yields. As a consequence of these developments, people in some parts of Garo Hills have not been able to continue their practice of shifting cultivation and have become fully dependent on the cultivation of orchards.

The cultivation of areca nuts, replacing forests with monocultures that leave no room for biodiversity, demands what Escobar calls an "extractivist" approach to the environment (Escobar 2016, 14). These orchard crops are in no way integrated with the traditional Garo community religion, and do not involve any of the restrictions and taboos that pertain to shifting cultivation. Enabled by the growing prominence of a transcendent divine, as propagated by Christianity, people throughout the hills seem to be more and more accepting of a secularised idea of "nature." Yet today's followers of the traditional Garo community religion also grow areca nuts, to which end they utilise the same cultivation techniques as Christians. As far as the cultivation of areca nuts goes, even the followers of the traditional Garo community religion have thus come to consider nature as a commodity.

The commodification of nature is even more obvious when it comes to the mining of coal. Garo Hills contains many rich deposits of coal located in the village domains. Coal is in high demand in neighbouring Bangladesh, and is transported by truck in large volumes across the border. The people who own the land, the residents of the village domains, mine the coal either on their own or with the help of outside entrepreneurs. Dependent on the size of a deposit, villagers themselves do the digging, or they do so with the help of outside labour. The mines are typically small, lacking in safeguards, and the media use the rather derogatory term "rat-hole" mining (McDuie-Ra and Kikon 2016) to describe them. Many coal deposits are located below the water table and water must be pumped out of the mines. This polluted water is let out into streams and rivers where toxic sediment accumulates. For such reasons, the National Green Tribunal of India demanded a ban on coal mining that was then imposed by the Indian High Court in 2014. In 2019, after intense legal battles, the ban was revoked (Sitlhou 2019). While the ban was pending it never resulted in a complete closure of the mines, but once it was lifted it further encouraged people to mine for coal and thus contribute to severe environmental pollution.

Agentive boulders

While Christian divinity is attributed with the ability to "numb" the spirits, and while secular modernity simply denies their presence, these are only effective to a certain degree. Among Garo, the spirits continue to manifest themselves, even though perhaps less fiercely than before. At times, spirits continue to inflict illnesses on people (irrespective of their religious orientation), the curing of which demands dedicated healing practices that directly tap into Songsarek cosmology (de Maaker 2013). Likewise, the spirits continue to have a presence in relation to the "planted boulders," with which they continue to be associated, even in places where the traditional Garo community religion has ceased to be practiced.

In January 2019, I visited Rongkinggre, a village located on the northern slope of the Nokrek range. People in Rongkinggre were all profoundly Christian, and for decades, the traditional Garo community religion had ceased to be practiced. One morning, our local host took us on a tour of the village. He wanted to show us various kinds of fruit trees and give us an impression of the economic condition of the villages' inhabitants. Rongkinggre has several wards, spread out over quite a large area, and our walk easily took 40 minutes. Each ward had many detached houses, with roofs of corrugated iron, and walls of woven bamboo. In between wards were stretches of greenery. Central to the village was a large forested area, which was initially introduced to us as a "forest reserve" (*reserp*). Our host required little encouragement to tell us that this was the place where the "planted boulders" were located, and when we showed interest, he readily agreed to guide us on a narrow path into the forest (it was January, the dry season, and thus no leeches).

The trees were towering high above us, their canopies breaking the light of the sun. To me, it was reminiscent of entering a cathedral. The forest was teeming with many different kinds of birds. Our guide went silent as soon as we entered the forest, as is appropriate in such a place. My companion and I were familiar with the local cultural conventions, and we kept quiet as well. Inside the forest were Rongkinggre's boulders: the *kusi*-boulder in the middle, with at least eight supporting boulders in a circle around it. Our host then explained, in a low voice, that these boulders had been planted a long time ago. And, he said, at the time of their planting, "people" had been killed for them. A hushed conversation followed, which linked the boulders to the founding of the village, which had, according to our guide, taken place more than two hundred generations prior. I had heard these kinds of renditions before. What stunned me was that our guide had made clear, the evening before, that the traditional Garo community religion had truly been abandoned in the village. We were assured that there were no Songsarek anymore. Yet apparently, given the respect with which he (and we) approached the boulders, and the fact that a forest was kept around them in the middle of the village, the spirits continued to be attributed agency (Figure 7.4).[7]

In anthropology, agency is a concept that was foregrounded by Gell (1998), in his theory on the efficacy of art. Gell emphasised that art objects are included in social relationships that can attribute them with qualities equal to social entities. Gell's ideas were at least partially inspired by Mauss's theory of exchange. Mauss showed, with reference to the Maori *hau*, that holds that, in the context of exchange, objects are seldom "dead matter," but rather attain their value from a strong association with a certain person or entity. This association renders an object "active" and capable of direct social interaction (Mauss 1990, 10). The *kusi*-boulders were, in the context of the traditional Garo community religion, explicitly agentive, since they were attributed the capacity to create a space of a village for the residents to reside in, and for its surrounding forest to be used for farming. While the spirits in villages such as Rongkinggre are no longer addressed in religious ritual, their presence continues to be acknowledged less explicitly. The boulders continue to be feared, and are attributed the capacity to mete out sanctions, while reminding people of the religious practices of their

Figure 7.4 The planted boulders of Rongkinggre.

Videostill by Erik de Maaker, 2019.

forebears. The respect that people continue to have for the boulders is evident in their inability to touch or move them. It should be remembered that even in Emangre, people did not move the boulders: these had toppled over "due to a thunderstorm." Moreover, people continue to show deference in relation to the boulders, as is clear from the avoidance of raising their voices in their vicinity.

The boulders thus continue to be "agentive," in that they demand respectful treatment. Moreover, as long as they remain planted, they enforce the people's maintenance of prime forest around them, and thus at a central location in their village. In other words, the spirits of the boulders demand that people conserve forest, encompassing trees, medicinal plants and a great variety of animals. Garo have adopted an "extractivist" approach towards much of their land and forest, but as long as these boulders are erect they cannot, or do not, extend this to the forests surrounding *kusi*-boulders. The spirits of the boulders, in other words, use their agency to protect at least these small patches of forest, and the many species that they provide shelter to. In giving moral directions, the boulders express normative and ethical dispositions, which motivate their actions towards people. This translates into an environmentalist perspective where prime forest matters, as a resource that should be valued.

Rather than being replaced, the relational ontology that is central to the traditional Garo community religion thus continues to exist next to an approach that allows for rather straightforward extractivism. In Emangre, once the boulders had toppled over, their influence was sharply reduced, allowing for hardwood trees to be planted in their vicinity. The changes in land use in Garo Hills over the last couple of decades have resulted in very substantial loss of prime forest, creating a serious threat to the biodiversity of one of Asia's richest hotspots. In addition, the persistence of villagers to mine coal (or have others do so on their behalf), results in severe pollution of water and land, which creates serious hazards to health. Clearly, such an exploitation of the environment is, at an ontological level, enabled by a nature/culture divide, which does not entail the kind of long-term reciprocal relationships that are central to the traditional Garo community religion. Garo extractivism is reminiscent of the OWW perspective, but only with respect to certain aspects of (economic) life, since it has not erased the earlier relational ontology. While the ontology anchored in the traditional Garo community religion has lost its earlier dominance, it continues to be apparent and even to prevail in certain contexts and in relation to certain (agentive) objects.

Conclusion

In this chapter, I have explored ontological transformations in rural Garo Hills. I have shown that even though the vast majority of the Garo identify themselves as Christians, that does not imply that the relational ontology underlying the traditional Garo community religion has lost its relevance. Christianity and the secular modernism associated with it have paved the way for the large-scale cultivation of areca nut palm trees, which brings about the destruction of prime forest, and creates a serious threat to biodiversity. Likewise, it has allowed for the extraction of coal, which creates severe environmental pollution. These extractivist usages of the land have transformed the rural economy, integrating people firmly with markets for the sale of their produce, and for buying food, while reducing their dependence on subsistence agriculture.

De la Cadena and Blaser have challenged the hegemony of extractivist approaches to the environment that proceed from a nature/culture dichotomy by arguing for a "pluriverse." This creates room for the acknowledgement of ontological alternatives to a hegemonic OWW (de la Cadena and Blaser 2018, 4). Typically, such alternative ontologies foreground how people are part of the relationships that they maintain with their environment. Whereas the traditional Garo community religion emphasises how people exist in relationships which they maintain with spirits that are immanent to the landscape, this is much less so in the context of Garo Christianity as well as the secular modernity that it has facilitated. Rather than these distinct ontological traditions being juxtaposed, as argued by Escobar (2016, 13–14), I have shown that in the Garo context these can, in certain respects, be complementary. Here the pluriverse encompasses both an ontology rooted in the traditional Garo community religion, as well as extractivist one. Or, from an environmentalist perspective, it can be argued that while Garo are the region's "first conservationists," they are also responsible for causing environmental degradation (Snodgrass and Tiedje 2008, 9).

I have shown that the prevalence of a specific ontological perspective can be situational. The planted boulders are "emplaced," at sites associated with the founders of a village domain, "forging a complex intertwinement of earth and humans" (Smyer Yü 2014, 484). They thus can impose the relational ontological values associated with the traditional Garo community religion. Even if the spirits are no longer addressed through religious rituals, these maintain the authority to demand certain human behaviour in the prime forest, and can trigger lethal sanctions if those demands are not adhered to. Allison refers to this as "a place based spiritually-generated perspective [which] places humans within networks to which humans and other living and non-living beings contribute and respond" (Allison 2015, 440). Yet elsewhere on a village domain, an "extractivist" understanding can apply, which allows for the replacement of forest with monocultures of areca palms, as well as the environmentally destructive mining of coal.

The distinct ontological approaches prevalent among Garo are being reproduced and reinterpreted in a variety of contexts. The presence of both relational and extractivist perspectives in a single environment creates ample ground for dilemmas and conflicts of interest. While some people make attempts to extend the area of land covered by prime forest, others may not agree to do so. Yet the planted boulders, as agentive entities, can continue to provide moral guidance regarding a relational engagement with the environment.

Notes

1 This research visit materialised in the context of my involvement with the drafting of a nomination dossier for the creation of a UNESCO world heritage site in Garo Hills (tentatively called the Garo Hills Conservation Area). I am grateful for the support of Daniel Ingty, Venybirth Marak, Rangku Sangma, the Meghalaya Forest Department

(Shillong, Tura, Baghmara), the Meghalaya Basin Development Authority (Shillong), UNESCO C2C (Dehradun), Leiden University, as well as the Himalaya University Consortium of ICIMOD (Kathmandu).

2 I acknowledge Jandi Marak for helping me to understand this term.

3 In earlier writings I have sometimes referred to these spirits as "deities," but "spirit" does more justice to their not being integrated in an institutionalized religion.

4 In 1999, Garo Hills had 141 square miles of Reserved Forest, managed by the Department of Forest of Meghalaya State. Most of these forest reserves were created by the colonial government between 1883 and 1885, with later additions in the two decades that followed (Sangma 1999, 1).

5 The organic ginger that people grow, for instance, could thus fetch a higher price than comparable non-organic crops. Unfortunately, the marketing of the swidden crops is currently not done in a way that allows for this additional value to be realized.

6 The reasons are multiple, but these are some of the more prominent ones, which, for reasons of space are not discussed in the main text: the increased access of the rural areas to markets, and the growing presence of the state as an agent of economic development; a rather sharp increase in population due to the gradual extension of basic medical care, resulting in lower infant mortality rates and increasing population pressure in the rural areas; a growing need for cash, since people came to depend on markets for goods which they would previously have produced themselves (such as clothing).

7 The boulders were placed by the founders of a village near to where they lived. In Rongkinggre, the forest area that has the boulders is therefore also referred to as the 'old village' (*songgitcham*). Typically, such former village wards would be abandoned when some of its inhabitants were struck by disease, died in labor, or met with fatal accidents. Such incidents are blamed on *marang* (death pollution), which can invisibly be contained in the earth. *Marang* is associated with blood (including that of child-birth), conflict, and violence. Because *marang* poses a lethal danger, people are keen to either avoid, contain, or purge it. Land, with its capacity to harbor *marang,* then has the agency to direct the behavior or influence the people who depend on the land for their livelihood.

References

Allison, Elisabeth. 2015. "Religion Inscribed in the Landscape: Sacred Sites, Local Deities and Natural Resource Use in the Himalayas." In *The Changing World Religion Map*, edited by S.D. Brunn, 439–459. Dordrecht: Springer Science+Business Media.

Barman, Santo. 1994. *Zamindari System in Assam during British Rule (A case study of Goalpara district)*. Guwahati: Spectrum Publications.

Bhuyan, Gangotri, and Tengnang Dawa Sangma. 2019. "Social Markers of the Garos: Looking at a Tradition Undergoing Change." In *Megalithic Traditions of North East India*, edited by Queenbala Marak, 209–223. New Delhi: Concept Publishing Company.

Bird-David, Nurit. 1999. "'Animism' Revisited: Personhood, Environment, and Relational Epistemology." *Current Anthropology* (1999): S67–S91.

De Castro, Eduardo Viveiros. 1998. "Cosmological Deixis and Amerindian Perspectivism." *The Journal of the Royal Anthropological Institute* 4 (3): 469. doi: 10.2307/3034157.

de la Cadena, Marisol, and Mario Blaser. 2018. *A World of Many Worlds*. Durham: Duke University Press.

de Maaker, Erik. 2006. "Negotiating Life : Garo Death Rituals and the Transformation of Society." Phd Thesis, Leiden University.

de Maaker, Erik. 2013. "Have the Mitdes Gone Silent? Conversion, Rhetoric, and the Continuing Importance of the Lower Deities in Northeast India." In *Asia in the Making of Christianity: Conversion, Agency, and Indigeneity, 1600s to the Present*, edited by R.F. Young and J.A. Seitz, 135–162. Leiden: Brill.

de Maaker, Erik. 2018. "On the Nature of Indigenous Land: Ownership, Access and Farming in Upland Northeast India." In *Indigeneity on the Move: Varying Manifestations of a Contested Concept*, edited by Eva Gerharz, Nasir Uddin and Pradeep Chakkarath, 29–48. New York: Berghahn.

Descola, Philippe. 2016. "Presence, Attachment, Origin: Ontologies of 'Incarnates.'" In *A companion to the Anthropology of Religion*, edited by Janice Boddy and Michael Lambek, 35–49. Chichester: John Wiley & Sons.

Descola, Philippe. 2013. *Beyond Nature and Culture*. Translated by Janet Lloyd. Chicago: Chicago University Press.

Escobar, Arturo 2016. "Thinking-Feeling with the Earth: Territorial Struggles and the Ontological Dimension of the Epistemologies of the South." *Revista de Antropología Iberoamericana* 11 (1): 11–32. doi: 10.11156/aibr.110102e.

Gell, Alfred. 1998. *Art and Agency: An Anthropological Theory*. Oxford: Oxford University Press.

Grim, John A. 2001. *Indigenous Traditions and Ecology: The Interbeing of Cosmology and Community*. Cambridge, Mass: distributed by Harvard University Press for the Center for the Study of World Religions, Harvard Divinity School.

Harvey, Graham. 2005. *Animism: Respecting the Living World*. London: Hurst & Co.

Karlsson, Bengt G. 2011. *Unruly Hills: a Political Ecology of India's Northeast*. New York: Berghahn Books.

Kurien, Amit John, Sharachchandra Lele, and Harini Nagendra. 2019. "Farms or Forests? Understanding and Mapping Shifting Cultivation Using the Case Study of West Garo Hills, India." *Land* 8 (9): 26.

Lambek, Michael. 2016. "What is Religion for Anthropology? And What Has Anthropology Brought 'Religion'?" In *A Companion to the Anthropology of Religion*, edited by Janice Boddy and Michael Lambek, 1–24. Chichester: John Wiley & Sons.

Law, John. 2015. "What's Wrong with a One-World World?" *Distinktion: Journal of Social Theory* 16 (1): 126–139. doi: 10.1080/1600910X.2015.1020066.

Li Murray, Tania. 2014. *Land's End: Capitalist Relations on an Indigenous Frontier*. Durham: Duke University Press.

Mackenzie, Alexander 1995 [1884]. *The North-East Frontier of India (originally published as: History of the Relations of the Government with the Hill Tribes of the North-East Frontier of Bengal)*. New Delhi: Mittal Publications.

Marak, Queenbala. 2018. "Rice from A•ba: Stories, Rituals and Practices of the Garos." *South Asian Anthropologist* 18 (2): 161–175.

Mauss, Marcel. 1990. *The Gift : the Form and Reason for Exchange in Archaic Societies*. London: Routledge Classics, 2002.

McDuie-Ra, Duncan, and Dolly Kikon. 2016. "Tribal Communities and Coal in Northeast India: The Politics of Imposing and Resisting Mining Bans." *Energy Policy* 99: 261–269. doi: http://dx.doi.org/10.1016/j.enpol.2016.05.021.

Meghalaya Basin Development Authority. 2017. *Atlas for Natural Resource Management in Meghalaya*. Shillong, Meghalaya: GIS LAB, Meghalaya Basin Development Authority.

Ogden, Laura, Billy Hall, and Kimiko Tanita. 2013. "Animals, Plants, People, and Things: A Review of Multispecies Ethnography." *Environment and Society* 4 (1): 5–24. doi: 10.3167/ares.2013.040102.

Robbins, Joel. 2004. *Becoming Sinners: Christianity and Moral Torment in a Papua New Guinea Society*. Berkeley, Calif.; London: University of California Press.

Sangma, Milton S. 1999. "Reserved Forests in the Garo Hills." *Journal of the North East India Council for Social Science Research* 23 (2): 1–2.

Sitlhou, Makepeace. 2019. "Return to the Rat Hole." *Foreign Policy*. https://foreignpolicy.com/2019/12/24/rat-hole-coal-mine-meghalaya-government-protections-workers-environment/, accessed March 10, 2021.

Smyer Yü, Dan. 2014. "Sentience of the Earth: Eco-Buddhist Mandalizing of Dwelling Place in Amdo, Tibet." *Journal for the Study of Religion, Nature and Culture* 8 (2): 483–501.

Snodgrass, Jeffrey G., and Kristina Tiedje. 2008. "Indigenous Nature Reverence and Conservation: Seven Ways of Transcending an Unnecessary Dichotomy." *Journal for the Study of Religion, Nature and Culture* 2 (1): 6–29.

Sponsel, Leslie E. 2012. *Spiritual Ecology*. Santa Barbara: Praeger.

van der Veer, Peter. 1996. *Conversion to Modernities: The Globalization of Christianity*. New York; London: Routledge.

8 Cordyceps, climate change and cosmological imbalance in the Bhutan highlands

Kinley Choki

Introduction

Bhutan is at the heart of eastern Himalayan biodiversity. Our mountains and rivers have respectively shaped diverse vertically elevated habitats due to the steep ecological gradients and the fertile valleys fed by numerous rivers. Pastoralism and farming are naturally complementary modes of subsistence in Bhutan. The geological forces have also strengthened our spiritual relations with mountains and rivers being not merely physical parts of the earth but also embodiments of different supernatural beings. Growing up in an urban family, my first experience conducting fieldwork in the highlands of Lingzhi took place in the fall of 2018. At the time of my arrival, the local story about the disappearance of a bird-like sacred rock on a cliff above the village of Gonyuel reaffirmed my sense of the Bhutanese highlands as a shared home of gods, humans and animals.[1] This was not an ordinary rock but one that Lingzhips—as people of Lingzhi are referred to—worshipped as a citadel of the powerful, territorial deity Sumi-jachung. Lingzhips interpreted this disappearance neither as an ordinary occurrence, nor as a natural event. Instead, they all agreed that it was a deliberate act perpetuated by Sumi-Jachung. They thus interpreted it as her communicating her displeasure with the Lingzhips. What must have upset her, the Lingzhips reflected, were the recent socioeconomic and cultural changes that had unfolded in the village. In Lingzhi, everyone knows that, once upset, deities could communicate, and exert punishment, by causing natural calamities, among many avenues of possible punishment (Figure 8.1).

One sunny afternoon, I sat down with a 39-year-old yak herder who was tying together several bunches of dry grasses to store for the winter. As we talked about yaks and grasses, climate change, and deities and mountains, he pointed to the cliff where the rock once was and lamented: "It was right there. But one day it just flew away." Not just him, but all Lingzhips I spoke to could clearly recall the inauspicious day on which the rock had disappeared and all insisted that the rock had simply flown away. Given how common this narrative was, I felt that I had touched upon something important. A herder in her late twenties told me thus:

> It has been three years since the sacred rock disappeared and the people were very disheartened when that happened. Ever since, people have been

Figure 8.1 Gong-yuel and Sha-yuel Chiwogs with the citadels of their deities Sumi-jachung and Sew-dratse standing majestically over the village.

> discussing about what has made the deity so angry as to make the sacred rock fly away.

When I asked why people assumed that the deities were angry, she added, "I think they are angry because we have started constructing a lot of new houses and for that we have started digging more ground and cutting down more trees than we should. In this way, we are harming the environment. This must have angered the deities because we live on their land." The 39-year-old herder added to this the observed decline of ritual and cultural observances in the village. He said:

> I think it is because we Lingzhips do not follow the culture and traditions that were followed by our forefathers. Since most of the rituals are not conducted well, compared to the past, it must have angered our local deities.

This rock's disappearance is just one of several inauspicious events in Lingzhi, including landslides and floods. Alongside the narrative of global climate change, Lingzhips invariably interpret all these destructive environmental events as the result of deities' anger. Several Lingzhips affirm that the people have made the deities furious with activities that do not abide by their restrictions and requirements. All such activities had begun with the onset of the fungus-harvesting economy followed by excessive extractive economic activities. As natives to Bhutan, we all understand well that we live in a densely knit network of interrelations between humans, deities and the environment. This is the relational ontology of our homeland, which is based on the mutual dependence of the living, the spiritual and the physical earth. It also means that the formless deities are also environmental beings who live co-constitutively with humans. In what follows, I wish to address three human-induced causes to the environmentally manifested anger of the deities.

First, I discuss the recent emergence of a prosperous cordyceps-centred economy that has generated vast wealth for many Lingzhips. However, the collection of cordyceps, or *yartsagunbub* in Dzongkha (The national language of Bhutan), comes with people's experience of karmic retribution. I then argue that the harvesting and auctioning of cordyceps is both locally invoked as the cause and consequence of the currently experienced cosmological imbalance in environmental and spiritual terms. Second, I recount my field observation about how the villagers' declining participation in rituals honouring the deities is displeasing to the mountain gods. The third contributing factor is the increasing number of "outsiders" who nowadays enter the highlands and who, even if often unwittingly, regularly transgress the rules, taboos and prohibitions laid down by Lingzhi's deities.

The Fungus economy and the ambiguity of wealth

"Cordyceps must be shivering in fear because many of us are going to search for them today," Kinza, a female yak herder in her late twenties remarks as she readies herself for a day of crawling through the highlands with her eyes cast on the soil, in search of the caterpillar fungus known in scientific annals as *Ophiocordyceps sinesis* but to Lingzhips as *yartsagunbub* that translates as "insect in winter and grass in summer." Later that day, when Kinza finds two cordyceps she murmurs a song while carefully unearthing them: "*Rang rang so soi dungyel* (everyone has their own difficulties in life)." She is referring to the unimaginable hardship a caterpillar must be experiencing when a mushroom starts growing from its head, in the process of which the caterpillar succumbs. This introduces us to a central ambiguity related to the fungus economy in the highlands, and which revolves around the question of whether a caterpillar is still a sentient being once a mushroom has sprouted from its head and thence—and this is where the apprehension lies—whether its collection and trade, and the vast wealth this has generated over the past decade and a half, is ultimately rooted in sin and thus accumulating karmic debts for those earning vast worldly riches out of it. Science insists that it isn't, in the sense that it establishes cordyceps as deceased insects, plain and simple. Not all Lingzhips are so convinced about this however, as Kinza's statement and mutterings reveal.

In this section, I discuss the possibility, as apprehended by Lingzhips themselves, that the wealth generated through cordyceps collection is both a cause and consequence of a cosmological disorder with deities disapproving of the conduct of the herders. I start with the cause. The cordyceps-boom, however, has not just transformed the material condition of yak herders, it has also spurred a host of other changes, including a changing epistemology of the insect/grass whose essence is now evaluated in a frame of the neoliberal market, thus introducing, if by "stealth" (Li 2014, 9), commodified views of nature.

Tshering Pelzom, a member of the research team, noted in her diary, "Even though it is only recently that the cordyceps collection has been legitimatized and is flourishing, the highlanders have excellent knowledge about it." This knowledge, however, is of a particular kind, which is derived from "market-science" that does not privilege cordyceps' role and function within the local ecosystem, but that is vested in considerations of market value. "There are three types of cordyceps and we can differentiate them by the color of their eyes," a younger herder who is also an ardent cordyceps collector, explains. "The yellow-eyed cordyceps fetch us the highest prices. The other two types are those with red and black eyes. They too sell well, but it is the yellow-ones that earn us the highest amounts." Besides eye-colour, another way in which the market value of cordyceps is evaluated is by the size of the mushroom (the "grass") that grows from its head, which, in terms of market value, should be short and small. As states another ardent cordyceps collector who is in his early thirties, "The shorter the fungus that grows from the caterpillar's head, the thicker and stronger is the cordyceps flesh, and it is the flesh we need. This is because the mushroom draws the nutrition away from the insect, thus weakening its healing powers. So, we get higher amounts when the mushroom is small."

It is only occasionally that yak herders talk about cordyceps in relation to the local eco-system, of which it is nevertheless a nourishing part. With the escalating material wealth in Lingzhi, people have started to look at the environment more in terms of profit. This conversion of what naturally exists in nature into profit-generating resources is a case of ontological occupation. Here, the economic perspective of cordyceps has clearly taken over its traditional or spiritual perspective of which karmic debt of uprooting and collecting cordyceps is a significant part of. This form of ontological occupation is not a physical occupation by non natives but one driven by change in peoples' values and social behaviours solely due to the extractive economic activities surrounding cordyceps. According to Escobar, a case of ontological occupation took place in the southernmost area of the Columbian pacific, around the port city of Tumaco, when the mangrove and humid forests were destroyed in the early 1980s for oil palm plantation and industrial shrimp cultivation. Escobar wrote that in the process of doing so, the relations maintained with and by the forest-world were being effaced (Escobar 2017, 245). In Lingzhi, such an ontological occupation seems to be emerging internally, although of course in intimate relation to the global market. In the process of uprooting and excessively extracting cordyceps from the land for economic reasons, the Lingzhips' relation with the highland-world is being disrupted. In addition to that, with the onset of the fungus economy, there has been an increase in the number of houses constructed and in the number of trees being cut down for such construction purposes: this is a major reason for the disruption in the negotiated relationship between the Lingzhips and their deities. In what follows, I will explain why and how.

It has long been argued that Tibetan attitudes to the natural environment have been shaped by ancient and indigenous beliefs and by ideas introduced later, particularly from India and China, and mainly by way of Buddhism. The common feature in all these influences is that the phenomenal world is held to be inhabited by a host of spirit powers and deities who are organised into a single "ritual cosmos" (Samuel 1993, 157) and must be ritually acknowledged in relation to most areas of human activity (Huber and Pederson 1997, 584). People in Lingzhi have long held similar strong views as the Tibetans. The way they perceived of their natural environment was strongly influenced by their faith in religion and beliefs in their local deities.

However, this has changed with the onset of the fungus economy, as today some Lingzhips look at the environment more in terms of profit. This again has been increasingly exasperating to the local deities because with people holding such economic views of the natural environment, they have been overexploiting the environment by cutting down trees, digging ground, drillings into rocks and increasingly over-picking the cordyceps leading to a drastic decrease in the growth of cordyceps and also a decrease in the quantity and quality of the region's grasses. All these extractive economic activities that harm the environment have disrupted the Lingzhips' centuries of maintained relationship with the local deities who are the prior inhabitants of the land. According to Allison, these enlightened beings who pre-date Buddhism have rights to the land that supersede those of the villagers (Allison 2009, 1169), which is why the extractive and harmful economic activities make the deities furious towards the people.

The now-popular neoliberal epistemology, for all its increasing dominance in local discourse, does not foreclose a still raging, and more existential, debate regarding the question "What is a cordyceps?"and in particular, "Is a cordyceps a sentient being?" Opinions on this are many and diverse in the highlands. According to Elizabeth Allison, the doctrine of reincarnation gives rise to the idea that any other being may have been one's mother in a previous life. This gives rise to a spontaneous feeling of compassion and reverence, and the desire to reduce suffering in the lives of other sentient beings (Allison 2009, 1169). Kinza, the female yak herder mentioned above invokes the rebirth doctrine that asserts that reincarnation for humans does not necessarily guarantee another human rebirth, but a being's existence in one of the six *gati* (realms), known as Bhava chakra, allow one to assert that cordyceps are sentient beings. She also explains: "Cordyceps are people who smoked in their previous life. They blew smoke from their mouths which is sinful, and now they must atone by experiencing a mushroom growing out of their heads." Kinza's view is not an idiosyncratic one. The association between cordyceps and smoking is commonly and variously invoked by Lingzhips. Says an elderly yak herder:"People who smoke tend to find more cordyceps compared to those who do not. This is because cordyceps themselves were cursed for smoking in their previous life as a human being."

On the issue of sentience, another herder expressed thus: "Collecting cordyceps is sinful, but since it gives our people a good source of income, no-one wants to call it a sin in public. It is said that those who collect a lot of cordyceps in this life accumulate so much negative karma that they become a cordyceps in their next life and must endure the pain of a mushroom sprouting from their heads." Other yak herders, however, insist that cordyceps are dead and that, thence, their collection does not result in the accumulation of negative karma: "Even as some people keep saying that collecting cordyceps is sinful, I don't think it is. They are already grasses when we pick them. They die by the third month of the Bhutanese calendar. There is no end to cordyceps; it is like us humans. We die and are then reborn. So do cordyceps. When we pick them, they are already dead, and will grow again."But even those Lingzhips who insist that cordyceps are not sentient beings allow room for ambiguity in their reasoning, as is exemplified in the following statement by a senior yak herder in the village:

> The head is fungus only, but when we look carefully it is an insect. We even see them crawling around when they have not become a fungus, when they are not dead yet. Some say the fungus grows from their heads, while some say it grows from their backs. When they are alive, they can move both forward and backward. If we think he is facing this side because he is coming towards us and we try to obstruct his path, then he will start crawling backwards. The fungus grows from the head because we can see its eyes. The legs will remain intact even if it is a fungus. It is already dead when we collect them. Then, I think there is no soul left. The main reason they turned into fungus is because they are dead. But, originally, they are insects, and just like us humans they will be reborn and grow again.

But whether cordyceps grow again, or do so in the same quantity and quality is something many Lingzhips are not so sure about any longer. While the question of whether cordyceps are sentient or not remains to be settled in Lingzhi, many herders allude to the view that cordyceps collection is rooted in sin, thus resulting in karmic debts. As such, the cordyceps economy is sometimes seen as a prime reason behind deities' current anger. They believe so because with the increasing collection of cordyceps, people have not only started to cut down trees to build large mansions with the help of the money accumulated from collecting and trading cordyceps, but they have now also started to look at the environment from an economic perspective. As noted, all of these cause a tension between the materialist and the religious practitioner, and people have started to prioritise material wealth above religion, thus eventually leading to religious decline. Such activities have not resulted in any favour to the Lingzhips in the recent years because along with unfavourable weather conditions and a decrease in the quantity and quality of pasture grasses, the growth of cordyceps has also drastically decreased in Lingzhi.

Cordyceps have injected near unimaginable wealth into the highlands, now firmly ranking yak herders, who collect the fungus from where their yaks graze, among the wealthiest strata of Bhutanese society; a sharp contrast with the relative poverty and hardship that was long ascribed to their largely nomadic lifestyle. It is hard to exaggerate the sudden wealth yak herders have accumulated, which some of them compared to having won the lottery, and the wealth's exuberant presence can easily be discerned from the large mansions that have replaced the earlier stone and clay houses (some of which now sit uncomfortably empty next to the extravagant new structures), the many more that are currently being constructed, the meticulously painted and decorated furniture and woodcarvings inside of them, the helicopter flights yak herders now regularly charter for purposes of shopping in Thimphu and Paro, and the landed properties they have acquired in the same valleys. Each new house is now built larger and more luxuriously than the next, the result of a clear sense of competition in wealth among Lingzhips, as some of them would occasionally confess to. Now a few of these houses resemble monasteries in their size and expensive decoration (Figure 8.2).

However, the construction of these mansions has required the increased cutting of trees for their construction, and this is a site of tension in the village. According to Allison, deities are most frequently found in places with massive old trees which the villagers will not disturb for fear that retribution, such as sickness or even death, will come to themselves or someone in the village (Allison 2009, 1169). In the case of Lingzhi, oral history tells that some of the trees in Cheybeysa (a village in Lingzhi) where the most luxurious houses are being built are the physical embodiment of the bodyguards of the Great Tibetan lama Zhabdrung Ngawang Namgyel. Cutting down these trees is considered a taboo in Cheybeysa. This particular belief portrays the tension between the materialist and the religious ways of viewing the land and its beings. In order to construct these houses, Lingzhips need to cut an increasing number of trees. There is, however, a tension given the spiritual and historical relevance of these trees, and the restrictions against cutting these trees that are patrolled by deities.

As put forward by Ura (2001), the perspective that some part of the environment, such as sacred sites, cliffs and lakes, are not accessible as resources to be exploited displays a deep environmental consciousness in Bhutan. Allison (2009, 1169) writes, "In addition to the Buddhist religious prohibitions on cutting down trees, villagers also experience the claim of deities, the prior inhabitants of

Figure 8.2 A newly built house in Cheybeysa Chiwog.

the land believed to inhabit various features of the landscape." Yet, we now see these logics beginning to disintegrate in Lingzhi where it has become seemingly inevitable for the herders to cut down trees to increase the number and size of locals' houses. In the process of doing so, people have also started to excavate more ground to construct foundations, and to break more rocks. All such acts are thought to invoke the vengeance of deities as they are increasingly harming the environment where both the deities and the people reside.

According to the yak herders, this is a prime reason for their deities' anger and therefore the cause of natural calamities and a general change in climate. During one of my interviews, a 50-year-old yak herder explained the change in the climate to me. He spoke about the change in the intensity and pattern of snowfall and rainfall, as did all my other respondents. When I asked him his opinion on the change in the climate, he added:

> It must be because of the activities that people indulge in these days like cutting down a lot of trees, which angers our deity. The people are harming the environment. The deities are avenging [the abuse of] the land and punishing the people for the harm caused to the environment.

What this indicates is that the way the Lingzhips experience and explain the unpleasant changes in the highland is related to the recent arrival of vast wealth through a flourishing fungus economy.

A decline in religious and cultural observances

Escalating material wealth has coincided with the emergence, and I will argue, convergence, with the perceived dwindling of culture, customs and traditions in the village, and which is a cause of rising tension, especially among older yak herders. This vast decline in culture, according to elderly Lingzhips, is one of the major reasons that have angered the local deities who have thus caused unfavourable changes in the climate.

In this section I will narrate this decline through the eyes of Tshewang, a senior herder in the village, for whom this is a major concern which he tries, through various means, to mitigate. "I have to impose fines on people who do not attend these rituals and festivals just to make people attend," Tshewang began his explanation. For the senior herder the origin of this decline was easy to trace to the onset of the fungus economy: "Ever since the introduction and legitimatization of cordyceps, people's interest and focus has greatly shifted towards cordyceps collection. Now that they have a better source of income, they spend most of their time and energy in collecting and selling cordyceps." This is one reason for the dwindling in culture and religion. "Local perspectives are primarily concerned with respect for the gods of the world. It is conceived of morally in terms of binding relationships between localized human communities and non-human occupants of the same localities and entails mutual obligations, "Huber and Pederson (1997, 584) write, however, this section shows that the Lingzhips are believed to no longer be fulfilling their obligations.

On the third day of a 3-day ritual conducted in Lingzhi, everyone gathered at the Basic Health Unit after a day of making offerings to their local deities. That evening, everyone gathered to sing and dance as a way of closing the ceremony. We joined them in a big circle to dance and sing their songs. As I looked around, I noticed that the BHU was just one room and I started to wonder how it accommodated all the people in the village so I asked an elderly herder who was sitting next to me how that was possible and he said,

> Not even half the people in the village are present today which is why we have been asked to gather here in the BHU. In the past, there was no such thing as making people pay a fine because people were naturally present. People did not have alternative sources of income so they did one work [activity] and were not as occupied as they are today. People were very much interested in mass gatherings in the village. Sometimes celebrations would be extended for days. Today, that is not the case. People have their own work and everyone is always busy doing one thing or another. Now the celebrations have been reduced to one day and people are fined if they are absent just to make them attend the celebrations and rituals. People do not attend willingly. People only think about their own work. They are not concerned that we should conduct rituals for the local deities who protect us and our land. In order to make people practice our culture and tradition, we have to impose penalties or fines.

Every year, Lingzhips conduct rituals for the local deities in which they make offerings and say prayers to reaffirm their mutual relationships. However, with the flourishing of the fungus economy many Lingzhips, many locals lament, have become primarily engaged in generating money, and this at the expense of their cultural and religious responsibilities. Secondly, and as noted, the cutting of trees and the digging of the soil that comes with the construction of large mansions denotes a digression from religious observances. Either way, Tshewang argues, the escalating material wealth has greatly affected people's interest and faith in their cultural and religious practices. So, when people do not attend these rituals, it is considered to be an act of disrespect towards the deities, potentially inviting their wrath. The cultures and traditions are still preserved but are not practiced well, Tshewang explained further. Not just he, but multiple respondents told us that whereas in the past, Lingzhips always looked forward to conducting rituals and ceremonies because everyone gathered during these occasions, nowadays they attend out of compulsion, to avoid having to pay fines. In the past, moreover, every year, there would be different families willing to sponsor the event, providing alcohol and meat for such rituals and occasions; today very few families come forward for this.

There are also some Lingzhips who have migrated to the region's towns and cities, but who still visit Lingzhi to collect cordyceps; however they are not always a part of the rituals and ceremonies that are conducted in Lingzhi because they spend most of their time elsewhere. After collecting cordyceps for a month, they walk back to their homes in the towns and cities. As such, they take from the environment without giving back to it in terms of ritual observances and offerings, and this, from the deities' point of view, is unacceptable.

Allison states, "Many non-western people ascribe agency and even intention, to the activity of nature or to spirits that direct seemingly natural acts" (Allison 2014, 441). Lingzhips have a similar notion. Allison gives an example of the Himalayan villagers who may attribute a landslide or a hailstorm to deities who are angry because of the violation of some social rules. Similarly, when we asked the Lingzhips about their perspectives on some of the changes in the village such as in the climate and the local environment, all their perspectives were directly or indirectly related to their faith in their local deities and the relationship people have with them. "From the human point of view, maintaining this relationship by means of correct, regular offerings ensures a stable environment which yields its bounty. The local deities must be carefully appeased, manipulated, or otherwise coerced, and any contamination thought to exist must be cleansed to restore an equitable balance between humans and the environment they live in and use" (Huber & Pederson 1997, 585). This is not a new concept for the Lingzhips, as they are very well aware of their responsibility towards the local deities as much as the deities are of their responsibilities towards the Lingzhips. The Lingzhips also realise the consequences of blaming the deities for the unpleasant changes taking place in Lingzhi. Unfortunately, as noted in the above narratives and opinions of the Lingzhips, the Lingzhips have failed to maintain faith in their religion and consistent dedication in their cultural practices. Hence, the angered deities are causing changes in the climate and landscape that are not benefitting the Lingzhips in any way.

More outsiders, more pollution

In the spring of 2019, we first visited Choga, a secluded yak-herding settlement inhabited by Tandin, a 28-year-old male yak herder and his family of five for more than 20years. When we reached Choga, Aum (aunt) Kinley waved at us using a cloth to make sure we could see her in the fog that had surrounded her tent. We greeted each other and sat down around the *bukhari* (fire pit). Tandin, the eldest son in the family, asked us if we had seen the lake and when we said no, he laughed and said, "Why not? It is right here. You might not have seen it because of the fog. Come I'll take you." We stood up and followed him. To our surprise, the lake was not even a minute's walk away. The moment we saw the lake, we froze because we were scared to take a step further. He said, "This lake is called Tsho-kar and the deity of this lake is called Tshokar Dungsokam. I grew up herding yaks around this lake, watching it grow smaller and smaller over the years." We became curious and wanted to learn more about the lake. As we walked back to the tent, Aum Kinley murmured in a lower voice, "It starts to drizzle and become cloudy around the lake when there are outsiders visiting," and when, out of curiosity, I asked her why, it was Tandin who responded,

> I grew up here listening to my late father tell me about the mountains and the lakes, specially this lake. There were times when the weather would

> suddenly become very cloudy and it would start to drizzle around the lake. My late father would say "I think there are going to be visitors at the lake" and he was always right. We believe that it is the doing of the deity to protect its lake from new visitors whose intentions the deity is not aware of. The weather was just fine this morning. It only started to drizzle and become very cloudy when you were about to arrive.

This shows that the deity of the lake can clearly recognise and differentiate outsiders from the Lingzhips, and as is protective of her territory (Figure 8.3).

Besides the emergence of the lucrative fungus economy, another palpable change in Lingzhi is the dramatic increase in the number of "outsiders" who visit the area. These outsiders usually fall into two groups; government officers and tourists. Most of these outsiders are not aware of the cosmological landscape and the presence and personalities of distinctive deities, including related prohibitions and taboos. In this section, I discuss how, in terms of maintaining a cosmic balance, this is problematic, and that many Lingzhips directly or indirectly explain the deities' anger as, in part, caused by outsiders not following prescribed rules and restrictions.

In thinking about this encounter, I found Tony Huber and Paul Penderson's (1997, 590) argument that, "Modern knowledge constitutes the environment as an ecological space, whereas in traditional Tibetan knowledge it is a moral space," helpful to think through the situation. In the narration above, it is said that the lake tends to hide itself under thick fog and drizzle whenever there are outsiders visiting, as the deity of the lake is unaware of the outsider's intentions. This very tendency reveals how the environment is seen as a moral space in Lingzhi. Tibetans and the Lingzhips are grounded in similar belief systems when it comes to their perception of the environment. Lingzhips believe their landscapes to be celestial beings hence the activities surrounding them are also considered to be activities involving the deities.

Figure 8.3 Lake Tsho-kar.

Lingzhips perceive that when someone from outside visits their yak-herding place, the person might bring some bad omen with him or her and cause sickness to their yaks, so they maintain the ritualistic practice of burning a particular incense, an act which they call *sutangni*. Tandin said,

> When people from outside visit any yak-herding place, the yak herders of that place will secretly burn some incense. They do so as a gesture of asking the deity to stop the outsiders from bringing any bad energy or sickness to their yaks. It is also a gesture of asking the deity not to cause their family or the outsider any harm in case the outsider displeases the deity unintentionally.

This they deem necessary as there are several past instances in which yak herders were affected by misdeeds perpetrated in by outsiders. This usually happened when outsiders camped and littered. This applies not just to Lingzhi, but has wider implication in the Bhutan highlands. Pommaret (2004, 56) argued that in "high altitude regions devoted mainly to yak herding, the deities are so sensitive that they cannot tolerate being polluted by any kind of "dirty smoke," and that is why dead bodies are never cremated but [are] exposed to the vultures." In Lingzhi as well, bodies are not burned but are exposed to vultures; even garbage is buried, never burned. However, when outsiders visit Lingzhi and camp on their lands, they tend cook meat or chilli over an open flame, producing smoke, and they often litter; something the territorial deities just cannot tolerate. In such cases, the deities tend to punish the yak herders who live nearby. Tandin explained thus:

> Once, I had to go to Juphugoenpa [a monastery in the area] and make offerings and ask the deity to stop punishing my family members for the misdeeds that outsiders conduct on our land. Henceforth, the deity stopped punishing my family. Before that, when outsiders visited the lake and littered near it, the deity would get upset and punish my family by causing sickness to my sister, my mother, or to our yaks. However, after I made offerings and asked the deity to spare us from the punishments for the misconduct of the outsiders, the deity has stopped punishing us.

As we sat around the *bukhari*, Tandin narrated to us an incident he still remembered vividly, and which further indicates the local deity's sensitivity and power.

> In 2013, there were three men who visited the lake. In the lake there are strange yellow fishes that we call *ser-nya*. They wanted to make a documentary about the fish so one of them caught a fish from the lake as the other two men video recorded the process. Nothing happened on that day but when these three men reached back to their base camp, one of them fell mentally ill while the other had his body half paralysed. They asked an astrologer to find the reason for their sudden illness and the astrologer told them that it was because they disturbed a fish of the lake. That was when they called me and requested me to make some offerings to the deity of the lake on their behalf. As requested, I made the offering but I am not aware of their health now.

Instances such as these are narrated with considerable frequency in Lingzhi. On this, Pommaret (2004, 56) wrote: “If strangers come to the deity’s territory, they have to obey certain rules so as to not incur his or her wrath.” Similarly, in Lingzhi, the taboos and norms are not only for the villagers who live in the deities’ territory but also for outsiders who often fail to abide by the deities’ restrictions. With the extension of the government into the highlands of Lingzhi, it has definitely helped the development of the village but in terms of the cosmological landscape, it is not doing any favour to the Lingzhips. The number of tourists’ visits has rapidly increased throughout the years. The number of official visits has also increased apace with the developmental activities in the village. All of these, however, come at a cost and create a palpable tension between deities and Lingzhips, with the former blaming the latter of not preserving the territory in the way expected of them.

Conclusion

I wish to conclude this chapter with another narrative that evidences the relational ontology that exists between the Lingzhips, their environment and their deities. This story was introduced to us by Tandin Tshering, who said, “The deity of this place (Choga) is Jo-gyem and, as we believe that the lakes have deities of their own, the deity of this lake is Tshokar Dungsokam. My mother told us of a strange incident. Please tell them mother…” Aum Kinley smiled at her son and said,

> It was a long time ago, when your father was alive. He was a particular man. One day I was burning firewood to prepare dinner when I heard a noise outside, I peeked out from the tent and saw him standing near the lake staring at it and he had his back facing me. He was supposed to herd the yaks back and when I saw him I thought he was back early so I hurried and put the water pot on the *bukhari* to make tea for him. I was waiting for him to walk in at any moment but he didn’t. I went outside to look for him and to my surprise, I saw him walking down from the other end of the lake with the yaks. There was no way he could have reached the other end of such a huge lake in no time and if there were other people walking by or around, they would always stop in for a cup of tea. It was strange and when I thought of it, I realized that the man I saw outside was wearing a black *gho* (Bhutan’s national dress for men) and your father never owned a black *gho*. I was shocked and a little scared so I hurried back inside. When I told your father about it after he arrived, he calmly said, “you must have seen Tshokar Dungsokam (the deity of the lake) today. It’s a good thing but don’t be scared about it or tell other people. He might harm you if you do. If you don’t, it’s only a blessing.” When I thought about it after your father passed away, I think the deity was trying to tell me that your father’s death was approaching. I believe that the deity has always been trying to protect us as long as we protect him.

After listening to Tandin and his mother talk about the deities, it is impossible not to recognise the co-existing relationship between humans and deities. It is as though the deities and human cannot be seen existing apart from or independent of one another. The deities and humans engage in each other's existence in so many aspects of their lives, even in life and death. The way the Lingzhips perceive of their environment is also directly connected to how they perceive of their deities which is evidence again of the relational ontology that exists between the Lingzhips, the deities and the environment. This relationship between them also indicates that the way the Lingzhips think and live their life is partially shaped by nature and the deities who reside in the natural landscapes. This has relevance to the concept of environmental imaginaries. As defined by political ecologists Richard Peet and Michael Watts, Environmental imaginaries are the place-specific and regionally variable conceptions of nature that shape the ways that people perceive, discuss, work and play in nature and reflect a community's values with respect to its environment (Peet &Watts 1996 as cited in Allison 2014, 441).

Similar conceptions shaped the activities the Lingzhips engaged in, especially those related to land use and the use of natural resources. In the past, the Lingzhips would strictly abide by the prohibitions and requirements set by the deities to avoid natural calamities in their village, and a major part of such norms included not harming the environment. This paper has discussed the ways in which the people have failed to abide by such norms, their consequences and the causes, as explained by the Lingzhips themselves, in which the idea that the way the Lingzhips think and behave is influenced by nature and its events is clearly indicated. In the introduction, I posed the question, "Why are the deities in Lingzhi angry." This chapter has answered by showing how in recent years, there have been significant social and economic transformations in the highlands of Lingzhi. With the escalating material wealth and increase in extractive economic activities there is a rising tension between the materialistic and religious outlooks, with, among others effects, has produced a decline in cultural and religious observances. This, along with the increasing number of outsiders who fail to live by the rules of the deities, according to many Lingzhips, is believed to be the major factor that angers their local deities who then causes unpleasant changes in the climate and bring about personal misfortunes.

Acknowledgements

I would like to express my gratitude to the informants who readily shared their knowledge and to our hosts in Lingzhi for their hospitality and generosity. I also thank Dr. Jelle J. P. Wouters for his insightful comments and guidance throughout.

Notes

1 This research was conducted as a group-fieldwork with a total of five members. In the remainder of the chapter, when I refer to "we" or "us" instead of "I," this is in reference to the other members of the group.

References

Allison, Elizabeth. 2014. "Religion Inscribed in the Landscape: Sacred Sites, Local Deities and Natural Resources Use in the Himalayas." In *The Changing World Religion Map*, edited by Stanley D. Brunn, 439–459.Dordrecht: Springer.

Allison, Elizabeth. 2009. “Gross National Happiness and Biodiversity Conservation in Bhutan.” In *Enspirited Places, Material Traces: The Sanctified and the Sacrificed in Modernizing Bhutan, 1162–1179*. Michigan: UMI Dissertation Publishing.

Escobar, Arturo. 2017. “Sustaining the Pluriverse: The Political Ontology of Territorial Struggles in Latin America.” In *The Anthropology of Sustainability*, edited by Marc Brightman and Jerome Leiws, 237–256. New York: Palgrave Macmillan.

Huber, Tony and Poul Pederson. 1997. “Meteorological Knowledge and Environmental Ideas in Traditional and Modern Societies: The Case of Tibet.” In *The Journal of the Royal Anthropological Institute*3 (3): 577–597. doi: 10.2307/3034768.

Li, Tania M. 2014. *Lands End: Capitalist Relations on an Indigenous Frontier*. Durham: Duke University Press.

Peet, R. and M. Watts (1996) Liberation Ecologies: Environment, Development, *Social Movements*. Routledge, London.

Pommaret, Francoise. 2004. “Yul and Yul Lha: The Territory and its Deity in Bhutan.” *Bulletin of Tibetology*, 40 (1): 39–67. Sikkim: Namgyal Institute of Tibetology.

Samuel, Geoffrey. 1993. *Civilized Shamans: Buddhism in Tibetan Societies*. Washington, DC: Smithsonian Institution Press.

Ura, Karma. 2001. *Deities and Environment: A Four-Part Series*. Thimphu: Center for Bhutan Studies.

9 Local knowledge of floods and coping strategies in downstream Mahakali River, Nepal*

Rashila Deshar, Dibas Shrestha, Sarina Maharjan and Madan Koirala

We, two environmental scientists and one hydrologist, write this chapter as a work of environmental humanities. It is our first experiment to reframe our quantitative findings into a qualitative text for the purpose of reaching out to a broader interdisciplinary readership. All of us are native to Nepal; however, professionally, two of us faculty members were trained in Japan for our doctoral degrees and one graduate student is an environmental scientist in the making at Tribhuvan University. This environmental humanist experiment is an eye-opening opportunity for us to work with more nuances, especially the social and cultural aspects of local environmental knowledge systems. As "environmental humanities engage with fundamental questions of meaning, value, responsibility and purpose in a time of rapid, and escalating, change" (Rose et al. 2012, 1), our attempt in this chapter is to demonstrate a set of local environmental problems caused by both natural forces and anthropogenic impacts. We also recognise that the humanist approaches to environmental studies have frequently been cast as "non-science" (Rose et al. 2012, 2; Heise 2017, 2). For this purpose, our study is intended to show the scientific and humanist values of the local environmental knowledge of resource preservation and safety preparation in the event of natural or human-induced flooding. We also intend to converse with existing perspectives on the topics of local environmental knowledge and environmental implications of customary and systemic social issues, such as caste system. Towards the end of the chapter we wish to share with our readers what we mean by "local knowledge." We do not wish to pre-define it; instead, it is most fair for our readers to walk through our research findings first before we offer our conceptual thoughts.

The geographical area of our case study is in the downstream of Mahkali River. Mahakali River is a transboundary river flowing in between India and Nepal,

* This work was undertaken by Central Department of Environmental Science, Tribhuvan University, as parts of the Himalayan University Consortium Transboundary Rivers of South Asia (HUC TROSA) Water Research Grant 2018–2019 through the TROSA project funded by the Government of Sweden. The HUC Secretariat is hosted by the International Centre for Integrated Mountain Development, headquartered in Kathmandu, Nepal. Views expressed are solely those of the authors.

which originates from Limpiyadhura in the Himalayas. The river bears different names in India and Nepal. Its upper reach in India is called the Sharda River. Sharda is also another name for Saraswati, the goddess of learning in Hinduism. Its lower reach in Nepal is called Mahakali River. The roots of the word Mahakali are "Maha" or "great" and "Kali" or "blackness" in Sanskrit. In Hindu religion, Kali is the name of a Hindu goddess known as the destroyer and the creator of life. She is worshipped by devotional movements and tantric sects as the Divine Mother and Mother of the Universe (Harding 1993, 25; McDaniel 2004, 5). In Hindu theology, she is also regarded as the divine protector and the one who bestows moksha, or liberation (Hawley and Wulff 1982, 152; Coburn 1984, 111–112). The goddess Kali is one of the fiercest, strongest and most complex deities in all of Hinduism. The violence, destruction and creative power of the river might have provoked the riparian people's spiritual imagination to name the river Mahakali, which expresses the similar violent characteristics of goddess Kali. It is common that our Himalayan ancestors often treated environmental flows as spiritual entities or bodies of deities as also shown the chapters by Charisma Lepcha, Kinley Choki, Thinley Dema, Rongnyoo Lepcha and Mongfing Lepcha, and Erik de Maaker in this book. The forces of nature continue to be deified and personified among folks in contemporary Nepal.

Anthropogenic context of floods from Mahakali River

Downstream region of Mahakali River of Nepal is one of the most flood-prone area during monsoon season between June to September. The geological conditions of the Himalayas and human alterations of environmental flows also contribute to the floods and making the tributaries of Mahakali River more susceptible to flooding. The Sharda Barrage and Tanakpur Barrage in the upstream of Mahakali River in India are the primarily human-induced drought in the winter season and floods in summer months in the downstream area of Nepal due to the seasonal closure and opening of the floodgates. Along with regional artificial riverbed mining and poor embankment structure management, global climate change also contributes to heavier monsoonal rains over a longer period and to more frequent opening of the floodgates of the barrages in the summer. Many downstream villages in Nepal below the barrages are exposed to imminent or immediate floods.

The Mahakali Treaty, the first bilateral water resources treaty signed in 1990 between Nepal and India, is the most controversial agreement between the two Himalayan nations (Gyawali and Dixit 1999; Shukla 2006; Subedi 2006; Lama 2019). As stated in the treaty, both countries have equal entitlement in the utilisation and management of the waters of the river without prejudice to their respective existing consumptive uses of the waters of the Mahakali River. India built Tanakpur Barrage on a stretch of Mahakali River to replace the aging Sharda Barrage constructed by British India in 1928 after getting permission from Government of Nepal. In this regard, India claimed that the barrage site and its protection works were wholly in Indian territory. Such mindset allows Indian imperialist monopoly over Nepal's water resources. As a result, the construction of the barrage and allied structures between 1983 and 1990 eroded about 0.36 square kilometre of land in Nepal (Dhungel and Santa 2009). Every year, when the water level in Sharda Barrage rises, water is released to avoid the burst of the dam. Most releases are conducted at the peak of the monsoon season resulting in substantial increase in the discharge of the Mahakali River. As a result, flooding in downstream is inevitable. It is, therefore, a human-induced natural disaster.

In this backdrop, we conducted our research in two communities, viz. Kutiyakhabar in Mahakali Municipality and Belauri in Bhimdutta Municipality of Kanchanpur District. The residents of Kutiyakhabar are predominantly the Sunar, who are also known as a part of Nepal's Dalit population. Belauri's residents are the Tharu native to the region for centuries. Both communities were chosen for our case study based on their flood history and on their local knowledge for the prevention of and the recovery from disasters. The community experiences flooding nearly every year. The floods are caused by both the heavy rainfall during monsoon season and the backflow of other streams connected to Mahakali River, viz. Jogbudha River and Doda River, which lie respectively in Kutiyakhabar and Belauri communities. People from both communities speak their native dialects as well as Nepali Devanagari that is why we need not to face problem in our communication with the community members. During our fieldwork, we inquired into what impacts floods have in the area and what the main local flood prevention and coping strategies are being utilised. Information was collected through interviewing key figures in the communities, focus group discussions and transect walks. Two focus groups were organised in each community to address above questions. The participating community members responded to our queries based on their experiences of floods. Overall, we practiced participant observation of community everyday life.

Sunar of Kutiyakhabar community

The Sunar from Kutiyakhabar in Mahakali Municipality, Kanchanpur District is an Indo-Aryan Nepali speaking group. Situated in the Nepalese multiethnic society with a heterogeneous caste hierarchy, the Sunar are treated as Shudras, also called Dalits or untouchables. Their caste was defined by their occupation as blacksmiths. The word Dalit does not refer to a specific caste of Hindu social classification. It is a rather general term grouping together those who live at the bottom of the society. The word is actually a modern coinage during the British India as the translation of the British classificatory phrase "depressed classes" of people whose social oppression is justified by the Hindu religiously defined out-caste status and/or by their service-based livelihoods, such as janitor and blacksmith. The etymological roots of Dalit are found in the Hebrew "Dal" meaning "broken" or "crushed" and in Sanskrit "dälit" meaning "downtrodden," "oppressed," or "exploited." Dalits, comprising 13% of the Nepalese population, include 26 separate subcaste groups; however, they are mainly divided into the two categories, namely the Hill Dalit and the Tarai Dalit (Bishwakarm 2016). Like in India, Dalits have been the most disadvantaged and marginalised community in Nepal, too.

The current national politics of Nepal indicates the persistence of the inequality that the Dalits are experiencing. They are not fully included in the political process let alone being empowered. As a result, very few Dalits participate in the central committees of major political parties. However, in recent years, Nepal's effort to work towards an inclusive democracy has made some difference for more political participation from and fair representations of the Dalits. Individual Dalits have steadily benefited from the governmental initiatives favouring their education and employment in the public sector as well as representation in the political institutions. For instance, the election of the Constituent Assembly in 2008 was a significant achievement in terms of Dalit representation. The Dalit representatives were able to develop a common informal platform, improving their bargaining strength (Khanal et al. 2012, 16). Despite the Dalits' struggling over the past seventy years for their rights and inequality, they are not yet fully integrated in political institutions and inequality still persists in Dalits' socio-economic, educational and political status (Bishwakarm 2016, 5).

In this national backdrop of Nepal, the Sunar living in the downstream region of Mahakali River continues to remain marginalised. Although they historically had no connection with Jogbudha River, they currently dwell in the vicinity of the river. It is because the caste-based water distribution practice in Nepal, not the Sunar's free will, has pushed them to the flood prone riversides as their residential area that is far enough so that they will not "pollute" the social space of higher castes. The Sunar have access to the water from the river but are exposed to the floods and contaminated water. In many ways, the Sunar being Dalits have to struggle for water when it comes to that of the distribution of water based on class and caste. In Hindu customs, Dalits are disentitled and not allowed to use taps and wells located in non-Dalit residential areas. Such social discrimination has contributed to their marginalised social and living space such as riverbanks far away from non-Dalit people. The name of the location of the Sunar community—Kutiyakhabar—also explains their involuntary social exclusion. "*Kutiya*" in local language means "hut" and "*khabar*" means "area" indicating the community of economically marginalised and/or low caste people. The Sunar's traditional environmental knowledge consists of not only their lived experiences with the riverine environment but also the ancient but continuously effective social and religious discrimination and marginalisation. It is in their socially segregated and environmentally near-inhabitable space that they have accumulated coping strategies and methods for survival from frequent flooding from generation to generation.

Tharu of Belauri community

The Tharu of Belauri community in Kanchanpur District is an ethnic group with a long history of living in Terai of the downstream region of Mahakali River. Terai is a narrow strip of land in the southern Nepal bordering with Northeast India. The word Tharu is derived from "*Sthavir*" meaning follower of Theravada Buddhism (Skar 1995, 35). There are several endogamous subgroups among the Tharu. In this study, the late anthropologist Dor Bahadur Bistha found that because the ancestors of the Tharu had migrated from Thar, a desert area in the south, they became known as the Tharu (Bisht 1994, 15). Among the different subgroups of Tharu, the Chaudhary and the Rana are the residents of Belauri community. Based on our fieldwork, 80% of the community members are the Chaudhary Tharu and the remaining 20% are the Rana Tharu. Mr. Ram Narayan Chaudhary, one of our interviewees, told us about the ancestral origin of the Tharu, "There is no proper written documentation about the origin of Tharu people; however, many traditional oral accounts are in existence. The Rana Tharu living in Kanchanpur District claim that their ancestors were a branch of the Rajput, who belonged to a high caste in Rajasthan, India." Both Chaudhary and Rana Tharus are considered the second lower caste in the caste system of Nepal. Despite the fact that written histories regarding the origin of Ranas in Kanchanpur are very few, their past has been remembered through local oral traditions.

What needs to be mentioned is that Tharu people became bonded labourers in a system known as *Kamaiya*. It was because after the unification of Nepal in the late eighteenth century, members of the ruling families received land grants in the Terai and were entitled to collect revenue from those who cultivated the land (Regmi 1994, 77). In 1854, Prime Minister of Nepal Jung Bahadur Rana enforced the so-called *Muluki Ain*, a General Code, in which both Hindu and non-Hindu castes were classified based on their habits of food and drink (Stiller 1993, 93). The Tharu were categorised as "enslavable" people together with several other ethnic minorities (Gurung 2005; Krauskopff 2007, 203). After the overthrow of the Panchayat system in Nepal in 1990, the Tharu ethnic association *Tharu Kalyankari Sabha* joined the umbrella organisation of ethnic groups, a predecessor of the Nepal Federation of Indigenous Nationalities (Skar 1995, 32). The government of Nepal in 2000, outlawed the practice of bonded labour prevalent under Kamaiya system, prohibiting anyone from the employing any person as a bonded labourer, and declared that the act of making one work as a bounded labourer is illegal.

For the Tharus, like many other peoples in Nepal, life begins with water. Our field finding coincides with that of Giesele Krauskopff, "Tharu are not forest dwellers but people of the swampy deforested lowland; they are farmers and fishermen" (Krauskopff 1999, 21). For the Tharus, water plays an important role for irrigation and ritual associated with marriage. For their subsistence livelihood they depend on rivers, lakes, oxbow lakes and wetlands. Tharu women use the hexagonal hand lift nets as their fishing tool (Figure 9.1). They have their own gods and their religion is animistic.

Flooding in Kutiyakhabar and Belauri communities

The Sunar at Kutiyakhabar community reside in Jogbudha River catchment, a tributary of Mahakali River. It is a place highly exposed to floods. In 2018, the community was flooded due to the incessant rains and additional water released from Sharada Barrage in India. The flood eroded riverbanks and destroyed crops, putting the nearby farmland and homes at risk. Delay in the construction of an embankment at Kutiyakhabar during the time made sleepless nights to the

Figure 9.1 A Tharu woman fishing with a hexagonal hand lift net locally known as *Thathiya.*

Photographed by Rashila Deshar 2019.

residents. One resident of Kutiyakhabar shared with us, "It is true that flood in the downstream are becoming more destructive not only because of heavy rainfall but also because in every monsoon season, water is diverted from barrages constructed by India." The Indian government constructed the barrage in contravention of international law that is supposed to ensure security and protection of land and people in both India and Nepal. However, opening the floodgates of the barrage during the monsoon causes devastating floods. The Sunar believed that the 2018 flood was mainly caused by the water released from Sharada Barrage. This anthropogenic factor actually makes the floods more destructive. Residents of Kutiyakhabar have been facing the brunt of floods every year whenever the barrage opens. One of them told us, "The flood can enter into the community any time, any season; so we are living in fear of floods all the time." Another resident also expressed his anger, "Despite millions of rupees are being poured for river training works by both government and non-government organization, the problems remain the same every year." It was disheartening to hear community members' accounts of the floods destroying properties, crops, and roads, killing humans and livestock, contaminating water sources, and eroding soil and river channels.

Global climate change is increasingly becoming a local knowledge as one of the causes for heavier rainfalls and longer monsoon season. The correlation of climate change and longer monsoon season in the Himalayas and South Asia is actively being confirmed by scientists (Loo et al. 2015, 818; Seth et al. 2009, 64). On the community level, natural disasters like floods are no longer the works of nature absent of human causal contributions and alterations. The Sunar experience the floods from both the monsoon and the floodgates of the barrages in India. In our research, the anthropogenic aspect of the flood they experience are threefold.

First, it is the accelerated melting of the Himalayan glaciers due to global warming. ICIMOD's latest commissioned scientific studies show that one of the most susceptible impacts of global climate change on the Himalayas is glacier melting that increases extreme events such as flooding, droughts, glacier melting and river flow, resulting in water scarcity when it is needed and too much water at other times (Roy et al. 2019, 107). Glacier melting increases flood discharge and, thus, contributes to the vulnerability of people especially those who live in downstream areas of rivers (Dhakal et al. 2019, 170). The Sunar is fully exposed to the harm of global warming.

Second, the transboundary governance of the Himalayan rivers continues to remain to be the lip service of the different states in the region, such as China and India. Environmentally, the natural courses of rivers are insensitive to human national boundaries and yet are unprecedentedly subjugated by humans in the modern times in the manner of dam building. Nations in upstream river regions are not concerned about their neighbours downstream when they build dams. The British set the examples in the Himalayas when they built the Sharda Barrage in the 1920s and the postcolonial India replaced it with the Tanakpur Barrage in the 1980s without serious consideration of the Nepalese living downstream Mahakali River. The Sunar has been under the exposure of human-induced floods for the last one hundred years.

Third, Nepal's own unforgiving caste system and politics have subjected the Sunar to the danger of floods. Like in India, the social space of the Sunar as a part of Nepal's Dalit population is a segregated space based on the Hindu notion of pollution that the untouchables are the "pollutants" to the "cleanness" of the higher castes. Cultural customs never work independently from modern political processes; instead, they are often unconsciously embodied in policymaking processes and inter-communal relationships. As discussed earlier, the Sunar have no historical roots to their current habitat along Jogbudha River and yet have to accept the ongoing spatial segregation that adds to their susceptibility to floods.

Protection of seed, food grains and cropland

In Belauri and Kutiyakhabar, farmers often experience a shortage of seeds due to failure and destruction of crops caused by regular floods. Traditionally, in Tharu community of Belauri, the grains and seeds are stored in artistic mud vessels (Figure 9.2a) and bamboo grain storage basket with mud plastering from outside. These baskets are kept in an upper storey of house (Figure 9.2b). However, in Kutiyakhabar, storage of grains and seeds in artistic vessels was not observed in Sunar communities. They store in normal vessels. The Tharu has traditional knowledge of making grain storage basket plastered inside and out with mud and cow dung. They have traditional knowledge in making such basket. In art and architect, Tharu people have a unique knowledge that is reflected in the basket. In the village, we found very few houses having mud vessels.

(a)

(b)

Figure 9.2 Mud vessels (a) and bamboo (b) grain storage basket with mud plastering practiced by Tharu.

Photographed by Rashila Deshar 2019.

A resident of Belauri told us, "The traditional knowledge of making seed storage mud vessel and basket has been transferred to me from my ancestor, however, the younger generation are not interested to learn such an art due to education and exposure to the modern world and there is no conservation and promotion effort done by government." He further explained to us that the basket has been raised above the ground to prevent the loss of the grains and seeds to flood, "We used to place the mud basket on the ground but now due to flood, we have the basket elevated high enough so that flood can't reach." With this old traditional practice, the Tharu communities are battling the foods by ensuring the safety of grain storage and preserving traditional seeds in addition to the commercially improved crop seeds.

As frequent floods compel both Tharu and Sunar communities to establish village based seed bank, a variety of main and alternate crop seeds adapted to flood prone areas are identified for being stored in the village seed banks for use in contingency situations. The main sources of seeds are farmer's own saved seeds and seeds from neighbours and relatives. Most farmers' seed selection process takes place before and after harvesting. In the past, the collected seeds were stored in community seed storage bin (Figure 9.3). The bin is made of bamboo with mud and cow dung plastering from outside. At present, the villagers have established a cooperative to conserve local seeds. This cooperative purchase, stores and distributes seeds within the community and sets up a mechanism for regenerating

Figure 9.3 Abandoned community bamboo seed storage basket.

Photographed by Rashila Deshar 2019.

Figure 9.4 Seed storage in aluminium beans.
Photographed by Rashila Deshar 2019.

seeds. Seed are now stored in aluminium beans (Figure 9.4). The cooperative distributes seeds without cost to flood affected households every year.

Farmland loss due to sand deposition is higher in both communities. As a result, farmers have started to cultivate sugarcane and watermelon because the sandy soil is more suitable for such crops. The study in Mohana-Macheli watershed of western Nepal carried out by D. Neupane and his colleagues (2019, 228) shows that cultivation of sugarcane in sand area is one of the mitigating measures of offsetting flood effects and preventing river bank erosion. Areas with sugarcane cultivated are better protected from being eroded in comparison with the areas without sugarcane cultivation. Places without mitigation measures applied suffer more erosion at the river banks.

Protecting houses and saving lives

People at Kutiyakhabar have developed their preventative and coping strategies to save lives and guard their houses and household belongings against flooding. In Sunar community, planting of trees and sow grass seed around their settlement is found as an effective method guarding against river cutting caused by the floods from Jogbudha River. The community has also developed methods to rescue human lives by building boats made from banana trunks and by making life preservers from dried, unpunctured gourds to help people and protect household belongings in the time of flooding. Community members take these initiatives in the absence of governmental help. A Dalit Sunar shared with us, "Our village in Kutiyakhabar lies in an India-Nepal borderland; so Nepalese government doesn't take care of us. During flood we don't get any support from Nepal government; we take the responsibility to rescue and saves lives of each other." He keeps a dried gourd as his life preserver in his home to be used in the event of flood. He thinks that it is best for him to take actions when the government is not taking responsibility to save and rescue the community during flooding. He added, "The community has received few life jackets provided by non-government organizations. This is not enough for our community. So we prepare rescue boats made of banana trunk and the floating dry gourds. We learnt to make them from our parents and grandparents." These specific techniques used in Kutiyakhabar show that local knowledge tends to be locally initiated, context specific, experientially practiced, verbally shared and transmitted by repeated action from one generation to another. They are closely related to subsistence and survival (Shaw et al. 2009, 32) and pertains to the adaptability of local knowledge to changing living environments (Dekens 2007).

The Tharu at Belauri community, on the other hand, have developed their own preventative and coping strategies to guard houses against flooding, such as raising the height of the plinth and building storage spaces in the upper section of the house for food, water, fuelwood, seed and other valuables during flooding. Among the two subgroups of Tharu, Rana Tharu make their house elevated by increasing the height of the plinth of the house. They have unique knowledge and skills of wall painting, straw work, mud work and wood carving. One Chaudhary Tharu from Belauri says, "Rana Tharu have built their houses by raising the platform since many generations. I didn't know why Rana Tharu build such houses. After several floods, we also started to elevate the plinth of our house and cowshed to protect our house and household belongings."

Rana Tharu community has a tradition since long past to protect their grains and other accessories from flood. From previous experience of flooding, other Tharu from the community tend to elevate their house and cowsheds. Some use raised platforms outside their houses to keep food grains, fodder, fuelwood straw and hays to protect from flood water. The houses, which are remarkable for their architect and spatial organisation, reflect the traditional knowledge of Tharu communities. Before a flood comes, livestock are transferred to the elevated safe places. Our study suggests that the building of the elevated houses and structures has become a normal practice. The studies conducted by Shitangsu Kumar Paul and Jayant Routray (2010, 495) in flood prone areas such as Bangladesh emphasise how villagers built their dwelling units on raised foundations or on earthen platforms so that a low magnitude flood cannot reach the plinth level. Similar practice of erecting shelters was found in the Tharu community, too.

In downstream region, Kutiyakhabar and Belauri communities have their traditional knowledge associated with their life from time immemorial. Local knowledge systems in these communities encompass dynamic, culturally specific knowledge, practices and beliefs. These systems evolve through adaptive processes based on the location specific learning-by-doing behaviour and intergenerational transmission. The local knowledge and practices of Rana Tharu show how traditional knowledge is passed from one generation to another and also spread within a community and how it often forms the people's identity.

Preventing water contamination and river channel erosion

Many community members of Kutiyakhabar and Belauri shared with us their experiences of suffering from contaminated drinking water due to the floods. Floods invade wells and contaminate sources of clean water for drinking. To prevent the contamination, folks increase the height of the wellhead supported usually by a brick platform. When we asked community members about the specialised design of the wellhead and its relation with traditional knowledge, they said that the elevated platforms of the well existed in the past but only for avoiding contamination. Now, the new design seals the well, increases more height to the platform supporting the wellhead and adds concrete walls to the three sides of the platform. These new features more effectively prevent water contamination and physical damages to the well. To add, the use of concrete, not a part of the old design, increases the durability and availability of the well (Figure 9.5).

Traditional bioengineering is enacted to minimise the riverbank erosion in Doda River and Jogbudha River in Kutiyakhabar and Belauri communities. Without this bioengineering intervention, the erosion of the riverbanks will lead to flooding and changing the direction of the river channel. Both communities use living bamboo as the bioengineering material to protect the riverbanks. Traditionally, bamboos are being used as living plants as well as construction material in different soil bioengineering techniques in many parts of Nepal. Upon consulting with a technical expert, the Tharu and Sunar communities have also adopted the strategy to develop a bamboo spur filled with

Figure 9.5 An elevated tube well in Shankarpur, Belauri Municipality.
Photographed by Rashila Deshar 2019.

sand bags placed diagonally along the river bank. This was further strengthened through planting local varieties of grass species, such as Napier grass (*Pennisetum purpureum*) and Bush Morning Glory (*Ipomoea carnea*), along with bamboo on the banks of the river. This is a technically feasible method developed from a traditional design and can be easily replicated with locally available resources and little involvement of specialists. These bioengineering structures help divert river currents, reduce the speed of undercurrents and also help limit river bank cutting. These traditional-cum-technical methods using local materials are implemented on Jogbudha River and Doda River. They are proved to be successful in controlling river cutting. To note, there is a high level of community support in such intervention, which is extremely heartening and remarkable as told by deputy mayor Mrs Chini Ojha Giri and Mr. Gyanendra Sunar, chairman of Mahakali Municipality of ward number 7. The effectiveness of the local knowledge as the source of people's coping strategies is, however, dependent on ecological and social conditions within which they must take into account (Figure 9.6).

Discussion—braiding together local knowledge with modern science and policymaking

During the course of our writing up this chapter, we were tempted to prefer "indigenous knowledge" over "local knowledge" as both an empirical and a conceptual phrase for our field narratives and conceptual discussions. If we did, it

(a) (b)

Figure 9.6 Community constructed bamboo spur and bioengineering in Doda River (a) and Jogbudha River (b).

Photographed by Rashila Deshar 2019.

would be legitimate, too, to do so because both the Sunar and the Tharu are classified by the Nepalese state as indigenous and minority peoples (Bhattachan 2003, 37; Subba et al. 2014, 20). From the perspective of indigenous studies, we find that the ancestors of the Sunar and the Tharu had their Indian roots. At the same time, both ethnic groups have been an integral part of Nepal since ancient times. As Nepalese, we also believe they are indigenous peoples; however, we do not wish to distract our readers to the wider, fruitless debates of how to define "the indigenous" in the global academic and policy communities. It is because thus far with countless specialists' contributions, UN has not been able to reach a universal definition but resorts to indigenous peoples' self-definitions, as discussed in the chapter by Zainab Khalid in this book. From the perspectives of environmental science and environmental humanities, it makes more sense to present the indigenous knowledge of the Sumar and the Tharu as local knowledge.

Our decision is based on our understanding of the existing definitions of indigenous knowledge and traditional ecological knowledge. We list two of the most influential definitions for discussion here. Paul Sillitoe defines:

> Indigenous knowledge in development contexts may relate to any knowledge held more or less collectively by a population, informing understanding of the world. It may pertain to any domain, particularly natural resource management in development currently. It is community based, embedded in and conditioned by local tradition. It is culturally informed understanding inculcated into individuals from birth onwards, structuring how they interface with their environments. It is also informed continually by outside intelligence (Sillitoe 2002, 9).

Fikret Berkes defines:

> TEK [Traditional Ecological Knowledge] is a cumulative body of knowledge and beliefs, handed down through generations by cultural transmission, about the relationship of living beings (including humans) with one another and with their environment. Further, TEK is an attribute of societies with historical continuity in resource use practices; by and large, these are non-industrial or less technologically advanced societies, many of them indigenous or tribal (Berkes 1993, 3).

In our research, we find that "local knowledge" or "local environmental knowledge" is more effective term for our chapter. The foremost reason for our preference is the environmental humanists' understanding of the indigenous in relational terms with the changing world, as it reflects a conscious effort not to regress into the colonial and the naturalist hyperseparation of the indigenous and the coloniser, and human and nature (Plumwood 2006, 128) as if the indigenous were a perpetually unchanging Other. The whole reason why the indigenous peoples are called "the indigenous" because of the violent encounters from the global colonialism, particularly relevant to Americas and Australia. The Himalayas became an indigenous home for many peoples whose remote ancestors came from elsewhere. The mobility of the indigenous people is a complex topic, which we do not wish to take away our central concern about how the Sunar and the Tharu implement their environmental knowledge in facing the current climate change and human-induced natural disasters. Like the UN's choice, we decide to give the indigenous knowledge an alternative nomenclature contextually specific to the Sunar and the Tharu. Thus, local knowledge is our first choice for two reasons. First, local knowledge is conceptually inclusive of indigenous or traditional knowledge. Second, it does not exclusively emphasise an unbroken continuity or isolation from the rest of the world. The notion of the local is relatively understood as place, meaning that when one place ends, another begins. Relative isolation is a topographical factor across the Himalayas; however, it does not mean that people do not migrate or trade with each other (Smyer 2017, 11–44). In the case of the Sunar and the Tharu, their environmental knowledge of rivers and floods since the advent of modern era is an accumulative body of knowledge in relation to colonial environmental practice of the British, the postcolonial India's hydraulic management, and the marginalising effects of Nepal's domestic environmental policies and Hindu caste-based customs. In a word, in this multifaceted context, local knowledge in this chapter is meant to capture the intertwinement of the global climate change, the anthropogenic impacts of India's management of the barrages and socio-religious limitations of Hinduism. It reflects the wisdom and resilience of the Sunar and the Tharu living under environmentally and socially adversary conditions.

With this said, due to geographical diversity and presence of many ethnic communities, Nepal is considered one of the societies with diverse traditional knowledge systems. In the case of flood, the knowledge concerning pre-disaster mitigation and coping strategies is accumulated from local people's experiences of floods, past and present. Different communities have their varied early warning systems and coping strategies developed from their human-nature relations situated in different ecosystems with different hydraulic patterns. These locally based environmental knowledge systems are proven to be effective in saving lives and properties and recovering from natural disasters. From the environmental humanist perspective, the qualitative understandings of local knowledge admittedly have little presence in the modern scientific and policymaking communities in contemporary Nepal. The intangible aspect of local knowledge often appears to be a challenge to the quantitative approaches of natural scientists. Like many other parts of the world, indigenous or traditional communities in Nepal have their environmental knowledge woven into their place-based living environments, religious beliefs, folklores and modes of subsistence. It is a type of integrated knowledge often situated in socially and politically marginalised contexts. It thus has very little chance to converse with modern science and co-innovate effective methods for sustainable human flourishing with the healthy environment. Modern scientific community often overlooks local environmental knowledge and wisdoms, while local communities are very much aware of modern science as a system of knowledge as well as an instrument of power sanctioned by the government and research institutions. However, based on our work, modern environmental science has much potential to collaborate with small scale societies and to translate their environmental knowledge into modern scientific terms making it available for public awareness and governmental policymaking.

We hope our study of the environmental knowledge of the Dalit and Tharu communities to be an initiative of environmental scientists, which provides insights to the policy arena of Nepal. The current state of affairs is that policy makers in Nepal have not yet taken the environmental knowledge of the Dalits and Tharu into account. For effective policymaking, modern scientists and policymakers are compelled to work together. We scientists play important role in presenting local knowledge to policymakers and inform the public of its potential valuable contributions to sustainable living. At the moment, both scientists and policymakers have much ground to cover before they can work together seamlessly. To be fair with our scientific community, we are proactive in our environmental research and making attempts to approach policymakers. The challenge for policymakers is that they have very little time to consider the value of modern scientific findings for their policymaking due to the complexities of intra-governmental politics. Speaking of the government, the culture of its centralised decision-making process is incompatible with knowledge-based policymaking methods. Modern scientific knowledge is thus largely excluded let alone the indigenous and local knowledge of the Dalits and Tharu communities. Our common challenge is how to overcome the barriers between scientists and policymakers, and between policymakers and local communities. The alliance between all of them is waiting to be materialised. We trust that, not too long from now, local stakeholders, scientists and policymakers will join hands for the capacity building of those communities living under disaster prone conditions. The ultimate goal of both modern science and governmental policymaking is the same when it comes down to the wellbeing of people and the environment. Therefore, national policies of Nepal should lend substantial support to the scientific documentation and promotion of community-based environmental knowledge for a sustainable future of Nepal.

References

Berkes, Fikret. 1993. "Traditional Knowledge in Perspective." In *Traditional Ecological Knowledge: Concepts and Cases*, edited by J. T. Inglis, 1–10. Ottowa: International Development Research Centre, Canadian Museum of Nature.

Bhattachan, Krishna B. 2003. "*Report: Indigenous Nationalities & Minorities of Nepal*". London: Minority Group Rights International.

Bisht, Bhagwan S. 1994. "*Tribes of India, Nepal, Tibet Borderland: A study of cultural transformation.*" South Asian Books, Missouri, Columbia.

Bishwakarm, Mom. 2016. "Democratic politics in Nepal: Dalit political inequality and representation". *Asian Journal of Comparative Politics* 2(3):261–272

Coburn, Thomas B. 1984. *Devī-Māhātmya – Crystallization of the Goddess Tradition.* Motilal Banarsidass, Delhi, India.

Dekens, Julie. 2007. *Local knowledge for disaster preparedness: a literature review.* International Centre for Integrated Mountain Development (ICIMOD), Kathmandu, Nepal.

Dhakal, Shobhakar, Leena Srivastava, Bikash Sharma, Debajit Palit, Brijesh Mainali, Rabindra Nepal, Pallav Purohit, Anandajit Goswami, Ghulam Mohd Malikyar and Kul Bahadur Wakhley. 2019. "Meeting Future Energy Needs in the Hindu Kush Himalaya."

In *The Hindu Kush Himalaya Assessment: Mountains, Climate Change, Sustainability and People* edited by P. Wester, A. Mishra, A. Mukherji, and A. B. Shrestha, 167–208. Cham, Switzerland: Springer.

Dhungel, Dwarika N., and Pun Santa. 2009. *The Nepal-India water relationship: challenges*. Springer Science and Business Media.

Gurung, Harka. 2005. "Social exclusion and Maoist insurgency". *National Dialogue Conference on ILO Convention* (169):19–20.

Gyawali, Dipak, and Ajaya Dixit.1999. "Mahakali Impasse and Indo-Nepal Water Conflict". *Economic & Political Weekly* 34(9): 553–564.

Harding, Elizabeth U. 1993. *Kali: The Black Goddess of Dakshineswar*. Nicolas Hays. Inc., York Beach, ix-xx.

Hawley, John S., and Donna M. Wulff. 1982. "*Sri Ramakrishna: The Spiritual Glow*". Motilal Banarsidass Publisher, Delhi, India.

Heise, Ursula K. 2017. "Introduction: planet, species, justice—and the stories we tell about them". *The Routledge Companion to the Environmental Humanities*, 17–26. Routledge.

Khanal, Krishna, Frits S. Gelpke, and Uddhab P. Pyakurel. 2012. *Dalit Representation in National Politics in Nepal*. Nepal National Dalit Social Welfare Organisation.

Krauskopff, Gisele. 1999. "A Marshland Culture: Fishing and Trapping among a Farming People of the Tarai". *HIMALAYA, the Journal of the Association for Nepal and Himalayan Studies*, 19(2):21–37.

Krauskopff, Gisele. 2007. "An 'indigenous minority' in a border area: Tharu ethnic associations, NGOs, and the Nepalese state". In *Resistance and the State: Nepalese Experiences*, edited by David N. Gellner, 199–243. New York: Berghahn Books.

Lama, Nabraj. 2019. "Re-negotiating the Mahakali Treaty in the changing geopolitics of Nepal". *International Journal of Scientific and Research Publications*, 9(1): 417–420.

Loo, Yen Yi, Lawal Billa, and Ajit Singh. 2015. "Effect of climate change on seasonal monsoon in Asia and its impact on the variability of monsoon rainfall in Southeast Asia." *Geoscience Frontiers* 6 (6):817–823.

McDaniel, June. 2004. *Offering Flowers, Feeding Skulls: Popular Goddess Worship in West Bengal*. Oxford University Press, New York, USA.

Neupane, Dinesh, Pradeep Gyawali, and Dinee Tamang. 2019. "*Study of river channel migration and identification of potential sugarcane cultivation area in the Mohana-Macheli watershed using remote sensing*". *ISPRS Annals of Photogrammetry, Remote Sensing & Spatial Information Sciences*, 4.

Paul, Shitangsu K., and Jayant K. Routray. 2010. "Flood proneness and coping strategies: the experiences of two villages in Bangladesh". *Disasters*, 34(2):489–508.

Plumwood, Val. 2006. "Concept of a Cultural Landscape: Nature, Culture and Agency of the Land." *Ethics & the Environment*, 11(2):115–150.

Regmi, Rishikeshab R. 1994. "Deforestation and rural society in the Nepalese Terai". *Occasional Papers in Sociology and Anthropology* (4):72–89.

Rose, Deborah B., Thom van Dooren, Matthew Chrulew, Stuart Cooke, Matthew Kearnes, and Emily O'Gorman. 2012. "Thinking through the environment, unsettling the humanities". *Environmental Humanities*, 1(1):1–5.

Roy, Joyashree, Eddy Moors, M. S. R. Murthy, S. V. R. K. Prabhakar, Bahadar Nawab Khattak, Peili Shi, Christian Huggel, and Vishwas Chitale. 2019. "Exploring Futures of the Hindu Kush Himalaya: Scenarios and Pathways." In *The Hindu Kush Himalaya Assessment: Mountains, Climate Change, Sustainability and People* edited by P. Wester, A. Mishra, A. Mukherji and A. B. Shrestha, 99–126. Cham, Switzerland: Springer.

Seth, Anji, Alessandra Giannini, Maisa Rojas, Sara A. Rauscher, Simona Bordoni, Deepti Singh, and Suzana J. Camargo. 2009. "Monsoon Responses to Climate

Changes—Connecting Past, Present and Future." *Current Climate Change Reports*, 5(2):63–79.

Shaw, Rajib, Anshu Sharma, and Yukiko Takeuchi. 2009. *Indigenous knowledge and disaster risk reduction: From practice to policy*. Nova Science Publishers, Inc, New York, USA.

Shukla, Deeptima. 2006. "India – Nepal Relations: Problems and Prospects". *The Indian Journal of Political Science*, 67(2):355–374.

Sillitoe, Paul. 2002. "Participant observation to participatory development: making anthropology work." In *Participating in Development: Approaches to indigenous knowledge* edited by P. Sillitoe, A. Bicker and J. Pottier, 1–23. London: Routledge.

Skar, Harald O. 1995. "Myths of origin: The Janajati Movement, local traditions, nationalism and identities in Nepal". *Contributions to Nepalese Studies*, 22 (1):31–42.

Smyer, Dan. 2017. "Introduction: Trans-Himalayas as Multistate Margins." In *Trans-Himalayan Borderlands: Livelihoods, Territorialities, Modernities* edited by D. S. Yu and J. Michaud, 11–43. Amsterdam: Amsterdam University Press.

Stiller, Ludwig F. 1993. *Nepal: Growth of a Nation*. Human Resources Development Research Center, Kathmandu, Nepal.

Subba, Chaitanya, Bishwamber Pyakuryal, Tunga S. Bastola, Mohan K. Subba, Nirmal K. Raut and Baburam Karki. 2014. "*A Study on the Socio-Economic Status of Indigenous Peoples in Nepal*". *Kathmandu: Lawyers' Association for Human Rights of Nepalese Indigenous Peoples (LAHURNIP)*.

Subedi, Surya P. 2006. "Hydro-diplomacy between Nepal and India, and the Mahakali River Treaty". In *Dynamics of Foreign Policy and Law: A Case Study of Indo – Nepal Relations*. New Delhi: Oxford University Press, 120–152.

Part IV

Transboundary environmentality and indigenous commoning

10 Indigenous irrigation system linking people, place and the planet

The practice of *Jamfwi* on the India–Bhutan borderlands

Anwesha Dutta and Shailendra Yashwant

Introduction

There are 56 rivers and hundreds of streams that flow into the Terai region known as The Dooars of Assam, along the India–Bhutan border, and that go on to meet the Brahmaputra River. Traditionally this land is home to the indigenous Bodo people (formerly known as Kacharis) who continue to practice a hundreds-of-years-old community-managed indigenous irrigation system known as *dongo* or *jamfwi*, that diverts water from the rivers to paddy fields via a series of earthen levees and canals. Our interlocutor, Krishna Bahadur Chhetri, President of Saralbhanga Dong Management committee, tells us:

> What is fascinating is that this system that is heavily reliant on voluntary and equal contribution of all communities, across religions, castes, [and] clans, has survived for hundreds of years, most of them turbulent, with a constantly changing demographic dynamics. The Bodos, the Nepalis, other *adivasi* [indigenous people] that are not Bodos, including the Bengalis and the Brahmos and even the Bhutanese from that side of the river—so many of them have come and gone and come back again, all of them seem to instinctively pick up where the last settlers left off to repair and revive the *dongo*.

Sidney Endle, in his 1911 treatise on the Kachari people, elaborated on the Bodo practice of a highly efficient but inexpensive "Public Works Department" of their own devising in the field of agriculture where the cultivation of both *kharif* and *rabi* crops were prevalent.[1] The irrigation system used water from numerous streams originating in the Bhutanese hills. This was done through the construction of canals comprising the indigenous *dongo* or *jamfwi* system. The earthwork embankments for diverting water from riverbeds into rice fields were carried out with much expertise and strength, that led Endle to state that "failing some very overwhelming convulsion of nature, it would seem to be hardly possible that a famine could take place in the Kachari Duars" (Endle 1911, 12) (Figure 10.1).

Saralapara Village, on the banks of the Saralbhanga River, in Kokrajhar, was once a small settlement nestled between the river and vast fields of rice that has grown into an unplanned small town, is still bound by the *dong* network. Here, the Bodos know the Saralbhanga, a tributary waiting to unite with the Bhrahmaputra

Figure 10.1 Dong channel flowing from Bhutan to India.
Photo by Shailendra Yashwant/Oxfam-TROSA.

River as Swarmanga, and consider the river to be a *dei*, an elder sister that sustains them. All of the tributaries of the Brahmaputra, a male river, are female and are referred to as older sister in popular lore.

Across the border, the hills of Bhutan are covered with lush forests, but on the Indian side there are vast tracts of dry plains with occasional patches of severely denuded forests. Not very long ago the forests were contiguous across the borders but internal migration, poverty and increasing demand for fuelwood changed the landscape drastically on the Indian side of the border. Mattam Mardi, Secretary of the Saralbhanga Dong Management Committee, recounted to us:

> *Dong* are primarily used to water paddy [fields] in Saralpara. They are most needed in the dry periods of May to August, during planting season. Over three thousand families of Saralpara are dependent on the *dong*, but only about five hundred families get the full benefit, while others get it in varying degrees, and some none at all. There has been a scarcity of water for last few years due to changes in climate. Sometimes we have to steal water from other people, even our neighbors.

Drawing on the narratives of community members gathered over the course of our ethnographic fieldwork, the following contribution situates itself within the vibrant discussions about the complex relationships between people, place and planet. In doing so, this contribution situates itself within the growing field of environmental humanities as, "an intrinsically interdisciplinary or even transdisciplinary research practice" (Peterson 2019, 4) with an emphasis on the inseparableness of humans and nature, and the need to bridge ecology and the humanities

(Rose et al. 2012, 3). We especially resonate with two basic precepts of the Australian Aboriginal worldview succinctly phrased by Mary Graham: "The Land is the Law" and "You are not alone in the world" (Graham 1999, 105–106). The former emphasises the fact that it is the land that creates and nourishes humankind. The latter highlights the kindred relations of humans with the land that is a living entity having a life of its own. Both precepts are efficacious for us to explain the Bodo community's relationship to the *jamfwi*, the system that long precedes the creation of political borders been India and Bhutan and other forms of regulatory regimes in relation to water supply, management and use. We thus find that the environmental-humanist approach is most relevant to our case study in which humans and nature mutually lodge in a logic of connection (Rose 2015, 5) or a web of mutual interdependence rooted in the earth.

Our focal point in this chapter is a conceptualisation of the *jamfwi* as an indigenous system of managing and sustaining a transboundary water commons on the borderlands of Bhutan and India. We expand on the notion of transboundary commons by conversing with the larger body of work on commons, indigenous practices, and borderlands. While the notion of commons usually suggests shared interests (Yong 2020, 205), the Saralbhanga River commons are paradoxically also sites of inter-ethnic conflicts and contestations over land and environmental resources. The transboundary commons, in this case, is thus constituted by the dynamically enmeshed processes of "environmental commoning" (Miller et al. 2020, 298) and "b/ordering" that co-produce borderland spaces for environmental actions (Miller 2020, 474). Through the narratives collected from our fieldwork, we then delve deeper into the interface between local knowledge and modern science and highlight how this particular practice, in fact, relies on the everyday human–nature and human–human relationships between neighbouring countries. We conclude by reasserting the importance of these commoning practices but also questioning when and why certain indigenous knowledge forms gain prominence, under what conditions and whose/what forms of knowledge gain prominence over others.

The conflict context of Northeast India borderlands

Northeast India, which is comprised of eight states, is both an Asian frontier and a borderland region. Its only terrestrial connection to mainland India is through a narrow corridor of land, only 27 kilometres wide at one point, which is known as "the chicken's neck." The region shares about 98% of its border with the neighbouring countries of China, Myanmar, Bangladesh, Bhutan and Nepal, and only 2% of its border with the rest of India. The region can be physio-geographically categorised as the eastern Himalayas. It is territorially part of sovereign India and yet geographically and ethnolinguistically embedded within a trans-Himalayan context. As the territorial foundations of many modern, post-colonial, trans-Himalayan nation-states rely on their imperial predecessors' maps, national borders often become barriers against the traditional overland trade routes. This is also the case with the India–Bhutan borderland. These processes, as Smyer Yü (2017, 26) describes, have reshaped the trans-Himalayas into what he terms "multistate margins" (Smyer Yü 2017, 26), meaning the collapse or dispersion of this once cohesive region through territorial slicing into various nation-states.

In this modern geopolitical context, unlike the comparatively well-studied borderlands of India, Bangladesh, Nepal, Pakistan, China and Myanmar, the India–Bhutan borderland is rather understudied. Although there is a history of ethnic violence on the Indian side of the border (see Dutta 2016, 481; Barbora 2015, 292), it has not affected the everyday social and economic exchange across the border.[2] Therefore, despite being a peripheral space, this is not a conflict zone characterised by attacks, famines, the denial of dignity, or the saga of partition. The border between Bhutan and India does not speak to issues of belonging, forced displacement and banality (Cons 2013, 38; Jackson 2013, 117). In fact, the Ahom kingdom that ruled the region for centuries before the British, had shared friendly trade relations with Bhutan since the 1300s (Ganguly 2006, 22). In Bogle and Manning's (1879, 59) account, "The Bhutanese, the people of Lhasa, and those of several other countries located to the northwest of the Brahmaputra carried out uninterrupted trade with Assam." British rule did not dampen these exchanges, and interviews carried out with Bhutanese nationals in Gelephu often reiterated these connections. One of my interlocutors said: "My grandfather used to come down to the border every summer and get rice, vegetables and *eri* silk in exchange for dry fish and wool. My grandfather also spoke fluent Assamese and his friends in Assam spoke Bhutanese."

However, the India–Bhutan borderland, where the practice of *jamfwi* is found, is not completely immune from conflicts. In fact, the village Saralpara, through which the Sarabhanga River flows from Bhutan into Assam, was affected by violent conflict in 2012 and in 2014. Demographically the area is inhabited by ethnic Bodo, Nepali, and *adivasi* populations. As it is surrounded by forests, Saralpara is also known as a forest village. Section 2(t) of the Forest Rights Act defines "forest villages" as follows:

> …forest villages means the settlements which have been established inside the forests by the forest department of any State Government for forestry operations or which were converted into forest villages through the forest reservation process and includes forest settlement villages, fixed demand holdings, all types of *taungya* [a sustainable system of forest management] settlements, by whatever name called, for such villages and includes lands for cultivation and other uses, permitted by the Government.
>
> (Forest Rights Act, 2006, 3)

These being interior forest villages, there is an absolute lack of basic amenities, with water scarcity being one of the key challenges.

On the diplomatic front, Bhutan and India share special bilateral relations in trade and development by virtue of the Indo-Bhutan Treaty of Friendship of 1949. It was updated in 2007 and called for the expansion of economic relations and cooperation in the fields of culture, education, science and technology. Unlike Nepal, whose 1950 treaty with India has been the subject of great political controversy and nationalist resentment for decades, the nature of Bhutan's relationship with India has not been affected by concerns over the treaty provisions. Collaborating with the Indian government, from 2003 to 2004, the Royal Bhutanese Army conducted operations against the Assamese insurgent group called the United Liberation Front of Assam (ULFA) that was operating out of Bhutanese territory. Overall, Bhutan and India are on good terms with each other.

It is within this backdrop that the practice of *jamfwi* is situated. We elaborate more about the ways in which people-to-people networks have transcended national borders and acted as early warning systems in case of flash floods on the Indian side. In fact, the relationship between local communities and external nongovernmental agencies help subvert the state-centric borderland narrative (Figure 10.2).

Jamfwi as a transboundary system of common pool resources

To begin with, the state of Assam, in general, does not have a great track record of providing irrigation facilities for farmers in its borderlands with Bhutan. Though Assam receives over 2 metres of annual rainfall, water continues to be a scarce resource due to the fast run-off caused by the sloping terrain of the Sorolpara area. Groundwater is largely inaccessible as well, and with high iron contamination, its quality also suffers. Notwithstanding such adversities, the survival of the water-intensive paddy cultivation in the region owes mainly to the indigenous *dongo* canal system. This indigenous water distribution system involves building temporary bunds or causeways on permanent stream or river courses originating from the eastern Himalayan foothills of Bhutan and flowing south to meet the Brahmaputra. The *dongo*, as a diversion system, redirects some of the water from the rivers and

Figure 10.2 Dong channel in Sorolpara.

Photo by Shailendra Yashwant/Oxfam-TROSA.

streams into fields and homesteads in the downstream villages. So, it makes the sense that a *dongo* network starts at the point of diversion from a river or water source, in this case, the Saralbhanga River. Larger *dongo* systems comprise subsidiary channels, around three to five feet wide, taking off from the main *dongo* channel that is seven to twelve feet wide. These subsidiary canals break off into *jamfwi* or field channels that supply water to the agricultural fields. The canals are usually opened for a particular period for a specific village so that the village residents water their fields or store some water in their backyard ponds.

Similar to the case of the *rajaki* water system in Hunza Valley described by Zainab Khalid in this book, when it comes to the sustenance and everyday upkeep, every household in each village is expected to send at least one member of the household to work voluntarily to maintain the *dongo* system and ensure uninterrupted water flow. In this sense, the *dongo* is a collective system managing water as a common pool resource as well as a transboundary common, given its borderland context. Each point in the intricate network of the *dongo* is managed by local community institutions, called *dong-bandh* committees, that have emerged over the decades to maintain these indigenous irrigation structures. These committees exist intermittently throughout the four districts of Bodoland Territorial Region (BTAD), Kokrajhar, Baksa, Udalgiri and Chirang. The *dong*, therefore, are a vital component of a dynamic hydro-social landscape managed by the indigenous commons system through everyday interactions.

Assam's agriculture department states that the Baksa district does better in autumn cultivation than the state average, and officials of the agriculture department credit this indigenous irrigation system for this. Yet, during our fieldwork, we found that some households obtained more water than the others, depending on the distance between households and the channel, the status of how many voluntary workers a household contributes, and the ability to pay fines in lieu of physical labour. This links to the problems associated with the use of common pool resources. Ostrom and her colleagues (1999, 280–281) suggest two distinct solutions to the management and distribution of commonly shared resources, namely restricting access and creating incentives. In our case, restricted access seems more plausible given the inevitable reliance on *jamfwi* as the primary source of water in the absence of other alternatives. Towards this end, common pool theorists have further argued that sharing of resources encompasses comprehending the "underlying rights and powers of access, use, management, exclusion, and transferability that are conferred through rules governing resources" (Agrawal 2001, 1650).

The responsibility of the maintenance of the *dongo* is realised through the notion of *shramdaan*, or collective labour. Tasks like dredging or clearing a canal, or repairing a broken dike are done by the farmers under the supervision of the core committee members (usually including the "captain", who is the head of the *dongo* system at the village level). Farmers must either go themselves or they must send an able-bodied man in their stead (women and children are not allowed to work). If a farmer fails to turn up, he must pay a fine set by the committee. Repeated absence and defaulting on penalty payments would result in the committee stopping water to his fields. Apart from contributing to *shramdaan*, individual households also have to pay a nominal fee, either in cash or kind, for the maintenance of the *dongo*. This is usually in proportion to the land owned by the farmer.

In recent years, however, increasingly erratic monsoon rainfall patterns and climate variability have led to severe and frequent flood disasters in the region. It is common knowledge that climate change, whether natural or human-induced, all leads to environmental impacts in negative terms. In the India–Bhutan borderlands, this has resulted in longer droughts, shallow river flows and repeated flash floods. Satyaraj Narzary, one of our interlocutors, recalled the floods of July 16, 2012, "When I woke up in the morning, there was no water. Nor were there any signs of flood. But around eight o'clock, the water started rising and before we realized what was happening, the whole area was flooded. About ten to twelve houses were washed away. Many families lost their cattle in the floods, their standing crop of paddy was destroyed, and considerable amount of land was lost due to erosion. We have not been able to farm on that land ever since." He said that the floods happened following heavy rains in the foothills of Bhutan when the Saralbhanga River broke a temporary embankment. "We had to run with our children and the few belongings we were able to gather. One contingent of the Army rescue team was deployed to rescue us with lifeboats but the force of water was such that the first lifeboat turned turtle in the middle of the river and the villagers ended up rescuing the soldiers," said Kamal Basumatari. "After that we made a raft with few tire tubes and bamboo to ferry women, children, and a few belongings across the river with great difficulty. I have never seen Saralbhanga so furious. The water was also very cold," he added.

The floods of July 2016 wiped out the entire town of Sarphang on the Saralbhanga river in Bhutan and wreaked havoc in Sorolpara. Both floods and droughts caused the loss of lives and property to people on both sides of border. Yet at the same time, for downstream villagers in Assam, the Saralbhanga River is their lifeline. The transborder complexity of the river puts the downstream areas inside the India border in a worse situation. After the flash floods in 2016, Bhutan tried to divert the river away from human settlements to mitigate the damage and also gave permission to stone miners to retrieve and sell stones and boulders from the dried-up riverbed. This resulted in a lot of disquiet among local farmers who were concerned that these activities would increase the flows of silt in the river. To note, small changes in the river's flow, course and siltation adversely affect those who live on the banks of the Sarabhanga. In fact, upstream damming by Bhutan and increased landslides in the mountains have already transformed the once perennial sources of water into dry rivulets in winter.

Finally, the transboundary nature of the commons warrants some attention since this digresses from the notion of commons being a local or regional phenomenon as seen in the existing research (Ostrom 1999, 496; Agrawal 2001, 1651). Often, the heterogeneity of social, cultural and economic preferences and actors with divergent interests from different nation-states complicates the transborder governance and local management of a transboundary commons (Mitchell 1999, 32). Recently, more and more public and research attention has been given to Himalayan-sourced rivers as international commons. Hirsch (2006, 105) puts forth the notion of "bioregional transboundary commons" (2006, 105) in the context of the Mekong River Basin. He describes the Basin as a transnational area defined by a natural boundary within which the commonality of the shared resource is based on the interconnected nature of the river system. Soliev and Theesfeld (2020, 64) offer solutions to the challenging dilemma of the transboundary governance of water commons in central Asia. They focus on the benefits rather than the quantity of water with "economic-development," "egalitarian-social," and "environmental" priorities (Soliev and Theesfeld 2020, 65).

Interestingly, what distinguishes our case of transboundary commons is the absence of any formal treaty between India and Bhutan that governs water sharing and the collaboration of heterogeneous actors beyond the indigenous communities in the form of civil-society organisations on the Indian side as well as on the Bhutanese side. The informal, relatively effective management of the commons systems like the *dongo* among different actors is also indicative of the roles of nonstate actors, with their shared borderland history, can play in the absence of the state. This occurrence, of course, resonates with Miller et al.'s (2020, 473) notions of transboundary commoning and b/ordering. The former pertains to "bringing people together across borders around a shared interest," while the latter "directs attention toward where borders are (re)drawn and how they are reinforced to include or exclude certain groups of people from environmental benefits and to expose or insulate sections of society from emerging threats and crises (Miller 2020, 474–475). Miller's transboundary commoning and b/ordering also speak to the transborder, ethnically diverse population of Bodo, Nepalese, Rabha, Garo and other *adivasi* peoples in the Sorolpara area. The transboundary commoning embedded in the *dong* system, in fact, transcends ethnic and national differences by focusing on water as a shared, transboundary common pool resource. In a recent Dong Bund Committee meeting, as one of our local interlocutors recounted, its Nepali president, supported by the majority of members, said, "Water is for everyone. No one community has an exclusive right over it by excluding others. Nobody here believes that. Committee members are selected based on their individual merit, not based on which community they belong to." (Figure 10.3)

Drawing on the *kuhl* system in Kangra, Himachal Pradesh

Before delving further into this specific case, mention should be made of similar indigenous or community-led irrigation systems in other parts of India, and discuss the literature thereof. Yet, it is also important that this section distinguish the transboundary nature of *jamfwi*, especially in the context of India, and how that leads to unique arrangements of transboundary governance practices. Harry Fischer (2017, 480–483) describes the changing nature of governance of the *kuhl* in Himachal Pradesh. *Kuhl* are an intricate systems of earthen irrigation canals that carry water from glacial streams to agricultural fields for wet rice cultivation. Fischer (2017, 484) further argues that communities have harnessed the state, leading to the state emerging as an integral force in sustaining access to water, as the *kuhl* represent a form of "collaborative engagement between communities

Figure 10.3 Sorolpara dong committee at work.

Photo by Shailendra Yashwant/Oxfam-TROSA.

and state institutions that might become increasingly common in mountain and other regions in the coming years" (482–483). Fischer thus makes the case for a better understanding of the state and external interventions in shaping the process of management of common pool resources. J. Mark Baker (2007, 199) argues that it is the decentralised and linked nature of these systems that makes them work, in the absence of a central power. He notes that what makes the *kuhl* in Kangra work is their long and interwoven nature. Baker uses an inductive approach derived from common-property-regimes theory and the concepts of "regionality" and "socio-cultural embeddedness." In this way, he proposes a counterpoint to the prevailing understanding of common property regimes, which he believes "cannot predict why and how a stressed regime will either persist unchanged, transform to endure, or collapse" (Baker 2007,11). Thus, Baker emphasises the extremely adaptive capacity of the *kuhl* regimes to cope with and recover from natural hazards and anthropogenic threats over time. However, this does not account for contexts of geopolitical instability, the transboundary commons, or contexts where the state is either absent or present through forms of military intervention for counterinsurgency activities, as mentioned above. More recently, A.K. Sharma and his colleagues (2015, 352) noted that the *kuhl* system has almost become defunct in current times mainly due to declining participation from communities.

Thus, local context, geopolitics and civil-society organisations play an important role in the management of common pool resources. Transboundary commons go beyond just management and distribution of water resources but also bring in the indigenous communities' relationships with the commons as a resource and as historical forms of transboundary engagements rooted in a shared past and culture, as noted above. What distinguishes the transboundary commons in the Sorolpara area from their Himachal counterparts is the adaptive strategies to encounter transnational complications through the involvement of civil-society organisations and the formation of friendship groups on both sides of the border. Moreover, the indigenous commons are cultivated through the inclusive sharing of local indigenous ecological knowledge and actions directed at redistributing both benefits and costs in an equitable manner. The undefined, fluid spatiality of the transboundary commons is illustrated by its fugitive resource, that is, water. Moreover, the Saralbhanga cannot be deemed an open-access resource due to the existence of village-level committees in India and those on the Bhutanese side, and rules that govern the river at the local level. Given the historical trajectory of the India–Bhutan borderlands, it would be apt to say that the very processes of commoning and b/ordering are being continually altered in relation to each other to coproduce transboundary spaces for the sustained use of the commons (c.f. Miller 2020, 474). This also digresses from Ostrom's (2009, 420) emphasis on "well-defined resources" with clearly known properties, which are consented upon by a fixed collective of resource users in the logic of collective action (Miller 2020, 473–475).

As already mentioned, Sorolpara is a reserved forest area where the residents are termed as "encroachers." The provision within the National Water Policy of 1987 states that "efforts should be made to involve farmers progressively in various aspects of management of irrigation systems, particularly in water distribution and collection of water rates" (Ministry of Water Resources, Government of India 1987, 11). Several Indian states including Assam enacted the act in support of exclusive legislation over people's participation in irrigation management: Participatory Irrigation Management (PIM). Water Users Associations have also been formed in the state for the implementation of major, minor and medium irrigation projects. However, these schemes do not extend to Saralpara because it is forestland. This has led to the stepping-in of civil-society organisations in recent years to circumvent the nation-state border by focusing on people-to-people networks in sustaining the resulting micro dams. This example shows that small, low impact, social water management projects actually work in areas where the state refutes civic services, and these projects are minimally invasive with minimum damage done to the environment.

WhatsApp as early warning system—people-to-people network

To understand the extent of the human-to-human network of the indigenous commons, the example of telecommunication is particularly worth noting. While doing fieldwork, Yashwant learned from local community members that in the last few weeks of June 2019, a series of WhatsApp messages were sent from Bhutan to India to warn their cross-border friends downstream along the Aai, Saralbhanga and Manas Rivers about cloudbursts, swollen rivers and possible flash floods affecting people in the Indian state of Assam. Bhutan is in a high rainfall zone and in the last 15 to 20 years, with alarming regularity, there have been increased incidents of cloudbursts, resulting in severe flash floods that have led to the destruction of everything in their path in Bhutan and then downriver in Assam. These messages were sent by officials via unofficial channels. If they had gone through the official channels, they would have involved the administrative head of the *dzonkhag* (district or local administration) in Geluphu, who would, in turn, pass the information to the officials in Bhutan's capital, Thimphu. The officials in Thimphu would then inform officials in New Delhi, the capital of India. They would, in turn, inform officials in Guwahati, the capital of the Indian state of Assam, who would pass the warnings on to Kokrajhar district headquarters. In the final stage, these messages would be relayed from there to villages along the India–Bhutan border. In most cases this circuitous channel would take too long, with information either critically delayed or unclear, and of little use to most riverbank communities in downstream Assam. Currently, the communities on the Assam side are relying on these "WhatsApp early warnings" routed through members of the Bhutan–India Friendship Association (BIFA) to friends in local NGOs in Kokrajhar town, who pass the information to their network. Messages are forwarded within minutes, giving the villagers precious lead-time to prepare and escape the wrath of the suddenly rising rivers.

This shows that people-to-people networks in the borderlands can play an important role in the community management of common pool resources, thereby avoiding ineffective official diplomatic channels. This allows us to refer to the theory of knowledge commons (Hess and Ostrom 2007, 41–43), which states that although the physical attributes of a resource invariably play an essential role in shaping the community and its decisions, rules and policies, available technology determines the limitations and possibilities of a particular commons. The technology also determines the ability to harvest or appropriate the resource units. The production, dispersal and containment of knowledge are continually evolving. Although, the larger discussion around knowledge commons is outside of the purview of this chapter, we would like to mention that in the absence of standardised channels of communication regarding the management of the commons, for example, release of water from Bhutan and warnings about flash floods, WhatsApp has emerged as a commons digital technology to counter the pitfalls of formal governmental networks. In this process, transborder community networks survive and subvert the constraints of international borders. They highlight the interdependence of the transboundary indigenous commons, the *dong* system and the socio-political networks that sustains this system. Moreover, transborder, community-based communication also showcases the traditional ecological and economic wisdoms that flow through the system, that is, a century-old irrigation system uses the "modern" telecommunication technology of WhatsApp as an early warning system for preventing the harm brought by natural disasters like floods. However, decades of militancy in this corner of India has led to a complex situation, where telecommunication infrastructure is seen as both a threat and an opportunity, making it a politically challenging decision to strengthen the communication in the area. There is no Indian telecommunication network in the area and most local communities use Bhutanese SIM cards and deal in Bhutanese currency.

"Already this year, the WhatsApp warnings from BIFA to the last mile family travelled within ten minutes of being sent out. The delay was due to the fact that the last family ran out of phone batteries, so when we didn't see the tick-marks going blue, I borrowed a motorcycle to alert the family. Mobile phones are warning systems, but you need motorcycle for sure, as there are always those without mobile phones," said Aniram Basumatary, who does not own a motorcycle but is saving up for one.

The floods of July 2017 led to the Bhutanese authorities putting an embargo on the building of micro dams by Indian farmers from the Saralpara area of Kokrajhar district, which would adversely affect over 5,000 farmers on the Indian side. This would lead to the blockade of the *jamfwi*, thereby impacting irrigation and agriculture in the area. Under these circumstances, a local NGO along with other civil-society organisations like the All Bodoland Students Union and Bodo Women's Forum met with BIFA officials. Together they met with the Deputy Commissioner of Sarphang District of Bhutan and after much deliberation Bhutanese officials agreed to allow the farmers to build micro dams and to help the farmers to continue to divert the water of Saralbhanga River for irrigation purposes, which they have been doing for centuries now. It has to be noted that this decision was made by an individual Bhutanese officer and was not a state decision which leads the farmers on the Indian side to continue to toe the line of uncertainty. A relaxed border also brings with it forms of instability and uncertainty especially in a terrain both geo-politically and environmentally vulnerable.

Conclusion

The case of the *jamfwi* highlights some of the best practices in community management of transboundary common pool resources. It is also pertinent to the broader argument connecting indigenous communities and their struggles to global politics, and helps with the reconceptualisation of existing linkages among knowledge, resources and human communities (Hegeman 2019, 134–135). Although, borderlands are zones of anxiety, they are ultimately porous zones. The India–Bhutan borderlands serve as an example of how local governments across borders can successfully manage such resources through people-to-people networks. Our case study also shows that the absent state leads to the stepping in of local and international NGOs and civil-society organisations, which work together with local communities in the resurgence of the traditional irrigation system. Building on these relationships between BIFA and the key local NGO, fourteen civil-society organisations from Bhutan and India, including the Bhutan Transparency Initiative and a regional NGO in Assam, along with Oxfam India's Transboundary Rivers of South Asia (TROSA) program, came together in June 2019 to strengthen people-to-people ties and help safeguard the rights of riparian communities upstream and downstream by supporting community-led cross-border ecosystem management and conservation practices. Transborder collaboration for early warning and the sharing of water resources between border communities of Bhutan and Assam serves in many ways as a model for peaceful relationships in a context where transboundary sharing of resources is associated with conflict and contestation.

Nevertheless, not all is well with the status quo, since so far, the governance of these resources has been community-based in the absence of formal governmental treaties on water sharing between India and Bhutan. The socio-ecological changes produced by infrastructure development in upstream Bhutan, the lack of a standardised communication channel, and periodic inter-ethnic violence has galvanised the indigenous communities in the borderlands to find innovative ways to use and manage the commons resource. This has led them to defend not just their subsistence-agriculture-based livelihood reliant on water, but also their indigenous knowledge in relation to irrigation. Local NGOs and community-based organisations in the area have received seed funding from INGOs like Oxfam and more recently from the Bodoland Territorial Council (BTR) for the twofold process of preserving local indigenous knowledge through documentation of oral histories of *jamfwi* as well as for carrying out maintenance work and construction of newer canals and also for furthering dialogue with Bhutan. These conjoined efforts at sustaining an ancient water system engender part of a situated transboundary commons (Miller 2020, 476). Local people perform an essential role in their community-based communing, constituting environmental solidarities, collective actions and shared interests (Yong 2020). This is revealing of the fluidity of connections, where both the community and the resource shape each other's lived experiences and belonging. At the heart of the transboundary indigenous commons is the existence of a cultural logic that surpasses more-than-economic values to natural resources, thereby connecting local indigenous communities with global environmental discussions around heritage preservation and natural resource management.

It is clear that the place-based transboundary indigenous commons and traditional environmental wisdom surpasses modern borderland anxieties. Relevant international diplomacy and aid programs are helpful for sustaining an irrigation system and battling the impacts of climate change, ethnic conflicts and floods. Humans and nature in this transboundary indigenous commons are dynamically interdependent and are collapsing the previous nature/culture divide. As mentioned in the opening part of this chapter, our case study resonates with the two basic principles of Australian Aboriginal ecological worldview that the land, resources and indigenous people are inherently connected and not independent from each other.

Notes

1 The *kharif* crops include rice, maize, sorghum, pearl millet/*bajra*, finger millet/*ragi* (cereals), *arhar* (pulses), soybean, peanut (oilseeds), cotton etc. The *rabi* crops include wheat, barley, oats (cereals), chickpea/gram (pulses), linseed, mustard (oilseeds) etc.

2 Gathered through interviews with villagers, shopkeepers, and forest officials on both sides of the border.

References

Agrawal, Anil. 2001. "Common Property Institutions and Sustainable Governance of Resources." *World Development*, 29 (10): 1649–1672.

Baker, Mark J. 2007. *The Kuhls of Kangra: Community-Managed Irrigation in the Western Himalaya*. Washington: University of Washington Press.

Barbora, Sanjay. 2015. "Uneasy Homecomings: Political Entanglements in Contemporary Assam". *South Asia: Journal of South Asian Studies*. 38 (2): 290–303.

Bogle, George, and Manning, Thomas. 1879. *Narratives of the Mission of George Bogle to Tibet: And of the Journey of Thomas Manning to Lhasa*. London: Trübner.

Cons, Jason. 2013. "Narrating Boundaries: Framing and Contesting Suffering, Community, and Belonging in Enclaves along the India–Bangladesh Border." *Political Geography* 35: 37–46.

Dutta, Anwesha. 2016. "The Politics of Complexity in Bodoland: the Interplay of Contentious Politics, the Production of Collective Identities and Elections in Assam." *South Asia: Journal of South Asian Studies* 39 (2): 478–493.

Endle, Sidney. 1911. *The Kacharis*. London: Macmillan and Company.

Fischer, Harry W. 2017. "Harnessing the State: Social Transformation, Infrastructural Development, and the Changing Governance of Water Systems in the Kangra District of the Indian Himalayas." *Annals of the American Association of Geographers*, 107 (2): 480–489.

Ganguly, Jaland B. 2006. *An Economic History of North East India, 1826–1947*. Guwahati, India: Akansha Publishing House.

Graham, Mary. 1999. "Some Thoughts about the Philosophical Underpinnings of Aboriginal Worldviews." *Worldviews: Global Religions, Culture, and Ecology* 3 (2): 105–118.

Hegeman, Susan. 2019. "The Indigenous Commons." *The Minnesota Review* 2019 (93): 133–140.

Hess, Charlotte, and Elinor Ostrom. 2007. "A Framework for Analyzing the Knowledge Commons." in *Understanding Knowledge as a Commons: From Theory to Practice*, edited by Charlotte Hess and Elinor Ostrom, 41–81. Cambridge: MIT Press

Hirsch, Philip. 2006. "Governing Water as a Common Good in the Mekong River Basin: Issues of Scale." *Transforming Cultures eJournal* 1 (2): 104–113.

Jackson, Michael. 2013. *The Politics of Storytelling: Variations on a Theme by Hannah Arendt*. Copenhagen: Museum Tusculanum Press.

Miller, Michelle A.2020. "B/ordering the Environmental Commons". *Progress in Human Geography* 44, (3): 473–491.

Miller, Michelle A., Carl Middleton, Jonathan Rigg, David Taylor. 2020. "Hybrid governance of transboundary commons: Insights from Southeast Asia." *Annals of the American Association of Geographers*, 110(1), 297–313.

Ministry of Water Resources, Government of India. 1987. "National Water Policy." Ministry of Water Resources website. Accessed September 29, 2020]. http://jalshakti-dowr.gov.in/sites/default/files/National%20Water%20Policy%201987%20English.pdf

Mitchell, Ronald B. 1999. "International Environmental Common Pool Resources: More Common than Domestic but More Difficult to Manage." in *Anarchy and the Environment: The International Relations of Common Pool Resources*, edited by Samuel Barkin and George Shambaugh, 26–47. Albany: State University of New York Press.

Ostrom, Elinor. 1999. "Coping with Tragedies of the Commons." *Annual Review of Political Science* 2 (1): 493–535.

Ostrom, Elinor. 2009. "A General Framework for Analyzing Sustainability of Social-Ecological Systems." *Science* 325 (5939): 419–422.

Ostrom, Elinor, Joanna Burger, Christopher B. Field, Richard B. Norgaard, and David Policansky.1999. "Revisiting the Commons: Local Lessons, Global Challenges." *Science* 284 (5412): 278–282.

Peterson, Jesse D. 2019. "Doing Environmental Humanities: Inter/transdisciplinary Research through an Underwater 360° Video Poem." *Green Letters* 23 (1): 68–82.

Rose, Deborah Bird, Thom van Dooren, Matthew Chrulew, Stuart Cooke, Matthew Kearnes, and Emily O'Gorman. 2012. "Thinking Through the Environment, Unsettling the Humanities." *Environmental Humanities*, 1 (1): 1–5.

Rose, Deborah Bird. 2015. *The Ecological Humanities. Manifesto for Living in the Anthropocene*, edited by Kathrine Gibson, Deborah Bird Rose and Ruth Fincher, 1–7. New York: Punctum Books.

Sharma, Ashwani Kumar, Sharma K.D. and Brahm Prakash. 2015. "Death of Kuhl Irrigation System of Kangra Valley of Himachal Pradesh: Institutional Arrangements and Technological Options for Revival." *Indian Journal of Agricultural Economics* 70 (902): 350–364.

Smyer Yü, Dan. 2017. "Trans-himalayas as Multistate Margins." in *Trans-himalayan Borderlands: Livelihoods, Territorialities and Modernities*, edited by J. Michaud and Dan Smyer Yü, 11–44. Amsterdam: Amsterdam University Press.

Soliev, Ilkhom, and Insa Theesfeld. 2020. "Benefit Sharing for Solving Transboundary Commons Dilemma in Central Asia." *International Journal of the Commons* 14 (1): 61–77.

FRA, "The Scheduled Tribe and Other Traditional Forest Dwellers (Forest Rights Act)." 2006. Accessed September 29, 2020, https://tribal.nic.in/FRA/data/FRAAct.pdf

Yong, Ming L. 2020. "Reclaiming Community Spaces in the Mekong River Transboundary Commons: Shifting Territorialities in Chiang Khong, Thailand." *Asia Pacific Viewpoint* 61 (2): 201–218.

11 Rajaki

An indigenous approach to commoning in Hunza, Pakistan

Zainab Khalid

As a case study of indigenous knowledge of water use in Hunza Valley, in the western Himalaya, I write this chapter as both an environmental scientist and as a person with indigenous roots in Badakhshan in Afghan Pamir and in the Himalayan foothills in Abbottabad. Hunza Valley, accessible from Badakhshan via the Wakhan Corridor, is nestled in the tall mountains of northern Pakistan. The valley is surrounded by snow-capped mountains and glaciers, and is moistened by streams, rivers and *nalla* or tributaries of glacier melt and rivers. The local people of Hunza call themselves Hunzakutz or Brusho, and they speak the Burushaski language. They claim that their paternal ancestors were five soldiers who fell ill and were left behind by Alexander the Great in the fourth century B.C. However, scholars have found that Burushaski is a language isolate with Indo-European roots (Čašule 2014, 1). With folklore and scholarly findings combined, it is safe to say that the ancestors of the Hunzakutz underwent a process of ethnogenesis with Himalayan indigenous people. This historical encounter of Alexander's soldiers with local people allowed the Hunzakutz to diverge into several subgroups that claim to be one big family living in different settlements in Hunza Valley. According to the study by Sidky (1994), the pre-Islamic Hunzakutz practiced shamanism and revered their mountain spirits in the ancient times. It is thus fair to regard the Hunzakutz as indigenous people of the western Himalayas regardless of the changes in their religious and cultural affiliations from shamanism to world religions ranging from Buddhism in the past to Islam now. Indigeneity is not a stereotypically unchanging concept and human reality, which I will discuss more fully shortly (Figure 11.1).

Over the years, much has changed in the valley. The Hunzakutz, who once relied solely on local resources for their sustenance, now depend heavily on imported household goods from the commercial centres of Pakistan and China. In spite of the external influences that have brought many changes, the indigenous knowledge of Hunzakutz regarding the earth and natural resources is deeply rooted in their customs. Nature inextricably grows inside culture. Water is very much a central element in the human–environment relation in the valley. The Hunzakutz maintain the knowledge of *rajaki*, a local mutual aid system for a sustainable use of glacier water. While abundant glacier water feeds into the rivers and streams in the Himalayas and beyond, the topographical conditions of the mountains make the water difficult for human communities to access. Since

Figure 11.1 Google Maps image (July 2020) showing an overview of Hunza Valley in Pakistan.

Figure 11.2 A *kuhl* broken by a glacier surge. It was lined with big pipes as an adaptive measure.

ancient times, the Hunzakutz have self-engineered aqueducts or pipes to channel water from the glaciers into their villages. These water channels, known as *kuhl* are managed through the mutual-aid system of *rajaki* (Figure 11.2).

To build an interdisciplinary and intelligible study of this local community-based water management system, I have taken up the taxing task to bridge the social and natural sciences in the framework of environmental humanities. Substantiated with narratives from community members in Hunza and relevant peer scholars' perspectives, my aim is twofold. First, I recount how local people practice *rajaki* as an ancient communal system of water use in today's challenging environment. Second, I explore the complex human/nature interaction in Hunza Valley, in which each influences and transforms the other in different historical times. Since the discipline of environmental humanities is new to me, I would like to ask my readers to temper their expectations of a rookie who is still toddling in the terrain of the field. The focus of my writing and your reading is on an indigenous tradition more than on my inchoate exercise of environmental humanities as an emerging field in which many of my peer authors in this book are also toddling along. Noel Castree notes that even though the edges of environmental humanities are "currently fuzzy" (Castree 2014a, 234), the scholarship of this new field generously works with "forms of perception, cultural framing, modes and measures of value, imaginative and creative works, and emotional and informative responses" (Castree 2014a, 234). Bearing his observation in mind, I truly see environmental humanities as a precious opportunity to practice interdisciplinary studies of the environment as it demonstratively takes into account socio-political, cultural, climatic and environmental contexts of the complexity of human–earth relations, local and global. In the arena of indigenous environmental studies, Deborah Bird Rose, Thom Van Dooren, and other pioneering scholars place the study of indigenous knowledge in the centre stage of their scholarship (Rose et al. 2012; van Dooren and Rose 2016). At the same time, one cannot ignore the overall marginality of indigenous voices in this globally propagated inclusive field. By having a critical acceptance of a new and enchanting field, I trust I will yield serious results from my case study of an indigenous knowledge within the still fuzzy boundaries of environmental humanities.

Rajaki

In summer, 2019, when I went to Hunza Valley, it was unusually hot. My local hosts and friends told me that the sudden spike in temperature is a recent change in the Himalayas. Some time into my stay, I turned the faucet handle and no water came out of it. Where had the water gone? My host informed me "The *kuhl* broke last night, probably due to landslide," and said that I should use water from the storage tanks outside. The large blue and white tanks are a common sight in Hunza Valley, and are filled with water that is usually used for dishwashing, laundry or showering (Figure 11.3).

The next morning, I noticed a vehicle with a loudspeaker driving through the streets; a man was calling for a communal gathering. I was informed that the *charbu*, the man appointed to deliver messages to people, was making a call for *rajaki*. My 60-year-old host told me, "*Rajaki* is when the people get together and

Figure 11.3 A common water-storage tank in Hunza Valley.

work. Traditionally, the *numberdar* [the village head] asks the *charbu*, who uses the loudspeaker, to call people together for *rajaki*. The landowners go with spades and rakes to clear the way for water but floods disconnect the water-delivery channels again and again. *Rajaki* is a centuries-old system. Before people had proper spades, they used handmade wooden tools and stones."

The village head is entrusted by the government as an unofficial government representative and serves a bridge between government officials and villagers. Every village has its own *numberdar* who conducts community meetings and settles local disputes. My host used the word "landowner" instead of "people" because, on one hand, historically all Hunzakutz owned land, therefore they were landowners by default and, on the other hand, currently, some folks have constructed guest houses/hotels on their land and, therefore, no longer have land for agricultural use. These people are less interested in participating in *rajaki* as compared to the ones who still own land and practice agriculture. The term does not distinguish between who can or cannot participate in *rajaki.* Everyone is invited and encouraged to go participate.

A middle-aged housewife further explained how *rajaki* works:

> *Rajaki* means to work together in a group for mutual help. In ancient times, Hunza was less populated and people were poor and destitute. There weren't

> many opportunities to earn. People had to do all kinds of manual work by themselves as bringing in labor for work was out of question. At that time, Hunza was ruled by the *mir* [ruler]. All the people were his subjects and worked for him. Everyone from the village participated in *rajaki*; otherwise they had to pay fine. Even today, this custom continues to be upheld. One person from each house is required to participate in *rajaki*. A few decades ago, the participation rate was higher, maybe due to poverty. People participated as an obligation. *Rajaki* is important to clean the water channel especially for the crops and trees. *Rajaki* is physically challenging. That's why everyone needs to do it together.

The Hunzakutz have managed to sustain *rajaki* as their communal mutual aid system for water accessibility and distribution but they are also facing changes in the face of urbanisation, globalisation and climatic changes. It is an example of how the environment and humans are bound together in the same existential space.

Commoning and community building

Rajaki, as described here in the accounts of my host and the housewife, is clearly a cultural institution meant for the fair and effective management and distribution of glacier water as a common pool resource. It affords the community to form an environmental and existential relationship with the glaciers in the manner of what is known as "commons," which conceptually overlaps with common pool resource and common property. In my case study, I find the notion of commons in the context of *rajaki* is better understood with Elinor Ostrom and her colleagues' delineation in which commons "refers to a resource shared by a group of people and often vulnerable to social dilemmas" (Hess and Ostrom 2007, 349). It can be either an open-access resource or a collectively owned property with limited access. In my findings, the commons in the form of human managed glacier water is a communally accessible common-pool resource and, therefore, commons and common pool resource are interchangeably understood. Its ownership is not necessarily defined legally but is most certainly a type of customary ownership due to the ancient human–environmental relationship in the area. In Ostrom's work, commons as resource pool is understood as "natural and human-constructed resources" (Ostrom et al. 1999, 278), which impresses me with two types of commons, one directly from the earth and another artificially conceived. To be clearer for my case study, my field findings support the notion of commons as a combined pool of natural and human-constructed resources. Nature is never resource until humans or animals start harnessing and consuming life-supporting sustenance and materials from it. Thus, glaciers in the Himalayas would not be a part of a commons if the multispecies relations with them were absent. In this human–environment relation, commons is predicated on the affordance of resources from nature and is materialised in humans' systematic organisation of the distribution and consumption of the afforded natural resources. In this sense, *rajaki* is a communally based common pool resource regime (Figure 11.4).

Figure 11.4 A *rajaki* site.

Regarding how the Hunzakutz perceive and utilise their commons, I also resonate with Ostrom's critique of Garrett Hardin's notion of "the tragedy of the commons" (Hardin 1968), one of the most referenced and debated works on the economic-environmental perspective. Hardin argues that the tragedy of a commons happens when its individual members act independently to utilise the available common resources based on their self-determined needs and wants without any social rules and regulations. This competitive use of a common resource leads to the overuse and inevitable depletion of it; it is a tragedy of the community as well as of the environment. Hardin's concept of commons is apparently an unmanaged, unregulated, open-access resource in the context of a modern society in the West. Based on her study with Roy Gardener of the Nepalese irrigation system, Ostrom argues that the community-managed common pool resource is more successful than those managed by the government or without management at all (Ostrom and Gardener 1993, 108–109) because the local farmers are able to devise their own rules and regulations depending on different variables relevant to their specific context. Therefore, the tragedy of the commons can be circumvented. To add to Ostrom's argument, in my knowledge, the Nepalese farmers hold their traditional knowledge of the environment and are fully aware of the benefits derived from the communal common resource managed for sustainable use.

Peter Linebaugh (2009, 79) further reconceptualises "commons" as "commoning" suggesting the word "commons" can be a verb signifying humans' active, organised use of common resources. J. K. Gibson-Graham, Jenney Cameron and Stephen Healy expand the meaning of commoning "as a relational process—or more often a struggle—of negotiating access, use, benefit, care, and responsibility. Commoning thus involves establishing rules or protocols for access and use, taking caring of and accepting responsibility for a resource, and distributing the benefits in ways that take into account the well-being of others" (Gibson-Graham et al. 2016, 195). This perspective creates more room for an engaging discussion regarding the understanding of commoning as a community action and its implication in my case study of *rajaki.* Its conceptual richness provides me with an opportunity to refocus on *rajaki* as a set of community actions in the Himalayan context.

The human–natural environment of Hunza Valley can be seen as having been a relatively enclosed commons for centuries due to its topographical barriers limiting visitors and influences from the outside world. Here, the environment as the common pool resource shapes people's lives and vice versa. People have cared for and benefitted from the commons and established a deep relationship with the environment. *Rajaki* as a commoning exemplifies the ways in which humans emplace themselves in a more-than-human world through their relational existence with the environment that is not based only on their human free will devoid of the physical earth in a pure philosophical and moral sense. The community initiatives and actions are synced with the material affordances from the earth and foster their understanding of the human–environment relationality in such a way that they bring concepts of harmony, sustainability and resilience into play.

In larger national contexts, the notion of community is often politically imagined to strengthen the citizenry bond (Anderson 1987, 6–7); whereas in Hunza Valley, community and the actions of community members pertain more to the sustainability of human society in environmental terms. In turn, their sustained relationship with the environment strengthens their cultural, ethnolinguistic and political identity. In this sense, community is not a political abstraction or imagination as has occurred elsewhere in the world, but is woven together with both human and environmental fabrics. The study by Gibson-Graham and their colleagues also affirms the human–environment relationality in a variety of commons, such as biophysical commons, knowledge commons, social commons and cultural commons (Gibson-Graham et al. 2016, 192–194). My experience of commons in Hunza Valley resonates with that of Gibson-Graham. The commons as the result of human–environmental interactions is not only a utilitarian resource but is also a human–nature bond with community rules and regulations ensuring the social equilibrium among humans and environmental sustainability.

The continuity of the Hunzakutz as a community thus rests upon the interdependence of commons and community. Both mutually reassure each other's existence. As discussed earlier, nature's affordances, for example, water and edible plants, are latent resources when humans do not recognise them. The emergence of the commons as common pool resources requires humans to identify and utilise natural affordances. Resulting from their practice of *rajaki*, the Hunzakutz hold an accumulated understanding of "sharing the common to increase productivity" as well as of their cultural identity rooted in the Himalayan environment. As a relatively isolated mountain community relying on their living off the land, culture cannot afford to be independent from nature. It is rather nature, through humans' commoning, that culture can survive and thrive. In other words, a biophysical commons is a physical property of the earth; however, it does not conceive itself as a commons; instead, it is the human residents who conceive of resources as a provision of nature and a common resource pool. In the case of the Hunzakutz, nature and culture, commons and community, or commoning and communal continuity are inextricably intertwined. The practice of *rajaki* by the Hunzakutz is a perfect example of what Stephen Gudeman describes: "Without a commons, there is no community; without a community, there is no commons"; and that "taking away the commons destroys community, and destroying a complex of relationships demolishes a commons" (Gudeman 2001, 27). Having said this, *rajaki* as a form of commoning is indeed keeping the fabric of human community tightly woven with nature's provision, which is systematically managed and distributed.

Rajaki under the impact of the anthropogenic infrastructure building

The emergence of environmental humanities parallels worldwide debates about the Anthropocene. Although this term is not new, coined in the 2002 by Paul Crutzen (2002), it has quickly become a high-impact phrase beyond the geosciences. Social scientists, humanists and the concerned public are equally interested in it. Since the term's inception two decades ago, the Anthropocene has evolved from a geological hypothesis to an interscientific and interdisciplinary thought-field of diverse perspectives, interpretations and, most visible of all, debates concerning the responsibility of humankind for the current environmental crisis as well as in building a sustainable future of the earth as a more-than-human world. Environmental humanists are among the most active contributors to the global discourse of the Anthropocene. Lorimer argues that the validity of Anthropocene as a new epoch may be contestable in the scientific world; however, it certainly prompts more critical re-evaluations of the established nature–culture dualism and compels environmental thinkers to rethink nature as "multinatural geographies" with a multiparticipant biosphere (Lorimer 2012, 2–3). In conversation with Lorimer, Castree regards the "multinatural geographies" as "emergent geographies" whose "implication is that we can no longer talk about a singular, asocial 'nature' to justify various management/conservation/remediation/preservation/restoration measures (be they large or small)" (Castree 2014a, 241). In these multinatural geographies, there has never been a world without the presence of humans. Nature is so humanised that it is "not natural" at all (Castree 2014b, 105) except as found in the nature of pre-human times. In my field observation, the Anthropocene, on the ground level, manifests itself as a variety of anthropogenic activities and impacts that alter the biophysics and biosphere of the earth and, therefore, the lives of humans and nonhumans. Thus, a primary feature of multinatural geographies of the Anthropocene is the ontological mixing of different participants or "interacting actants" who are both humans and nonhumans, and both biotic and abiotic, inclusive of mountains and water bodies.

There is no doubt that humans have greatly altered the physical world, especially since the dawn of the Industrial Revolution. The once secluded Hunza Valley is now also undergoing transformations mainly due to the construction of interstate infrastructures for the New Silk Road initiated by China. Known as National Highway 35 in Pakistan and the China–Pakistan Friendship Highway in China, the Karakoram Highway was jointly constructed in 1974 by Pakistan and China via Hunza Valley. It is currently undergoing an upgrade as a part of the China–Pakistan Economic Corridor linking Kashgar of Xinjiang to Abbottabad and eventually to Gwadar Port operated by China Overseas Port Holding Company since 2013. The highway has certainly brought convenience for international trade. At the same time, from the environmental humanist perspective, it is an example of how nature is humanised in one of the most geologically active region of the earth. Both the human and the environmental costs are quite high. In early 2010, the section of the highway in Hunza Valley was closed due to massive landslides that created Attabad Lake and killed several dozen people and displaced many villages. The traffic flow was fully renormalised 2 years later after additional tunnels and bridges had been built. Nature speaks the language of lethal disasters when humans humanise nature on a massive scale. In this process of the humanisation of nature, will the indigenous lifeways remain sustainable in the face of globalisation, urbanisation, modernity and interstate capitalism?

In my fieldwork, older folks shared with me their memory of Hunza Valley before National Highway 35 had been built. They either used horses and mules for transportation or they walked. With limited arable land, the Hunzakutz diversified their staple crops as much as they could, growing crops such as wheat, millet and barley. They raised more cattle for labour and dairy foods. As they depended on nature's provisions, they gained impeccable knowledge about the surrounding natural environment. A 90-year-old man told me about life in Hunza before the construction of the highway (Figure 11.5):

> The houses were made of stones and mud. There was a big pond in the middle where people used to gather before sunset. Everyone (men and women) used to come out and participate in games and *tamasha* (a dance ritual with traditional instruments). In Hyderabad and Aliabad, there were four tribes. They all practiced *rajaki*. These tribes were extremely united. *Rajaki* was announced by the *charbu* (announcer) in the evening for the next day. If someone was absent, they were fined. It was mandatory for every household to participate in *rajaki* except widows and orphans.

This vivid account gives a glimpse of the recent past of an indigenous people living in a relatively isolated Himalayan valley. They lived in a close-knit community deeply rooted in the biophysical earth surrounding them. Their lifestyle reaffirms the environmental humanist perspective discussed thus far—culture grows roots in nature and nature saturates culture via human bodies, tools and materials for shelter. The Hunzakutz were bonded with their cultural history, language and religion and with nature. In this nature–culture mutual embodiment, nature in the forms of air, water and soil is the silent actor shaping the *Hunzakutz* indigeneity and their customs and livelihoods.

In Hermann Kreutzmann's study of the socio-economic transformation of Hunza Valley after the Karakoram Highway was built between Pakistan and China, he deems *rajaki* a type of forced labour (Kreutzmann 2000, 111). Based on my fieldwork, he may have mistaken the communal obligations embedded in *rajaki* as "forced labour." In fact, *rajaki* is described by the Hunzakutz as

Figure 11.5 The author in discussion with local villagers.

community volunteer work. The community workflow of *rajaki* can be seen as a loop stringing together the annually designated family in charge of the operation of *rajaki*, the community recipients of the channelled water and the portion of the water for drinking and agricultural irrigation. What I would agree with in Kreutzmann's work is that the rapid modernisation in the valley is eroding traditional practices of resource management. As previously mentioned, nonparticipating villagers to *rajaki* are monetarily fined. This communally agreed penal measure may sound coercive. In fact, it is largely the externally introduced pressure at work. For instance, many young and middle-aged males are leaving the valley for urban jobs in far-way places, while the modernisation process has not yet been able to solve the water accessibility issues. The shortage of needed labour to maintain *rajaki* is pressuring every household. This is an anthropogenic factor (Figure 11.6).

Gendered *rajaki*

The Anthropocene in its anthropogenic expressions shifts the ground of gender dynamics in a traditional society besides having its geological and the ecological impacts on the physical earth. Traditionally, the gendered division of labour for the operation of *rajaki* gravitates towards men. It is a men's business starting from its committee membership to the maintenance work on the *kuhl*. The maximum participation of women is food preparation for men. In recent years, the participation and the sense of leadership among women have grown significantly, not

Figure 11.6 A typical underground cistern.

just in the sustainability of *rajaki* but also in agricultural work. The male-centred trend of outmigration to cities is the primary force of this gendered change. Those who stay behind are mostly women, children and the elderly. Women are naturally becoming significant participants in *rajaki* and other community affairs. Like men, they are fully aware of the threatening fact that if there is no *rajaki*, there is no water for everyone. Now, in every *mohalla*, or a neighbourhood in the same alley, women are organisers in the operation of *rajaki*. They also form community-based organisations to better finance farming work, such as *tanzeem*, a community-based financial organisation owned and run by women. More and more Hunzakutz women are renting fields to grow potatoes and fodder, processing sun-dried fruits like apricots, making jams, extracting oil from apricot seeds and walnuts, and selling their products in the market. Women are no longer only staying in the domestic space but are becoming viable financial providers to their households. As they are making cash incomes, they have a say in terms of where to send their children to school and university.

Shall I consider this growing self-empowerment of the Hunzakutz women in the context of the Anthropocene as a type of feminism? I very much want to connect the current changes taking place among the women in the valley with the ongoing discussions of feminism and ecofeminism in the scholarly world; however, my hesitation is that feminism studies is overall a complex field focusing on power, the body and the sexuality of women mostly situated in Western gender politics. There is a growing subfield known as Islamic feminism that may speak to the experiences of the Hunzakutz women; however, the diversity among Muslim societies across the world prevents me to see a unified understanding of what Islamic feminism is and Muslim women's social experiences in vastly different ethnolinguistic and national contexts. In her study of Islamic feminism, Lila Abu-Lughod finds that the central point of debates is the politics of the private and public domains of Muslim women regarding their rights, clothing and bodies caught in the modernity-tradition dichotomy (Abu-Lughod 1998, 3–7). As Islamic feminism is centred on the question of Muslim women's equality and emancipation (Abu-Lughod 1998, 6; Al-Sharmani 2014, 83; Sullivan 1998, 216), the participating scholars in this feminist study are inevitably caught in divergent perspectives. Some of them reject feminism "as a form of Western colonialism and othering of non-Western Muslim women" (Al-Sharmani 2014, 84; Barlas 2008, 15). Resonating with Gayatri Chakravorty Spivak's subaltern study, Abu-Lughod asks, "What are we to think of white men saving brown women from brown men? Or white women saving brown women from brown men, as would be the case with those dedicated women missionaries who tell their tales in *Our Moslem Sisters*" (Abu-Lughod 1998, 14).

What is clear in the discursive trend of Islamic feminism studies is the topic of "new" Muslim women. I do see new changes among the Hunzakutz women; however, I prefer to situate my understanding of it in the Himalayan environmental context rather than in the urban politics of Islamic feminism led by Muslim and non-Muslim scholars in their privileged social positions in European and North American academic institutions. In the social realm of the Hunzakutz women who take the lead in promoting *rajaki* in the absence of men, the issue is not how they want to emancipate and modernise themselves from the dominance of men. Men are being taken away by the highways and jobs in far-away places. The issue is the survival of family and community under the anthropogenic pressures with fewer men present for work. It is a combined issue of the anthropogenically impacted natural environment and traditional social structure of labour and production. *Rajaki* as a community mechanism of commoning is being endangered if women do not take action in the absence of men. Entire villages can be at stake if *rajaki* dies out. In this backdrop, I would not call the new gendered division of labour in the Hunzakutz commoning effort as an Islamic feminism but would regard it as a new trend of community organisational and leadership change that demonstrates the capability of women no less than that of men. Like men, women are equally capable in the public sphere of their society. In this transformative gender dynamic, the changing environment is one of the motivating factors. To be clear, I am not advocating environmental determinism here; instead, like many geoscientists and environmental scholars, I fully recognise the destabilising weight of the anthropogenic pressures on the survival and the wellbeing of indigenous communities in the Anthropocene.

Indigeneity of *rajaki* sustained in a world religion

I started out this chapter with my regarding of the Hunzakutz as indigenous people. The notion of indigeneity is one of the most complex and complicated ideas about which scholars and politicians have attempted to reach a clear, universal definition, but in vain. The UN has made attempts but decided to leave it to the indigenous peoples around the world for their self-definitions (UN 2009, 5). What is commonly circulated in scholarly texts is the recurrent association of indigeneity with the dichotomy of tradition and modernity, conqueror and conquered, and pre-colonial inhabitants and colonised subjects (Corntassel 2003, 79; Popov-Gosart 2012, 97). In this global context, on one hand, the meaning of indigeneity is etymologically based, coming from the Latin *indigena* or native born, suggesting an indigenous people as the "original inhabitants of a country" (Coates 2004, 11). On the other hand, when the colonial history and postcolonial debates are added to the definitional research, the meaning of indigeneity leans more towards a given native human society as a formerly or currently colonised subject; therefore, the politics of colonialism and marginalisation takes the precedence in discerning who the indigenous peoples are.

Either way, if I try to frame the indigeneity of the Hunzakutz in these ongoing debated approaches, the Hunzakutz may lose their self-claim as an indigenous people of the Himalayas, as their paternal ancestors are said to be the five soldiers of Alexander the Great that were left behind in 330 B.C.E. and having come from ancient Macedonia, a geographical area covering current Greece and six Balkan states, including Bulgaria, Albania, Serbia and Kosovo. Perhaps, their maternal ancestors can quality Hunzakutz as an indigenous people, but who were the foremothers of the Hunzakutz? Did they come from elsewhere or were they Himalayan natives? Are my questions headed in the right directions for reasonable, if not convincing, answers? As aforementioned, I think that the idea of "pure" indigeneity does not speak to human realities. In this front, I concur with L. K. Csáji's work that we shall not only dwell on the established ethnonyms, but also need to take into account changing religious, linguistic, political and cultural realities (Csáji 2018, 111–135). The Hunzakutz are Himalayans. We shall not take away their rootedness to the Himalayas; instead, we recognise and respect their changing histories grounded in the Himalayas.

Therefore, I do wish to offer my conviction of them as an indigenous people based on my fieldwork and other scholars' findings. One older woman told me:

> There was a tradition called *brumkhani*. It is extinct now. Our elders used to take *shisher* (fried bread) and butter to the *kuhl* and pray there. This is a mountainous region and we have beliefs about angels and fairies residing in the mountains. The elders offered prayers to make the water abundant and favorable for us. After these prayers, they let the water flow in the *kuhl*. Times have changed now and these traditions are not followed.

This local narrative coincides with Homayun Sidky's study of pre-Islamic Hunzakutz shamanism and beliefs in mountain spirits. The mountain spirits in the native language are called *pari*, and they speak through *bitan*, their human spokesmen. The remnant of this pre-Islamic tradition was still in existence in the time of Sidky's fieldwork in the 1990s. In his research, there were three historical turning points that led to the extinction of the Hunzakutz mountain-cultural tradition. The first one was the introduction of Islam in the fourteenth century during which the indigenous, mountain-based beliefs were discouraged but permitted. The second historical point was the end of the nineteenth century when Islamic authorities began to prohibit the activities of *bitan*. This assessment echoes what David Lorimer wrote nearly a 100 years ago, "Wherever Islam has gone, its God has overshadowed indigenous shamanistic and polytheistic beliefs and practices" (Sidky 1994, 71; Lorimer 1929,, 511). The last historical point was associated with two modern events. One is the abolition of the "mirdom" (Kreutzmann 2003, 220, the traditional Hunzakutz governing system) and another took place in the late 1970s when the relative isolation of Hunza Valley was fully breached by the Karakoram Highway. Since then, Hunza Valley has attracted tourists and international traffic of goods and ideas. So, it was not too long ago when the Hunzakutz mixed their indigenous beliefs with Islam (Sidky 1994, 71–93). The former receded into their memories as the latter fully saturated the cultural ethos of the Hunzakutz. When religious conversion on the individual basis reaches a demographic tipping point, it becomes an event of cultural reorientation and remaking; however, it does not take away the indigeneity of the people undergoing the transformation.

Thus, based on my fieldwork and Sidky's findings, I argue for the Hunzakutz's Himalayan indigeneity. The pre-Islamic mountain culture of the Hunzakutz obviously had a custom of revering mountain spirits and gods. It may have formal differences from other Himalayan native expressions of mountain-spirit reverence; however, in essence it is similar to the mountain cultures of the Lepcha, the Tibetans and the Bhutanese elaborately presented in the chapters by Charisma Lepcha, Rongnyoo Lepcha and Mongfing Lepcha, Thinley Dema and Kinley Choki in this book. This human–mountain–spirit theme is a Himalayan indigenous universal. In the manner that the indigeneities of the Lepcha, the Bhutanese and the Tibetans are mixed and syncretised with world religions like Buddhism and Christianity, the indigeneity of the Hunzakutz has its own historical path intertwining with the cultures, languages and beliefs of the Greeks and the Balkans, and finally transformed by Islam.

Admittedly, as a monotheistic religion, Islam discourages mountain spirit worship; however, its overall theological cosmology affirms the earth as a sacred creation of God. In Islamic theology of nature, the human–nature relation is spoken through the creative act of God. In this sense, the earth is sacred (Nasr 2003, 94–95). Humankind is a representative of God on earth and, therefore, holds the responsibility of stewardship for the earth. In this regard, Islamic theology of nature emphasises *mizān* or balance for the complex ecosphere of the earth (Hancock 2015, 5). Its unitive understanding of the earth requires all Muslims to respect the earth in unison. This is where I find Hunzakutz's current human–nature relations are well expressed through *rajaki* as a commoning in which the participants are the stewards of the earth and of the wellbeing of their community. Although the pre-Islamic mountain spirits are no longer involved in the enactment of *rajaki*, the unitive theology of Islam and the community's cultural resilience continue to uphold their method of commoning carried from pre-Islamic times.

My recapturing of pre-Islamic Hunzakutz beliefs, their Islamic transformation, and the continuity of *rajaki* is meant to highlight the changing nature of indigeneity but not to idealise any "pure" indigenous practices. In my experience of practicing environmental humanities with the Hunzakutz, the understanding of indigeneity has to come from one's awareness of both human cultural changes and environmental conditions. Both humans and human ideas migrate, encounter, interact and transform each other. It is easier for scholars to differentiate the natives from the settlers in the historical contexts of the Americas, Australia, New Zealand and South Africa; however, it is not so smooth for those of us inquiring into the indigenous roots of peoples in the Eurasian context. Hunza Valley, along with Kashmir, northern Afghanistan, Tajikistan, Kyrgyzstan and southern Xinjiang, is a part of the Himalaya region that has gone through cultural transformations mostly due to the invasion of larger empires and civilisational systems as well as due to the introduction of Buddhism and Islam in different historical periods. Burushaski, as the native tongue of the Hunzakutz, is a language isolate but has its Indo-European roots (Čašule 2014, 1). Beneath the uniform practice of Islam, the Hunzakutz have the memory of their ancestral shamanism establishing human communication with mountain spirits. At the same time, Islam unites their village-based communities to sustain their human/nature relations in the manner of *rajaki*. Hunza Valley is undoubtedly a part of what I earlier referred to as the Muslim Himalaya. In its case, indigeneity signifies both place-based native roots and the indigenisation of externally introduced cultures and beliefs. Both are grounded in the immediate environment of the Himalayas.

The future of *rajaki*

Thus far, I have made an attempt to lay out the indigenous roots and the current practice of *rajaki* among the Hunzakutz in the anthropogenically impacted Himalayas. I have also discussed the changing gendered division of labour under the circumstances of modern infrastructural development and outmigration and how different religious beliefs and practices in different historical times have shaped human environmental practices in Hunza Valley. I hope my findings and perspectives will offer an understanding of the Muslim Himalaya in the field of environmental humanities. In my research, one of the most intriguing findings is the word *kuhl* (water channel). The people of Hunza claim it as a Burushaski word; however, studies conducted by Baker (1996) and Fischer (2016) show that communities in the Hindu Himalaya also use the same word for their irrigation water channels. Based on the regional ethnohistory, *kuhl* is a Burushaski word but has obviously migrated to other parts of the Himalayas. It has become a trans-Himalayan loan-word regardless of topographical barriers and human political boundaries. It is then reasonable to say that *kuhl* is both indigenous and indigenised when the history and cultural practices in Hunza Valley are seen in relational terms. The question is—does the indigenised also eventually become the indigenous?

The history of *rajaki* shows that it consists of both indigenous and indigenised cultural ethos as discussed earlier. The search for the artificial construction of a pure ethnic origin is perhaps a necessity in the arena of ethnic nationalism (Smith 2009, 8), but it is largely irrelevant when we address local sustainable living as a regional and a planetary question. The alliance between the Lepcha (the indigenous) and the Bhutia (the indigenised) in contemporary Sikkim (Arora 2006, 57) is exemplary of how local people of different ethnic origins collaborate with one another to sustain human–environmental wellbeing by fighting together against external forces of change that are altering the local environment and eroding local cultures. Likewise, *rajaki* is an ethnohistorical kaleidoscope reflecting the diverse cultural, linguistic and religious components of the past and the present. It is a bridge between ancient and modern Hunza and has demonstrated its relevance for human survival and environmental wellbeing regardless of its cultural form having shifted from mountain spirit cult, Buddhism and Hinduism to Islam and modern secular ideas. What if we remove *rajaki* from Hunza altogether? The absence of *rajaki* will definitely make the life of Hunzakutz unsustainable in the Himalayas. The reality is that the Hunzakutz are endeavouring to keep their indigenous practices intact and relevant in spite of the destabilising forces coming from the outside world. As an ancient method of sustaining the collaborative existence of humans and nature, there must be reasons why *rajaki* has existed thus far.

Currently environmental policymakers, development specialists and earth scientists are becoming more and more open to native environmental knowledge, especially those community-based and environmentally sensitive methods for collaborative management of natural resources. Given the currently intense and frequent geohazards manifesting as both natural and anthropogenic events in the Himalayas, one kind of environmental knowledge is not enough to ensure sustainable living. Instead, traditional collaborative models of resource management are steadily finding alliance with modern methods of community organisation and modern scientific knowledge of the environment in bigger pictures like climate change and overall environmental changes in the entire ecosphere of the earth. The openness of *rajaki* as a communal system of natural resource management is capable of offering opportunities to Hunzakutz, development workers and scientists to practice collaborative commoning for a sustainable future.

References

Abu-Lughod, Lila. 1998. "Introduction: Feminist Longings and Postcolonial Conditions." In *Remaking Women: Feminism and Modernity in the Middle East*, edited by Lila Abu-Lughod, 3–32. Princeton: Princeton University Press.

Al-Sharmani, Mulki. 2014. "Islamic Feminism: Transnational and National Reflections." *Approaching Religion* 4 (2): 83–94.

Anderson, Benedict. 1987. *Imagined Communities: Reflections on the Origin and Spread of Nationalism*. London: Verso.

Arora, Vibha. 2006. "The Forest of Symbols Embodied in the Tholung Sacred Landscape of North Sikkim, India." *Conservation and Society* 4 (1): 55–83.

Baker, Mark. 1996. "Changing Contexts, Steady Flows: Patterns of Institutional Change within the Communal Irrigation Systems (Kuhls) of Kangra Valley, Himachal Pradesh, India." *Himalaya, the Journal of the Association for Nepal and Himalayan Studies* 16 (1–2): 13–18.

Barlas, Asma. 2008. "Engaging Islamic Feminism: Provincializing Feminism as Master Narrative." In *Islamic Feminism: Current Perspectives*, edited by Anitta Kynsilehto, 15–23. Tampere: Tampere Peace Research Institute.

Castree, Noel. 2014a. "The Anthropocene and the Environment Humanities: Extending the Conversation." *Environmental Humanities* 5: 233–260.

Castree, Noel. 2014b. *Making Sense of Nature*. London: Taylor & Francis.

Čašule, Ilija. 2014. "Names of Plants of Mediterranean and Balkan Origin in Burushaski." *Himalayan Linguistics* 13(1). DOI: https://doi.org/10.5070/ H913223702

Coates, Ken S. 2004. *A Global History of Indigenous Peoples: Struggle and Survival*. New York: Palgrave.

Corntassel, Jeff J. 2003. "Who is Indigenous? 'Peoplehood' and Ethnonationalist Approaches to Rearticulating Indigenous Identity." *Nationalism and Ethnic Politics* 9 (1): 75–100. doi.org/10.1080/13537110412331301365

Crutzen, Paul J. 2002. "Geology of Mankind." *Nature* 415: 23.

Csáji, Laszlo Koppany. 2018. "Ethnic Levels and Ethnonyms in Shifting Context: Ethnic Terminology in Hunza (Pakistan)." In *Ethnonyms in Europe and Asia: Studies in History and Anthropology*, edited by Zsuzsanna Zsidai, 111–135. Hungarian Historical Review.

Fischer, Harry. 2016. "Harnessing the State: Social Transformation, Infrastructural Development and the Changing Governance of Water Systems in the Kangra District of the Indian Himalayas." *Annals of the American Association of Geographers* 107 (2): 480–489. doi:10.1080/24694452.2016.1232614

Gibson-Graham, Katherine, Jenny Cameron, and Stephen Healy. 2016. "Commoning as a Post Capitalism Politics." In *Releasing the Commons: Rethinking the Futures of the Commons*, edited by Ash Amin and Philip Howell, 192–212. London: Routledge.

Gudeman, Stephen. 2001. *The Anthropology of Economy: Community, Market and Culture*. Massachusetts: Blackwell Publishers.

Hancock, Rosemary. 2015. "Islamic Environmentalism in Great Britain." In *Muslims and Political Participation in Great Britain*, edited by Timothy Peace, 103–123. London: Routledge.

Hardin, Garrett.1968. "The Tragedy of the Commons," *Science* 162: 1243–1248.

Hess, Charlotte and Elinor Ostrom. 2007. *Understanding Knowledge as a Commons: From Theory to Practice*. Cambridge: The MIT Press.

Kreutzmann, Hermann. 2000. "Water Management in Mountain Oases of Karakoram." In *Sharing Water: Irrigation and Water Management in the Hindukush-Karakoram-Himalaya*, edited by Hermann Kreutzmann, 90–115. Oxford University Press.

Kreutzmann, Hermann. 2003. "Ethnic minorities and marginality in the Pamirian Knot: survival of Wakhi and Kirghiz in a harsh environment and global contexts." *The Geographical Journal*, 169(3): 215–235.

Linebaugh, Peter. 2009. *The Magna Carta Manifesto: Liberties and Commons for All*. Berkeley: University of California Press.

Lorimer, David L.R. 1929. "The Supernatural in the Popular Belief of the Gilgit Region." *The Journal of the Royal Asiatic Society* 507–536.

Lorimer, David L.R. 1935. *The Burushaski Language; Vol. I: Introduction and Grammar*. Oslo: Instituttet for Sammanlignende Kulturforskning.

Lorimer, Jamie. 2012. "Multinatural Geographies for the Anthropocene." *Progress in Human Geography* 5: 593–612. doi:10.1177/0309132511435352

Nasr, Seyyed Hossein. 2003. "Islam, the Contemporary Islamic World, and the Environmental Crisis." In *Islam and Ecology: A Bestowed Trust*, edited by Richard C. Foltz, Frederick M. Denny and Azizan Baharuddin, 85–106. Cambridge: Harvard University Press.

Rose, Deborah Bird, Thom van Dooren, Matthew Chrulew, Stuart Cooke, Matthew Kearnes, and Emily O'Gorman .2012. "Thinking Through the Environment, Unsettling the Humanities." *Environmental Humanities* 1: 1–5.

Smith, Anthony D. 2009. *Ethno-symbolism and Nationalism: A Cultural Approach*. London: Routledge.

Ostrom, Elinor, and Roy Gardener. 1993. "Coping with Asymmetries in the Commons: Self-Governing Irrigation Systems Can Work." *Journal of Economic Perspectives* 7 (4): 93–112. doi:10.1257/jep.7.4.93

Ostrom, Elinor, Joanna Burger, Christopher B. Field, Richard B. Norgaard and David Policansky. 1999. "Revisiting the Commons: Local Lessons, Global Challenges." *Science's Compass* 284: 278–282. doi: 10.1126/science.284.5412.278

Popov-Gosart, Ulia. 2012. "Indigenous People: Attempts to Define." In *Biomapping Indigenous Peoples: Towards an Understanding of the Issues*, edited by Susanne Berthier-Foglar, Sheila Collingwood-Whittick, and Sandarine Tolazzi, 87–116. Amsterdam: Brill/Rodopi.

Sidky, Homayun. 1994. "Shamans and Mountain Spirits in Hunza." *Asian Folklore Studies* 53(1): 67. doi: doi.org/10.2307/1178560

Sullivan, Zohreh T. 1998. "Eluding the Feminist, Overthrowing the Modern? Transformations in Twentieth-Century Iran." In *Remaking Women: Feminism and Modernity in the Middle East*, edited by Lila Abu-Lughod, 215–242. Princeton: Princeton University Press.

United Nations. 2009. *State of the World's Indigenous Peoples*. New York: United Nations Publication.

Van Dooren, Thom and Deborah Bird Rose. 2016. "Lively Ethography: Storying Animist Worlds." *Environmental Humanities* 8: 1.DOI: doi.org/10.1215/22011919-3527731

12 Transboundary environments, militarisation and minoritisation

Reimagining international relations in the Himalaya from Ladakh, India

Alexander E. Davis

Introduction

As the preceding chapters in this volume have shown, the Himalaya is an environmentally crucial global region, with enormous human and nonhuman diversity. Its ice caps moderate the global climate, and its rivers provide water to around half of humanity. Some estimates suggest that it is warming at twice global averages. More people, including troops and tourists, are taking to the mountains than ever before (Davis et al. 2020). The Himalaya is experiencing numerous forms of environmental and cultural destruction, many of which are driven by the behaviour of its states, such as India, China and Pakistan. State authorities have militarised their borders, facilitated by massive infrastructure builds, that enable an influx of troops and tourists to the region.

It is relatively clear that the environmental crisis in the Himalaya is shaped significantly by the international political context. The Himalaya, though, appears in most, indeed nearly all, international relations (IR) scholarship on the region, solely as a site of contestation between India, China and Pakistan (Malone and Mukherjee 2010; Raghavan 2019; Joshi 2011; Karackattu 2013; Saalman 2011; Singh 2008; Wang, 2011). Some postcolonial work in IR has looked at India and China as civilisational "twins," with many shared inheritances, and looks to forge solutions in the state-to-state relationship on this basis (Ling et al. 2016). This happy narrative, though, neglects the ongoing forms of slow violence and environmental destruction that India and China are unleashing on the Himalaya's minoritised people (Nixon 2011; Davis et al. 2020). It romanticises the relationship between the two states in the face of much contradictory evidence as to their actual behaviour in the region. In focusing so much on conflict and cooperation, IR has neglected far larger and more complicated questions about how the militarisation of the Himalaya, and the placing of lowland cartographical norms on its peaks and ice caps, affects its minoritised peoples and their more-than-human homes (Davis et al. 2020).

Supposedly "rational" state actors are seeking mastery over the Himalayan environment, enabled by technological developments and driven by their border disagreements. They are taking to the mountains, building massive new road, rail and airport connections, extracting resources, damming transboundary rivers and militarising the region. These Himalayan geopolitics take place on shifting mountains, inhabited by minoritised peoples, many of whom speak endangered languages, whose postcolonial states behave like colonial governors as they fight over the borders drawn by their imperial forebears. This takes place against a backdrop of climate change and slowly melting ice, which is only accelerated by the geopolitical contest. These issues are all fundamentally international. They are geopolitical. And yet, the study of IR's failure to go beyond this part of the story facilitates a likely catastrophic ending.

In order to explore ecological and geopolitical entanglements in the Himalaya, this chapter looks at the historical shifts of the colonisation, decolonisation and militarisation of Ladakh. I look at how political authority was organised in relation to environmental factors, to reimagine the Himalayan geopolitics as not being about India–China–Pakistan conflict, but centred on its more-than-human ecology. Doing this, though, requires putting IR in dialogue with environmental humanities. As Arturo Escobar (2016, 15) put it, the crisis the world faces is the crisis of "a particular world or set of world-making practices, the world that we usually refer to as the dominant form of Euro-modernity (capitalist, rationalist, liberal, secular, patriarchal, white, or what have you)." The field of IR, with its Eurocentric (Hobson 2011) and imperial (Thakur, Davis and Vale 2017) theorisation of the world, has itself been party to this world-making practice (Cox 1981). In this sense, IR's mainstream theorisations seek to explain the world by flattening it, boiling it down to its constituent parts—states competing for power—and naturalising this competition. IR's "state as actor" model—in which the international state is the only agent in international affairs—tells us that borderland peoples and ecologies do not matter. As such, an environmental humanities approach to IR seeks to destabilise international boundaries, while recognising the importance of the physical environment, the ecologies that underpin it, Indigeneity and particularly the cultures and ideas minoritised peoples with deep histories of living on a particular piece of territory.

Escobar (2016, 12), however, calls for a "pluriverse" or "a world where many words fit." This is defined by "a relational ontology," wherein "nothing preexists the relations that constitute it" (Escobar 2016, 18). This suggests immense and deep connectivity between peoples, politics, cultures and ecologies. When we read the history of borders and geopolitical conflict in Ladakh, we can see these constant entanglements. These entanglements, the connectivity between humanity, politics and ecology, define the international politics of the Himalaya in ways not yet unpacked by the IR discipline and its theorisation.

Putting international relations and environmental humanities in dialogue

Environmental humanities is an interdisciplinary field, which tends to incorporate environmental sciences and knowledges with qualitative, historical and cultural approaches to understanding the environment (Bird and Robin 2004). IR, however, has often relied on its own body of theorising, primarily looking at foreign policy elites and decision-making, intergovernmental organisations and the security of the international state. There is, however, space opening up for dialogue. Recently, studies emphasising climate change, the Anthropocene and environmental crisis have begun to appear relatively regularly in the IR literature, particularly within the branch known as "critical IR" (Harrington 2016). This broad "branch" of IR encompasses feminist, postcolonial, poststructuralist and "green" approaches. It seeks to understand the ways in which international affairs perpetuate global hierarchies of inequality, race and racism, gender and environmental destruction, with the endpoint in mind of critiquing and changing them. It is here that IR can find resonance with environmental humanities. It is particularly clear when we look at the Himalaya.

State-centric thinking, and security-centric thinking, to which IR is deeply prone, is anathema to thinking environmentally. It sees nature, air and water as being of short-term instrumental use to humans and their governing bodies. It rationalises the use and abuse of the planet, and seeks to protect it solely to maintain this exploitation. As Dyer (2017, 86) puts it, an ecocentric perspective on international affairs "implies a rejection of the split between domestic and international politics, given that arbitrary boundaries between nations do not coincide with ecosystems." Humans, moreover, are ecologically connected with one another, in ways which have global and local environmental importance (Dyer 2017, 86).

The fracturing of IR as a distinct discipline on its own makes possible a closer dialogue with environmental humanities (Cello 2017). Prior to the 2000s, environmental issues in IR were primarily discussed from neoliberal institutionalist perspectives, in which international organisations and states were seen as producing differing levels cooperation over international environmental issues. This approach, moreover, emphasises social scientific rationalism and is broadly positivist in nature. It offers very much a top down approach to understanding international environmental challenges. In the Himalaya, there are important forms of transboundary governance, such as the International Centre for integrated Mountain Development (ICIMOD), which do engage across borders. However, green IR has begun to identify a range of substate actors, including local and Indigenous social movements resisting particular infrastructure projects.

In order to accomplish its goals, environmental humanities must engage with the study of the "political" and the "global." Environmental crisis has obvious global, systemic causes, not least of all global warming, as well as particular local effects (Sörlin 2017). The goal here, then, is to connect postcolonial critiques of IR, which destabilise the international state and zoom in on Indigenous rights and agency, coupled with the literature on the "green IR" and the environment. This approach can speak to environmental humanities, and reveal to us the "international" elements of environmental crisis in the Himalayas.

Much of the work in IR which attempts a dialogue with the emergent literature on the Anthropocene has been "big picture" in nature, rather than granular, or engaging with a particular place. It has been theorised as "global" and systemic. In their manifesto calling for new investment between IR and the physicality of the planet, Burke et al. (2016, 500) stated that "Global ecological collapse brings new urgency to the claim that 'we are all in this together'—humans, animals, ecologies, biosphere. To survive, we must ask questions that are intimately connected to capitalism, modernity, and oppression." They also called for greater interdisciplinary dialogue, so as to "engage with a world of melted ice caps and permafrost, flooded cities, oceans so acidic they cannot support life, and the loss of the Amazon's rainforests" (Burke et al. 2016, 501). Environmental humanities though, is very much grounded in the physical environment of a particular place. Although abstract theoretical discussions about global environmental crisis are useful and necessary, for understanding the importance of IR and environmental humanities entering a dialogue with one another, ultimately, we need to connect across global and local scales. While a backdrop of planetary crisis understandably motivates scholarship, this means connecting the international with particular environments. The Himalaya, because of its extreme human and nonhuman diversity, is a particularly important place to begin this dialogue.

Security and transboundary environments

Perhaps the key mistake that IR analysts and theorists have made when engaging with environmental "threats" is its tendency to "securitise" the environment. To securitise something is to frame an issue or an object as a threat to someone's (often the state's) security (Trombetta 2008). This approach sees environmental destruction as being important, or worthy of protection, because a matter of national or human security. Framing something as a "security" issue turns it into a "life-and-death" threat and animates a military or militarised policy response. This makes it impossible to see humanity as connected to its environment. In the case of the Himalaya, presenting the environment as a security threat leads to further militarisation (Davis et al. 2020).

Aside from militarisation, securitising the environment can often lead to deeply undesirable policy results. This contest between states like India, China and Pakistan for Himalayan water resources, though, is ultimately unwinnable (Gamble 2019a; Davis et al. 2019). Today, there is a common refrain that future wars will be fought over water. This reflects the idea that we should care about the environment because it will create security threats. Resource scarcity, environmental degradation and the possibility of climate refugees are treated as important because they may generate conflict between states and therefore as "security" threats.

One of the best-known commentators on the possibility of water wars is Brahma Chellaney (2013), who has looked at water as a possible source of conflict in the future. Although he looks at the correct issues for our purposes, Chellaney's approach is ultimately anathema to environmental humanities. Chellaney has urged, as the solution to the crisis, more international cooperation over water resources by state level actors, and the "reasonable" exploitation of resources for development. However, Chellaney's (2013) analysis emphasises the region's environment but simplifies its complex issues into state-based environmental arguments. He blames all the mountains' ills on China, arguing that grassroots activism has checked India's development projects and prevented environmental destruction in the Indian Himalaya. He suggests that there is environmentally sustainable dam building in the region, and that this can be expanded (Chellaney 2018). Militarisation, competitive development and dam-building in the Himalaya, though, are already leading to water crises, without the need for an actual "war." The people most affected by these water crises, those living in the Himalaya—are not given adequate space to reject them. This is true in India as well as in China. Anti-dam activism may have slowed down some dams, but numerous dams have been built despite local protest (Gergan 2017; Huber 2019). The idea that wars might be fought of water has a securitising effect on the Himalayan watershed, even as it urges more state-to-state cooperation. Ultimately, we must undo this kind of thinking.

The call for more cooperation between states in the Himalaya, though obviously necessary, is insufficient to save the Himalayan environment. International agreements, when made solely between states and neglecting local agency or subnational administrations, frequently treat rivers solely as resources, rather than as watersheds or ecosystems. Using the example of the Lancang/Mekong river, which has its headwaters in Tibet, Fox and Sneddon (2007, 256) argue that transboundary river agreements fail to advance sustainable ecosystem governance because they securitise the environment. They further argue that "projects and policies that underpin this securitization are directly responsible for declining human and ecological security in both basins" (Fox and Sneddon 2007, 248). This is because the agreement uses the term "watercourse" to describe the river, rather than the term "watershed." Agreements based on a "watercourse" are successful in dividing up water resources between sovereign actors for exploitation.

They are also, however, based on inaccurate understandings of river basin ecology (Fox and Sneddon 2007, 249). River management agreements tend to place the control of water solely in the hands of states. When we are talking about geographically large and culturally diverse states like India and China, this is particularly ineffective. The watershed emerges from a region that is inhabited by people out of step of the mainstream of each of these societies. Even international cooperation, based solely on state-based actors, treats water security as a threat to be managed, and divides up its resources for rational use and extraction by state actors. This does little to address the overall health of the Himalayan watershed. I will return to this theme below when examining the Indus Waters Treaty between India and Pakistan.

The key for putting IR in dialogue with environmental humanities, then, is not subordinating the environment to the security of the state, or of people, but treating it a constitutive element of and agent in transboundary affairs. The Himalaya has an enormous shared ecology, but limited international cooperation and environmental governance to match it. Rethinking the international nature of the region, the role that its states are playing in the conflict, and the way that these states, and their militaries and engineers, relate to the Himalaya's ecologies, Indigenous peoples and its watershed, can help us to understand and unpick the environmental and geopolitical crisis the region faces.

Environmental state-making in the Western Himalaya

In IR, we often talk about the "rise of Asia," or of India and China. The Himalaya is literally Asia rising. It rises each year by about 5 cm, as the Indian continental plate crashes into the Eurasian plate, as it has done for the past 50 million years. But it then loses some of this height through erosion, with the rocks grating against one another. This process produces sand and silt, which makes the rivers particularly fertile, enabling the population density we see today in the South, South East and East Asia (Gamble 2019b). It also means that the mountains are unstable and prone to earthquakes (Gergan 2017). The Himalayas are geologically young, then, feeding both the geopolitical and geological instability. This sense of the earth being alive, though, should give us a sense of the role of ecology in the geopolitics of the Western Himalaya.

The stark environment of the region has had a profound effect on political organisation in the Himalaya. The region known as Ladakh is a high-altitude desert in the Western Himalaya. It sits on the South-West edge of the Tibetan plateau. It is nestled between Western Tibet and Southern Xinjiang (China), Gilgit-Baltistan (Pakistan), Kashmir and the Indian state Himachal Pradesh. To its immediate north is the uninhabited Aksai Chin, which is claimed by India but administered by China. According to the most recent Indian census (2011), there are 274,000 people living there. Today, it is made of two districts, largely run by hill development councils: Kargil, which is predominantly Shi'a Muslim, and Leh, which is predominantly Buddhist. The district of Leh has a population of 147,104 and Kargil has a population of 143,388. As Smith (2012, 1511) put it, the necessity of this formulaic description reveals to us the ways in which culture, language, religion and politics have combined to produce the territorialisation of the region. Some of its border regions, though, are largely uninhabited, with the exception of state militaries. This includes many of key contested territories, such as Siachen Glacier, Chulshul, the Galwan valley, Demchok, Pangong Tso and nearby Aksai Chin.

During the colonial period, the Himalaya was governed as a patchwork of small polities and princely states, which were influenced by nearby, plains-based empires, without being ruled directly (Gamble 2019a). With its sparse population and high altitude, Ladakh has been a target for expanding empires. The prevailing narrative of state-making in the region, though, is that the British, were the last empire to conquer these mountains. In the mid-nineteenth century, they defeated the region's previous rulers, the Sikhs, who had invaded the highlands from their base in Punjab. Ladakh had been under the suzerainty of the Sikh empire, under the armies of Raja Gulab Singh. Rather than ruling the region directly, the British created the Princely State of Jammu and Kashmir (1846–1947) from the Sikh Empire's ruins. Its Maharajas, the Dogra, were Hindus, based in the lowland city of Jammu, and ruled over a very diverse people who lived in its fragile environments. Most of the population of Jammu and Kashmir Princely state was located in its capital Srinagar. The realm included Ladakh; the majority-Shi'a, Tibetic-speaking, high-altitude Gilgit; the majority-Sunni, mid-altitude valley of Kashmir; and the majority-Hindu lowland city of Jammu (Rai 2018). Ladakh took up much of the land mass within the state, but was lightly populated.

The region was subjected to a series of invasions, with armies moving through and claiming suzerainty over the area. Militaries moved through, and fought over it, but lowland populations would never stay put in place, due to the region's harsh terrain, high altitude and cold winters. Although taxes were applied to its trade, the people of the region were largely left alone. In 1842, following the Dogra–Tibetan conflict, the Treaty of Chulshul (1842, in Sharma 2019, 322–323), stated that

> We will carry on the trade in shawl, pashma and tea as before by way of Ladakh; and if any one of the Shri Maharaja's enemies comes to our territories and says anything against the Rajah we will not listen to him, and will not allow him to remain in our country, and whatever traders come from Ladakh shall experience no difficulty from our side.

Ladakh, though influenced by external rulers and subjected to imperial geopolitical desires from the Qing, British and Sikh empires, was taxed, but its peoples were largely left alone in their day to day lives.

Imperial understandings of geopolitics were important underpinnings of why Ladakh and Himalaya became a borderland. In the nineteenth century, Lord Curzon framed India as a strategically vital empire which was under threat from various other imperial powers. British control of the Indian Ocean was relatively complete, and so the only threat to their Indian empire was the territorial border, the Himalaya. In particular, in his framing, it was mountain passes and valleys which needed to be defended. It was through these "grooves" that various other empires had made their way to Ladakh (Abraham 2014, 112). The British, then, sought to maintain territorial control over the region, seeing it as a potential weakness to their hold over South Asia. Their goal, then, was to create a linear border in the region, formalise control and then establish trading links across it. Within the British colonial "science" of border-making (Goettlich 2018), rivers, ice caps and watersheds were seen as ideal borders. However, high mountain passes between Ladakh and Tibet had checkpoints, where trade and pilgrims could pass easily. The notion of bordering the region in a linear fashion was an imposition of lowlands cartographic norms on the Himalaya's stark environment, dynamic and alive environment (Gardner 2019).

Depictions of Ladakh as harsh, dangerous and impossible to explore pervaded British colonial descriptions of the region as they sought to survey it (Gagné 2017, 225–230). As Gardner (2019, 2020) has argued, the inability of the British to survey and establish control of the region is one of the reasons the current border disputes were left as blank spaces on the map. By the time the Indian and Pakistani states inherited the territory, there were still missing borders from, due to the enormous difficulty in surveying these mountains (Gardner 2019).

The region's environment, the histories of imperialism and its role in Central Asian trade routes, combined with the intensification of colonial notions of map-making, race and identity, helped to produce the precarious position that Ladakh finds itself in today. Although thought of as idea places for borders, Himalayan rivers, mountains and ice caps rise and recede with the seasons. Sometimes, the changes are less subtle. For example, an eighteenth-century treaty defined the Gandak River as the border between India and Nepal. But the river slowly shifted course, and the village of Susta found itself subjected to an international border dispute (Jamwal 2017). Where the nineteenth-century cartographers struggled without the technology to map high valleys, contemporary scientific knowledge presents another issue for border makers: their science recognises how often this seismically active and ice-carved region transforms. Borders marked by lakes, ice and rivers move with the seasons.

The borders are alive.

The "Decolonisation" of Ladakh

The hurried departure of the British rulers of South Asia left the region's states with a series of cartographic difficulties, and its locals with three state's militaries to live alongside. Amidst the creation of India and Pakistan, there were calls for Jammu and Kashmir to become an independent, multicultural state (Bhat 2017). These, though, were given short-shrift by the region's newly independent states. Ladakh found itself at the crossroads of two interrelated geopolitical disputes. Today, it remains stuck between the India–Pakistan dispute over Kashmir and the border disputes stemming from the idea that the Himalaya is the "natural barrier" between India and China.

The British Raj was partitioned on the basis of the "two nation theory," a colonial narration of the many diverse identities and cultures that inhabited South Asia. This theory suggests that there were always two distinction "nations" of South Asia: a pluralist nation, perhaps with a Hindu majority, and a separate nation of Muslims. The Muslim league emerged out of the Indian National Congress (INC) with the concern that Muslims would be oppressed by a Hindu majority in India. India's identity was based on a rejection of the two-nation theory and claims to able to represent all the cultural and religious groups of South Asia. Pakistan relies on the theory for its foundational identity. The territory of Kashmir, with its Muslim majority, with a Hindu Maharaja who acceded to India, was a major test for this theory. Holding all pieces of Muslim majority territory was central to the founding idea of Pakistan. Likewise, holding some piece of Muslim majority territory was crucial to India's desire to reject the two-nation theory (Varshney 1991).

This means that culture was directly tethered to territory and environment at the moment of state-making. This is not a useful organising principle, however, for a region with sparsely or unpopulated mountain peaks and rivers being thought of as borders. Although they had made territorial claims, the British, the Dogra and Qing China had only limited capacity to actually govern the region. The dramatic geography of the Himalaya, and the lack of infrastructure, had made it too difficult to exert sustained influence. This meant that Chinese, Pakistani and Indian territorial claims had little to do Himalayan peoples cultures and identities, and there was limited contact still between the plains and the mountain peaks that were being imagined as its borders.

Ladakh's status, as well as its borders, were uncertain. In 1949, Chhewang Rigzin (1949, in Sharma 2017, 324–327), the president of the Buddhist Association of Ladakh wrote to Jawaharlal Nehru, to ask for self-determination within the Union of India. He argued that "[w]e are a separate nation by all the tests—race, language, religion, culture—determining nationality. The only link connecting us with the other people of the state being the bond of common ruler." It carried on that:

> If the Indian National Congress could persuade itself to recognize the Muslims of India as a separate nation although they had so much in common with the other elements of the Indian population the Government of India should have no hesitation in recognizing what is patent and incontrovertible fact in our case.

The letter emphasised the role of Ladakh as a crossroads, noting that "The Tehsil of Leh has Tibet and China among its neighbours and the town of Leh is the nerve centre of Central Asian trade" (Rigzin, 1949, in Sharma 2017, 324–327). The letter concluded with a request for self-determination: "Has not India repeatedly declared that it stands for the right of self-determination for all nations, and are we not a nation whose right of self-determination it should uphold and to whom it should extend the protection it seeks?"

The formalisation of the line of control between India and Pakistan split not only the Kashmir valley, but the Buddhist and Shia Muslim areas as well. The militarisation of the India–Pakistan border meant that connections between Gilgit and Ladakh were cut off. These families and cultures remain cut off today. The Indian and Pakistani armies, given the continual rise and fall of tensions in the region, began to stay in the region, fixed in place along disputed borders.

Ladakh, though, was not just a target of India and Pakistan. China's annexation of Tibet in 1950 changed the international circumstances of the Western Himalaya further. Starting in 1959, refugees from western Tibet came to settle in Ladakh following an uprising. Today, according to the Central Tibetan Relief Committee, there are 3,500 Tibetan refugees settled in Ladakh (Central Tibetan Administration 2019). Without an obvious or agreed-upon border, and local peoples being outside of each state's "core citizenry," India and China came to shadow one another in the Himalaya (Guyot-Réchard 2016). Each of these diverse and heterogenous states could claim to be the legitimate ruler of a particular piece of territory, and that drove their mutual anxiety and geopolitical tension.

With the Kashmir conflict entrenched, the regional environment came to be securitised. The Himalaya became subjected to the "national interests" of the Indian, Pakistani and Chinese states, for its environment, role in trade routes and its role of the region in the headwaters of the Indus River (Raghavan 2010, 101–146). With India–China tensions over Ladakh rising, India–Pakistan tensions over the Indus river were being dealt with through the Indus Waters Treaty of 1960 (Gilmartin 2015).

The treaty is generally seen as an example of healthy international cooperation between India and Pakistan. It covers the channels of the Indus, Jhelum, Chenab, Ravi, Beas and Sutlej rivers, and governs who can use which waters for what purpose (Indus Waters Treaty 1960). The Indus Waters Treaty takes a narrow view of a river and prioritises the rights of states to water over the overall health of a watershed. With state's water rights prioritised, and watersheds reduced to channels, the treaty does little to prevent militarisation of the river, and indeed encourages resource extraction. Once a river is seen as a resource to be managed by the state, the people that live alongside it and the ecologies that produce it, become invisible. The treaty prioritises the needs of Delhi and Islamabad, over the needs of the river, the watershed, or the people that live near it. Treating the river as something to be extracted from has gradually added to tensions over the region. This treaty has not been revisited since 1960 to be updated with advances in environmental sciences and our understanding of the Himalayan watershed (Fox and Sneddon 2007).

The 1962 war between India and China shattered a period of positive relations between India and China. Chinese soldiers simultaneously invaded Ladakh and Arunachal Pradesh to the east. Just as Chinese troops looked as though they had broken open the road from Sela Pass to Assam, they unilaterally announced a cease fire. They had, however, ousted India from Aksai Chin (Guyot-Réchard 2016, 231). India's foreign policy shifted in tone after its 1962 border war with China and with the death of Jawaharlal Nehru in 1964. Small memorials to this war are today dotted around the villages of Ladakh nearest the border. Indira Gandhi's foreign policy towards South Asia, in terms of India's border politics, rearticulated the Nehruvian vision of anticolonialism as more fundamentally tied to territorial integrity (Singh 2015). This meant the confirmation of what Sankaran Krishna has called India's cartographic anxiety. India's borders became increasing militarised and securitised (Krishna 1994). With the militarisation of the India–China border to Ladakh's north, the stand-off over Aksai Chin and the Siachen Glacier, Ladakh-Tibet connections, which went back centuries, were shut off as well.

The Kargil War, Union Territory status and renewed India–China conflict

In 1999, the region was the centre of a war between India and Pakistan. Pakistani troops crossed the Line of Control and settled on the outskirts of the Muslim majority city of Kargil. India, China and Pakistan each having nuclear weapons has even further frozen the conflict. With the cost of escalating conflicts so high, it became impossible for the armies to move forward. But this has not resolved the original border disputes. It has only become worse. Prior to the Kargil conflict, Indian and Pakistani soldiers would vacate their posts in winter, due to the extreme cold at high altitude. In 1999, though, Pakistani troops took advantage of India's army vacating their posts and occupied 130 winter posts that India had left, crossing some 10 km into the Indian territory. "Operation Vijay" took place in May–July 1999, as the snows receded and the Pakistani army was evicted from Ladakh. Much to the relief of those fearing escalation between two nuclear-armed neighbours, India avoided crossing the Line of Control in this operation (Ganguly and Kapur 2012). From this, the Indian Army learned not to withdraw for winter. The armies became even more frozen in place, with even the extremely high-altitude Siachen Glacier permanently militarised.

After Kargil, the number of Indian troops stationed in the region rose from 3,000 from the earlier 3rd Infantry Division, to 20,000, entrenching the army further in the lives and livelihoods of Ladakhis. Since the war, the Indian Army, and the Pakistani and Chinese armies has massively increased its presence in the region. Press reports suggest that the Chinese are able to deploy 80,000 troops to Aksai Chin and Ladakh within three days. India have 6,000 of the Indo–Tibetan Border Police (ITBP) on the Line of Actual Control with China at any given time, and have reportedly been considering raising this number by 8,000 (Press Trust of India 2018). At the time of writing, India and China were further militarising the region after violent clashes in 2020, and these troop numbers have gone up.

As we will see, the army and tourism, has dramatically increased the population of the region.

Throughout the 70 years of conflict, there were numerous efforts to solve the India–Pakistan Kashmir issue. In most of these negotiations, Pakistan has been willing to recognise much of Ladakh as part of India (Yusuf and Najam 2008). Ladakhi Buddhists, though, came to resent that their politics and representation was dominated by Hindu–Muslim tension. This led, through the 1980s, to a demand for Ladakh to gain more autonomy within India. This demand split the population of Ladakh along religious, regional and cultural lines. In this sense, the international politics of the region has had a profound effect on its culture.

Sara Smith has described the politicisation of religious identity in Ladakh through the 1980s. In 1989, the Leh Buddhist Association (LBA) announced a social boycott of the Muslim community, with the aim of drawing attention to their situation to the national government (Smith 2012). In response, governance was partially devolved, through the creation of the Ladakh Autonomous Hill Development Council (LAHDC), based in Leh, which was formed in 1995. This was in response to demands for the creation of Union Territory. A corresponding council was formed in Kargil. This allowed greater autonomy from Jammu and Kashmir. Governance had been decentralisation to a point, but Ladakh remained constrained by its border conflicts. The Ladakh Union Territory Front (LUTF) formed in 2002. The INC did not agree to the demand, however, and so the LUTF merged with the Hindu nationalist Bharatiya Janata Party (BJP) (Sharma 2019). Through this polarisation, the BJP began to win the support of local Buddhist groups by supporting the desire for greater autonomy. The longstanding demand for Union Territory status of Ladakhi Buddhists, was met by the Modi government in August of 2019. This decision sparked with celebrations in Leh and protests in Kargil. The decision dissected the polity of Jammu and Kashmir, while simultaneously centralising the power of the Indian central government over the region.

Shortly after the reorganisation of Ladakh, the India–China tension in the region reached surged. In June of 2020, fighting between Chinese and Indian soldiers over disputed Western Himalayan borders left 20 Indian soldiers dead, 20 seriously injured and an unconfirmed number of fatalities on the Chinese side (Dwivedi 2019). India and China have engaged in a troop build-up in Ladakh, particularly at Pangong Tso, Chulshul and the Galwan Valley (Hutcheon et al. 2020). The tensions have been the worst the region has seen for several decades, with reports of "warning shots" being fired in September 2020 (BBC 2020). Negotiations between India and China have been slow, though the two armies have recently disengaged at Pangong Tso. However, a likely result at the time of writing appears to be the year-round militarisation of the India–China border, just as has happened in Kargil and Siachen.

Cultural and ecological entanglements

With the surge of troops to the mountain and the recreation of the colonial idea of Ladakh as a harsh and contested borderland, I will now move on to look at how this environmental state-making has transformed Ladakh's culture and environment. With all of the military conflict, much of Ladakh has been transformed into an army base. The military presences has shaped the culture of the region. Gagné's (2017) ethnographic work on the region has shown how Ladakhis have taken in the conflict and the militarisation. Ladakhis tend to speak about the heavy military presence in Leh as a necessary evil (Gagné 2017, 236), or as something that prevents the region from becoming part of China or Pakistan. Local ecological knowledge has become used for defence of India. After the 1962 conflict, the Ladakh Scouts, a regiment of the Indian Army, were created. Joining the military has become a key source of employment. The ITBP are likewise deployed at high altitudes because of their experiences and their bodies. Indigenous peoples of high altitudes of the Tibetan plateau are far better able to deal with extreme cold and lighter levels of oxygen in the air (Petousi and Robbins 2016). In this sense, our bodies are deeply intertwined with the environments that our ancestors have inhabited. The bodies of Tibetans and Ladakhis are being used by the regions states to enable the further militarisation of high altitudes. Indian army officers began to describe Ladakhis as ideal soldiers for the region, because of their skills at mountain climbing, their resistance to extreme cold (Gagné 2017, 234) and their bodies are acclimatised to the altitude. Ladakh, though, was still presented by the Indian army as a harsh borderland not unlike the ways that British cartographers had done in the nineteenth century.

One major element of this militarisation has been a boom in infrastructure development. Prior to the Kargil war, infrastructure in the Western Himalaya was slow. One Indian engineer put this down to the mountains being full of "geological surprises" (Gergan, 2019). This curious turn of phrase treats the mountains as though they are alive, and in need of mastery and domination. Infrastructure in border regions is built by the Border Roads Organisation (BRO), who call themselves the "mountain tamers." When infrastructure is built in the region, it has often been slowed by unexpected water flowing out of the rocks whenever digging starts. One road, the Rohtang tunnel connecting Manali to Leh, has taken four decades but is expected to be completed this year (Dhillon and Chhina 2020).

Chinese engineers, however, appear to have mastered building roads and high-speed rail networks across Western Tibet (Bhutia 2016). This infrastructure boom has further facilitated the movement of troops and tourists. It fuelled Indian anxieties about China's ability to deploy troops to contested regions like Ladakh. India is trying to catch up, by completing 61 strategic border roads in the region by 2022 (Singh 2019). This level of control from the plains had previously impossible due to the unwillingness of expanding empires to actually stay in the region. The proximate cause of the recent fighting seems to have been infrastructure building on both sides. Building roads close to the border facilitates troop deployment, which provokes fear on the other side. The anxiety produces further infrastructure building.

The year-long militarisation of high-altitude regions is an extremely undesirable result not just for the environment, but also for the soldiers. Already, thousands of soldiers (Gao 2016) have died in the Himalaya, not from fighting, but the extreme environmental conditions. Landslides and asphyxiation are common, particularly at year-round occupied high-altitude sites such as the India–Pakistan stand-off at Siachen Glacier. The prioritising of security above all else, though it may capture how the region's states are behaving, does not capture how the politics of the region have worked historically, or how they work now.

The Indian Army made the region safe for tourism, but is also part of the attraction. Memorials to the Kargil war are among its most popular sites for tourists. The wild success of the Bollywood film *Three Idiots*, though, brought even more people into Ladakh, particularly Indian domestic tourists (Lundup 2013). Today, a visit to Ladakh enables Indians from cities like Delhi and Mumbai to experience clean air, to visit the sites of one of India's most iconic movies and to revel in the glories of India's military. *Three Idiots* features a closing scene of Kareena Kapoor chasing after Aamir Khan at Pangong Tso, a contested lake on the India–China border (Parviz 2019). It is marketed as an adventure tourism destination, where Indians from the plains can experience frontier life. Today, though, some of the most popular tourist sites, aside from the region's natural beauty, are war memorials, largely devoted to the Kargil war with Pakistan in 1999. These memorials both engage with local culture, while also emphasising the role of the Indian military.

In this sense, militarisation and tourism complement one another in important and interesting ways. The Kargil War has made the region a site of nationalist sacrifice. Just south of the city of Leh, past its military run airport, is the Indian Army's "Hall of Fame." The folding of local cultures into the conflict manifests physically. Out the front of this war memorial is a statue of the Buddha, flanked by howitzers on either side. A similar war memorial to the Kargil conflict exists in Dras, on the road between Srinagar and Kargil. Both of these are extremely popular and successful tourist sites for Indian tourists. The 2020 fighting between India and China will only add to this situation (Figure 12.1).

Himalayan states have also come to see the region's rivers as a source of "clean" electricity through hydroelectric projects. This, though, comes with enormous risks in a geologically unstable region. Amelie Huber (2019) has argued that these dams place the costs of development on local populations. Dam construction has often been met with local protest and when protests are ignored, social marginalisation increases (Gergan 2020). China's construction of dams on Himalayan rivers has also worried India about shared water resources (Gamble 2019b). All of the Himalaya's states are concerned about access to water resources and electricity generation. However, the rush to claim water resources leads these states to treat the watershed as something from which to extract value. This makes environmentally sound governance, which looks after the health of the watershed, nearly impossible.

Ladakh's ecology and its traditional ways of living off the land, are threatened by this surge of people into the region, as well as the backdrop of global warming. Schmidt and Nüsser (2017) portray the precarity of relying on slowly shrinking glaciers for water. Glacial retreat in Ladakh has been slow, yet any loss of glaciers would be quickly devastating for its traditional farming techniques. In many summers, ice melt is the only form of water for irrigation in the region. In response, locals have built ice stupas and artificial glaciers in the winter, to replace lost glacial melt for the spring (Nüsser et al. 2019). These forms of local adaption and knowledge cannot easily be transferred across the transboundary region of the

Figure 12.1 The Buddha flanked by Howitzers. Indian Army Hall of Fame, Ladakh, India. Photo by Alexander Davis.

Western Himalaya, though, because the interconnected communities of the region have been so thoroughly shut off by the border disputes. This comes despite the increase of connections between Ladakh and the rest of India. The region's climate is changing in other ways as well. Rainfall in Ladakh has increased quite dramatically in recent years. The average rainfall in Ladakh during summer has risen from 30 mm in 2009 to over 140 mm in 2019. Flash floods struck Leh in 2010, killing 255 people. (Dwivedi 2019). Although this might be a replacement for lost glacial melt, traditional homes in Ladakh are not built to withstand rain.

Security-centric thinking does not capture the reality of the situation in the Himalaya. Ecocentric thinking, though, captures the entanglements between geopolitics and ecologies of the region and reveals to us the complexity and precarity of the current situation. If India, China and Pakistan continue to see the watershed as a target for competitive resource extraction, this will only accelerate environmental degradation. The Himalayan environment will ultimately have the final say in this situation, as its ecology cannot sustain intense militarisation, climate change and hydropower extraction in the long term.

Conclusion

In November 1, 2019, a statue of Black-Necked Cranes was unveiled in Leh to celebrate the shift to Union Territory status. The cranes were selected to win and to reflect the support of the Leh Buddhist community for the change. The symbolism of the Black-Necked Crane, however, has multiple layers, some of which were likely not intended by the organisers. The birds migrate from Tibet to India during the winter, and are a key symbol of Himalayan Buddhist communities. They have become increasingly rare. In Arunachal Pradesh, they have been a cultural spark to mobilise against dam construction by local peoples (Khandekar 2016). The future of Black-Necked Cranes, then, is threatened by militarisation, dams and the destruction of the Himalayan environment, even as they remain a symbol for political changes (Figure 12.2).

When we read international politics in the Himalaya, we cannot just emphasise the region's states and the possibility of war. The Himalayan environment has always been deeply entangled in its politics. IR's ignorance of ecology and history and its assumption that the state is a rational actor, and the only one that matters in international affairs, obscures these entanglements. The connectivity between culture, ecology and geopolitics, also tie IRs and environmental humanities together. We cannot understand one of these factors without the others. Escobar's relational ontology captures this shift in a way that traditional IR thinking cannot.

Looking at the ecological entanglements with geopolitics in Ladakh tells us the ways in which the environment of the region is alive in the conflict. With Ladakh's glaciers slowly shrinking, its local people trying to adapt, understanding the Western Himalaya as a transboundary region is critical. The Himalayan

Figure 12.2 Black-Necked Crane Statue to celebrate gaining Union Territory status. Leh, Ladakh.

Photo by Alexander Davis.

mountain range has always played a crucial and active role in the region's politics. This, though, is unacknowledged by the regions states and IR's theorisations. The region's states need to recede from the mountains, allow greater local autonomy and build environmental protocols and governance that incorporate subnational administration, and listen to local voices and scientific knowledge. The history and experience of Ladakh shows the need to do this, and how flawed and damaging a rationalist, statist framing is for understanding the Himalaya. Lack of acknowledgement of both the regions fraught environment and its borderless history may underpin these ongoing tensions, but recognition of both suggests a way forward. If the highest peaks and rivers were de-territorialised, it would avoid much senseless loss of life, environmental destruction, and allow social and cultural exchange for the regions local communities.

References

Abraham, Itty. 2014. *How India Became Territorial: Foreign Policy, Diaspora, Geopolitics*. Stanford: Stanford University Press.

BBC. 2020. "China says Indian troops fired 'provocative' shots in border dispute" *BBC News* website, 8 September. Accessed October 26, 2020. https://www.bbc.com/news/world-asia-china-54067400.

Bhat, Safeer Ahmad. 2017. "Jammu and Kashmir on the Eve of Partition-A Study of Political Conditions." *South Asian Studies* 32 (2): 285–295.

Bhutia, Tshering Chonzom. 2016. "Tibet and China's 'Belt and Road' Will Tibet become China's bridge to South Asia under the Belt and Road Initiative?", *The Diplomat* website, August 30 Accessed February 11, 2021. https://thediplomat.com/2016/08/tibet-and-chinas-belt-and-road/.

Bird, Debora and Libby Robin. 2004. "The Ecological Humanities in Action: An Invitation" *Australian Humanities Review* website. April 1. Accessed October 26, 2020. http://australianhumanitiesreview.org/2004/04/01/the-ecological-humanities-in-action-an-invitation/.

Burke, Anthony et al. 2016. "Planet Politics: A Manifesto from the End of IR." *Millennium: Journal of International Studies* 44 (3): 499-523.

Cello, Lorenzo. 2017. "Taking history seriously in IR: Towards a historicist approach." *Review of International Studies* 44 (2): 236–245.

Central Tibetan Administration. 2019. "Sonamling Tibetan Settlement, Ladakh" *Central Tibetan Relief Committee* website. No date given. Accessed November 22, 2019. http://centraltibetanreliefcommittee.org/doh/settlements/india/north/sonamling.html.

Chellaney, Brahma. 2013. *Water, peace, and war: Confronting the global water crisis*. Maryland: Rowman & Littlefield.

Chellaney, Brahama. 2018. "A global environmental threat made in China." *Australian Strategic Policy* website, August 27. Accessed February 22, 2020. https://www.aspistrategist.org.au/a-global-environmental-threat-made-in-china/.

Cox, Robert W. 1981. "Social Forces, States and World Orders: Beyond International Relations Theory." *Millennium: Journal of International Studies* 10(2): 126–155.

Davis, Alexander E. et al. 2019. "Melting Opportunities: Managing Climate Change and Conflict in the Himalaya". *The La Trobe Asia Brief, volume* 3.

Davis, Alexander E., et al. 2020. "International relations and the Himalaya: connecting ecologies, cultures and geopolitics." *Australian Journal of International Affairs*, online first.

Dhillon, Gagandeep Singh and Man Aman Singh Chhina. 2020. "Leh at the end of the tunnel." *Indian Express* website, October 3. Accessed October 17, 2020. https://indianexpress.com/article/india/atal-tunnel-rohtang-pass-leh-ladakh-indian-army-6584674/.

Dutta, Amrita Nayak. 2019. "74 deaths in 3 years puts spotlight back on avalanche danger for Indian Army." *The Print* website, December 12. Accessed February 22, 2020. https://theprint.in/india/74-deaths-in-3-years-puts-spotlight-back-on-avalanche-danger-for-indian-army/334238/.

Dwivedi, Rama. 2019. "Traditional houses that served generations of Ladakhis are unable to cope with climate change." *Scroll* website, September 18. Accessed October 27, 2020. https://scroll.in/article/937195/traditional-houses-that-served-generations-of-ladakhis-are-unable-to-cope-with-climate-change.

Dyer, Hugh C. 2017. "Green Theory." In *International Relations Theory* edited by Stephen McGlinchey, Rosie Walters and Christian Scheinpflug, 84–90. Bristol: E-IR Publishing.

Escobar, Arturo. 2016. "Thinking-feeling with the Earth: Territorial Struggles and the Ontological Dimension of the Epistemologies of the South." *Revista de Antropología Iberoamericana* 11 (1): 11–32.

Fox, C. A. and Chris Sneddon. 2007. "Transboundary river basin agreements in the Mekong and Zambezi basins: enhancing environmental security or securitizing the environment?" *International Environmental Agreements*, 7: 237–261.

Gagné, Karine. 2017. "Building a Mountain Fortress for India: Sympathy, Imagination and the Reconfiguration of Ladakh into a Border Area." *South Asia: Journal of South Asian Studies* 40, 2: 222-238.

Gamble, Ruth. 2019a. "Unwinnable contest: The competition for resources." *East Asia Forum Quarterly* 11 (2): 30-31.

Gamble, Ruth. 2019b. "How dams climb mountains: China and India's state-making hydropower contest in the Eastern-Himalaya watershed." *Thesis Eleven* 150 (1): 42–67.

Ganguly, Sumit and S. Paul Kapur (eds). 2012. *India, Pakistan and the Bomb: Debating Nuclear Stability in South Asia.* New York: Columbia University Press.

Gao, Fei. 2016. "Days After Surviving Avalanche, Indian Soldier Dies". *Glacier Hub* website, February 16. Accessed October 21, 2020. https://glacierhub.org/2016/02/16/days-after-surviving-avalanche-indian-soldier-dies/.

Gardner, Kyle. 2019. "Moving Watersheds, Borderless maps, and Imperial Geography in India's Northwestern Himalaya." *The Historical Journal* 62 (1): 149–170.

Gardner, Kyle. 2020. "The elusive watersheds of Ladakh: Explaining India and China's missing border." *Observer Research Foundation* website, August 26. Accessed September 17, 2020. https://www.orfonline.org/expert-speak/elusive-watersheds-ladakh-explaining-india-china-missing-border/.

Gergan, Mabel Denzin. 2017. "Living with Earthquakes and Angry Deities at the Himalayan Borderlands." *Annals of the American Association of Geographers* 107 (2): 490-498.

Gergan, Mabel D. 2019. "Geological Surprises: State Rationality and Himalayan Hydropower in India." *Roadsides* 1: 35-42.

Gergan, Mable Denzin. 2020. "Disastrous hydropower, uneven regional development, and decolonization in India's Eastern Himalayan borderlands." *Political Geography* 80, online first.

Gilmartin, David. 2015. *Blood and Water: The Indus River Basin in Modern History.* Oakland: University of California Press.

Goettlich, Kerry. 2018. "The rise of linear borders in world politics." *European Journal of International Relations* 25 (1): 203–228.

Guyot-Réchard, Bérénice. 2016. *Shadow States: India, China and the Himalayas, 1910–1962.* Cambridge: Cambridge University Press.

Harrington, Cameron. 2016. "The Ends of the World: International Relations and the Anthropocene." *Millennium: Journal of International Studies* 44, (3): 478–498.

Hobson, John M. 2011. *The Eurocentric Conception of World Politics: Western International Theory, 1760–2010*. Cambridge: Cambridge University Press.

Huber, Amelie. 2019. "Hydropower in the Himalayan Hazardscape: Strategic Ignorance and the Production of Unequal Risk." *Water* 11 (3): 414–436.

Hutcheon, Stephen; Mark Dorman and Alex Palmer. 2020. "High stakes in a Himalayan hot-spot", *ABC News* website, July 10. Accessed February 11, 2021. https://www.abc.net.au/news/2020-07-10/high-stakes-in-china-india-himalayan-border-hotspot/12417248?nw=0.

Indus Waters Treaty. 1960. *United Nations* website. Accessed February 24, 2020. https://treaties.un.org/doc/Publication/UNTs/Volume%20419/volume-419-I-6032-English.pdf.

Jamwal, Nidhi. 2017. "As a river changed its course, a village on the India-Nepal border became disputed territory." *Scroll* website, March 19. Accessed September 12, 2020. https://scroll.in/article/831576/as-a-river-changed-its-course-a-village-on-the-india-nepal-border-became-disputed-territory.

Joshi, Shashank. 2011. "China and India: Awkward Ascents." *Orbis* 55 (4): 558–576.

Karackattu, Joe Thomas. 2013. "India–China Trade at the Borders: challenges and opportunities." *Journal of Contemporary China* 22 (82): 691–711.

Khandekar, Nivedita. 2016. "Arunachal hydropower project halted to save black-necked cranes." *The Third Pole* website, May 9. Accessed November 22, 2019. https://www.thethirdpole.net/en/2016/05/09/arunachal-hydropower-project-halted-to-save-black-necked-cranes/.

Krishna, Sankaran. 1994. "Cartographic Anxiety: Mapping the Body Politic in India." *Alternatives: Global, Local, Political* 19 (4): 507-521.

Ling, L. H. M. et al. 2016. *India China: Rethinking Borders and Security*. Ann Arbor, MA: University of Michigan Press.

Lundup, Tashi. 2013. "Contemporary Ladakh Culture, Commodification and Tourism." Institute of Peace and Conflict Studies, Issue Brief no 238.

Malone, David M. and Rohan Mukherjee. 2010. "India and China: Conflict and Cooperation." *Survival: Global Politics and Strategy* 52 (1): 137–158.

Nixon, Rob. 2011. *Slow Violence and the Environmentalism of the Poor*. Boston: Harvard University Press.

Nüsser, Marcus et al. 2019. "Socio-hydrology of artificial glaciers in Ladakh, India: assessing adaptive strategies in a changing cryosphere." *Regional Environmental Change* 19: 1327–1337.

Parviz, Athar. 2019. "Photo essay: Tourists and trash at Pangong Lake." *The Third Pole* website, June 26. Accessed February 21, 2020. https://www.thethirdpole.net/en/2018/06/26/photo-essay-tourists-and-trash-at-pangong-lake/.

Petousi, Nayia and Peter A. Robbins. 2016. "Human adaptation to the hypoxia of high altitude: the Tibetan paradigm from the pregenomic to the postgenomic era." *Journal of Applied Physiology* 116: 875–884.

Press Trust of India. 2018. "ITBP seeks more units in Ladakh to check Chinese transgressions." *Economic Times* Website, July 13. Accessed February 10, 2020. https://economictimes.indiatimes.com/news/defence/itbp-seeks-more-units-in-ladakh-to-check-chinese-transgressions/articleshow/47573860.cms?from=mdr.

Raghavan, Srinath. 2010. *War and Peace in Modern India: A Strategic History of the Nehru Years*. New Delhi: Palgrave Macmillan.

Raghavan, Srinath. 2019. "The Security Dilemma and India–China Relations." *Asian Security* 15 (1): 60–72.

Rai, Mridu. 2018. "Kashmir: From Princely State to Insurgency." *Asian History* website. No date given. Accessed November 20, 2019. https://oxfordre.com/asianhistory/asianhistory/view/10.1093/acrefore/9780190277727.001.0001/acrefore-9780190277727-e-184.

Rigzin, Chhewang. 1949. "Excepts from the Memorandum Submitted by Shri Chhewang Rigzin; President Buddhist Association Ladakh to the Prime Minister of India on behalf of the People of Ladakh in 1949". in Raghav Sharan Sharma *The Unfought War Of 1962: An Appraisal*: 324–327. Routledge: London.

Saalman, Lora. 2011. "Between 'China Threat Theory' and 'Chindia': Chinese Responses to India's Military Modernization." *The Chinese Journal of International Politics* 4: 87–114.

Schmidt, Susanne and Marcus Nüsser. 2017. "Changes of High Altitude Glaciers in the Trans-Himalaya of Ladakh over the Past Five Decades (1969–2016)." *Geosciences* 7 (27): 1–15.

Sharma, Aditya. 2019. "Service, Sentiment and Samarthan: How Buddhist Ladakh Made the Loudest Noise for a Change." *News* 18 Website August 9. Accessed November 21, 2019. https://www.news18.com/news/politics/service-sentiment-and-samarthan-how-buddhist-ladakh-made-the-loudest-noise-for-a-change-2263891.html.

Singh, Swaran. 2008. "India-China Relations: Perception, Problems, Potential." *South Asian Survey* 15 (1): 83–98.

Singh, Sinderpal. 2015. *Domestic Identity Politics and Foreign Policy from Nehru to the BJP*. London: Routledge.

Singh, Rahul. 2019. "India pushes to complete 61 strategic roads on China border by 2022", *Hindustan Times* website, February 7. Accessed February 11, 2021. https://www.hindustantimes.com/india-news/india-pushes-to-complete-61-strategic-roads-on-china-border-by-2022/story-mN9y4D98UOOu72me9RrmTI.html.

Smith, Sara. 2012. "Intimate Geopolitics: Religion, Marriage, and Reproductive Bodies in Leh, Ladakh." *Annals of the Association of American Geographers*, 102 (6): 1511-1529.

Sörlin, Sverker. 2017. "Scaling the planetary humanities, Environmental globalization and the Arctic." In *The Routledge Companion to the Environmental Humanities*, edited by Ursula K, Heise Jon Christensen, and Niemann Michelle, 433–442. London: Routledge.

Thakur, Vineet, Alexander E. Davis and Peter Vale. 2017. "Imperial mission, 'scientific' method: an alternative account of the origins of IR." *Millennium: Journal of International Studies* 46 (1): 3-23.

Treaty of Chulshul 1842. "English translation of the Persian text of the treaty signed at Leh on second of Asuj 1899 Bikrami (September 1842) between the Government of Maharaja Gulab Singh and the Government of Tibet". In *The Unfought War Of 1962: An Appraisal*, edited by Raghav Sharan Sharma, 322–323. Routledge: London.

Trombetta, Maria Julia. 2008. "Environmental security and climate change: analysing the discourse." *Cambridge Review of International Affairs* 21 (4): 585-602.

Varshney, Ashutosh. 1991. "India, Pakistan, and Kashmir: Antinomies of Nationalism(s)." *Asian Survey* 31 (11): 997–1019.

Wang, Vincent Wei-cheng. 2011. "'Chindia' or Rivalry? Rising China, Rising India, and contending perspectives on India-China Relations." *Asian Perspective* 35: 437–469.

Yusuf, Moeed, and Adil Najam. 2008. "Kashmir: ripe for resolution?" *Third World Quarterly*, 30(8): 1503–1528.

13 Symbiotic indigeneity and commoning in the anthropogenic Himalayas

Dan Smyer Yü

This chapter is written as an intertextual conversation with my peer contributors in this book regarding the changing lived experiences and understandings of indigeneity, commons and commoning concerning both the local and the planetary significances of Himalayan water contextualised in the Anthropocene and in the bordered reality of modern Himalayas. It also enters the broader discussions on the notions of transboundary commons and commoning shown in the works of Amin and Howell (2016), Miller (2020), Ostrom (1990), Linebaugh (2008), Brandshaug (2019) and Gibson-Graham, Cameron and Healy (2016). As a text of environmental humanities, this chapter answers environmental humanists' call for a "thicker" notion of humanity (Rose et al. 2012, 2) regarding what it means to be human on earth in the integrated sense of the creative forces of the earth, human–nature relations, human-to-human encounters and sustainable living.

Grounded in both local and planetary discourses of environmental ethics and sustainability, the thicker notion of humanity in this chapter is conceived from affective human–earth relations and articulates what I call "symbiotic indigeneity," "new-animist earth" and "transboundary water commoning." On the one hand, these three terms are coined to underscore the livingness of the earth and human affective bonds with it, and to re-emphasise an observable environmental fact regarding how the presence or the absence of human–earth affective bonds is directly correlated with the health or the illness of given local environments and of the neighbouring and far-away habitats that are interlinked through environmental flows such as rivers and weather patterns. On the other hand, the three terms are conceptually instrumentalised for recognising nature's innate environmental freedom as the imperative basis for contemporary international water commoning endeavours geared towards fair shares of hydraulic resources afforded from the same river systems originating from the Tibetan-Himalayas. In brief, environmental freedom in my upcoming conceptual discussion refers to the creative forces of nature on the scale of Deep Time. Environmental freedom indicates a gift of nature rather than a modern endowment of humankind. Synonymous with the earth, nature in this case is independent from humankind as a living planetary being that only began to enter relations with humanity after the emergence of humans in more recent time. In the last section of the chapter, my elaboration of environmental freedom culminates in an argument that commoning, as a collective action with nature, humans and nonhumans together, signifies the free movement of environmental flows and the niched movements of humans and animals. It is argued for as an inherent maternal principle of the earth and, therefore, an ethical basis of sustainable commons and commoning.

Symbiotic indigeneity

Unlike in the "Neo-Europes," (Crosby 2004, 2), such as the Americas and Australia, where the differences between natives and newcomers were historically shown in their distinct cultures, languages and religious beliefs, the historical sense of indigeneity in the Himalayas conveys migration, interaction and hybridity within the Himalayan region or between the Himalayas and the surrounding regions. Ethnolinguistic differences are relative, not absolute in the Himalayas, as shown in the geographical spread of the Indo-European, Tibeto-Burman, Austroasiatic and Dravidian language families. While ethnic differences are maintained through national identities or ethnic minority statuses, religions, such as Buddhism, Christianity, Hinduism and Islam, often become common cultural denominators of inter-ethnic groupings in the Himalayas as revealed by the popular designations of the Buddhist Himalaya, the Hindu Himalaya and the Muslim Himalaya. The familiarity of differences in proximity is a characteristic of Himalayan indigenous diversity. Many chapters in this book show the hybrid and intertwined sense of indigeneity among the Indo-European and the Tibetan–Burman language groups. For instance, the Sunar and the Tharo in Nepal are both classified as indigenous people; however, they also retain their oral histories of migration from ancient India (Deshar et al., this book). The Hunzakutz retain the oral history of their ancestral connection to the soldiers of Alexander the Great in addition to their claim to being indigenous to the Hunza Valley (Khalid). The Bengali Muslim chardwellers retain their Hindu past in their customary festivals (Das). While most Garos continue to live in their ancestral homeland, many of them have converted to Christianity (de Maaker). Likewise, many Lepchas have converted to Buddhism and Christianity; however, the memories of their indigenous past have survived their religious conversion (C. Lepcha, R. Lepcha and M. Lepcha). The sense of indigeneity in the Himalayas continues to evolve and interface with nonnative cultural, economic and political forces of change beyond the pre- and early-modern eras, especially in the currently anthropogenically impacted Himalayas or the New Himalayas as discussed in the introductory chapter.

Situated in both the historical and the anthropogenic contexts, this chapter proposes the current form of indigeneity in the Himalayas as what I call "symbiotic indigeneity." It refers to a set of hybrid indigenous environmental knowledge and cultural practices as the outcomes of pre-modern and modern encounters, interactions, counteractions and intertwinements of ethnolinguistic communities, larger civilisational systems, world religions, empires, colonial forces, modern nation-states, transnational corporate economic development, as well as the intervention of the global market economy and its expanding digital web that link individuals, communities and nations horizontally with cyber-transboundary connectivity. The list of the regionally and globally interfaced indigenous realities is long and exhaustive. Like our Arctic environmental humanist colleagues' understanding of the New Arctic, the New Himalayas has no peoples whose current life styles and cultural consciousness can be said to be "purely" indigenous. Even if people do not travel, things and ideas travel to people and transform or hybridise with their worldviews, livelihoods, social ethos and relations with the earth. The New Himalayas leaves no room for romanticising and imagining a pristine, untouched utopian landscape accessible only by explorers and scientists and neglecting to mention the "touch" of the people who already live there. Perhaps, in its initial encounter with Europeans between the seventeenth and the nineteenth century, the Himalayas could be seen from outsider's eyes as a topographical utopia (Smyer Yü 2015, 204), an actual physical place with natural barriers, such as frontiers, edges, gaps and in-betweens, limiting access to outsiders (Smyer Yü 2015, 41–42). Perceptually, it coincides with human mythical or ideal worlds of abundance as shown in the British colonial vision of trans-Himalayan commerce as "a perfect god-send to the starving peasantry of Ireland and of the Scotch Highlands" (Hodgson 1874, 89).

In contrast, the old Himalayas or the pre-modern Himalayas undoubtedly had a pristine quality of abundance, and they kept a relative distance from the greater world. However, the New Himalayas is not a topographical utopia. Flights, trains, buses and private vehicles are all readily available for those who wish to go in or come out of this majestic but highly stressed region. My proposed symbiotic indigeneity is meant not to encourage an unnecessary fantasised indigenous purity but to recognise the importance of discerning and reconceptualising the remembered indigeneity of the past, the lived indigeneity of the present, and the intertwinement of indigeneity and exogeneity as a third form, a symbiotic indigeneity. I feel compelled to connect my practice of environmental humanities in the Himalayan region with the ongoing worldwide conceptualisations and debates of indigenous knowledge in symbiotic terms, not as stand-alone, isolated cultural systems.

The authors in this book mentioned above are exemplary in illustrating the lived senses of symbiotic indigeneity particularly in the Tibet–Sikkim, Bhutia–Lepcha, Tibet–Bhutan and Hindu–Muslim nexuses. The indigenous histories of Bhutan and Sikkim demonstrate how indigenous memories and practices find their resilience, their new niches and their continuity in indigenised external political, economic and religious systems. In my own specialisation in the Tibetan-Himalayan Buddhist studies, the indigenisation of Tibetan Buddhism in Bhutan and Sikkim shows "a geomythological reorientation" (Smyer Yü and Sonam 2019, 7–10) of different indigenous landscapes and their resident deities and ancestral spirits that has engendered what is known as shamanic Buddhism (Samuel 1993). It is a commonly recognised pattern in the historical spread of Buddhism in the Himalayas and the Tibetan Plateau, that a process of converting not only local people but also deities and landscapes to Buddhism was involved. Besides its religious and spiritual impacts on local people and local physical environments, Tibetan Buddhism also engendered the cultural and political effects of "the taming of the mind, of society, of the environment or even of the country" and of "the subjugation of the spiritual inhabitants of the country" (Balikci 2008,

88). In this geomythological reorientation, the native Himalayan landscapes were subject to a perceptual shift by which they absorbed imported layers of exogenous meanings. The ecologically situated human dwelling places and natural surrounding environments were ontologically and spiritually given both pragmatic and transcendental meanings. The geomorphically embodied indigenous mountain deities and ancestral spirits were reoriented to be Dharma protectors and, thus, were imbued with Buddhist sacred status as *gne-sa* (sacred sites). Buddhism and indigenous pantheons, through their embodiment in mountainous landscapes, have become ever more symbiotic in modern times not merely in existential and spiritual domains but also in modern political, economic and environmental arenas. Kinley Choki's and Thinley Dema's case studies of how the globally linked economic wealth simultaneously triggers Bhutanese perceptual changes and evokes deeper indigenous histories concerning the cosmological gravity of their sacred mountains are revealing examples of the indigenous–non-indigenous interface and the felt collective challenges from it.

Concerning the transformative change of Himalayan indigenous lifeworlds and spiritual matrix, the case of the Lepcha indigeneity is particularly noteworthy. The interface of Tibetan Buddhism and the native shamanic tradition shows how the pre-Buddhist Lepcha indigeneity survives in adapted forms under the umbrella of a world religion. Both earlier and recent studies of the Tibetan Buddhist–Lepcha nexus show that Tibetan Buddhism does not completely diminish Lepcha shamanic practices. Instead, the Lepcha form of Tibetan Buddhism bears a syncretic character of "the lama-shaman duality" signifying Tibetan Buddhism's concurrent domination of and coexistence with the Lepcha shamanic tradition (Gorer 1938, 181; Balikci 2008, 4). A noteworthy example is Máyel Lyáng, a hidden place of abundance believed to have physical existence in an unknown location on Mt. Khangchendzong. Mythically, it was transferred from the Lepcha indigenous topographical utopia to the Tibetan tantric perceptual field of Beyul, a hidden spiritual world whose revelation is contingent upon the outer, inner and innermost spiritual states of a person. In other words, it is a religiously sanctioned mystical and sacred geography designated to the preservation of Buddha's teachings. It is only accessible by *terton*, or treasure revealers (Mullard 2011, 11). Appearance-wise, the Buddhist topographical descriptions of this tantric utopia are similar to those of Máyel Lyáng as a hidden place. They are thus perceptually persuasive enough to conflate Beyul with Máyel Lyáng (Rigzin 1998, 76). Essence-wise, the indigenous narratives of Máyel Lyáng focus less on the Buddhist sacred sense and more on the diversity and the abundance of flora and fauna in the company of a multitude of spirits and supernatural beings. This hidden place situated in Mt. Khangchendzong, is the cultural and spiritual centre of the Lepcha land. Thus, it is not a nonphysical place. Qualitatively, it is a place where earthly ecology converges with the shared spiritual sphere of its residents as shown in the chapter by R. Lepcha and M. Lepcha in this book. Although modern Lepchas have lost access to it, it is nevertheless remembered as a geomythical but actual place with animals, plants and Máyel dwellers—human-like spiritual beings and deities—and as the place of Lepcha ancestral origin (Balikci 2008, 72–74; Gorer 1938, 237; Little 2018, 341–352). The primary cause of the modern Lepchas' inability to access to Máyel Lyáng is the loss of their mother tongue (Balikci 2008, 74) and their land, as well as their cultural fragmentation into different worldviews and lifestyles. Their experience is one in which exogenous religious, cultural and linguistic elements are replacing their indigenous counterparts or co-evolving into symbiotic forms.

Ecologically, over the last 350 years, starting from the Namgyal Dynasty to becoming a protectorate of British India and then a state of modern India, the political changes applied to Lepcha territory have brought radical environmental transformations. British India's conversion of the forested land in Darjeeling to tea plantations wiped out the ecological niche of the forest-dwelling Lepchas in the region, their indigenous forestry knowledge rendered irrelevant to the modern cash-crop economy. British India's importation of migrant Hindu workers from Nepal increased economic productivity, but it transformed the religious and cultural characteristics of the Lepcha and their land, as did European Christian missionisation. Sir Herbert Hope Risley, a British colonial administrator, wrote in 1894:

> They [the Lepchas] are above all things woodmen of the woods, knowing the ways of birds and beasts, and possessing an extensive zoological and botanical nomenclature of their own. Of late years, as the hills have been stripped of their timber by the European tea planter and the pushing Nepalese agriculturist, while the Forest Department has set it face against primitive methods of cultivation, the tribe is on the way to being pushed out.
>
> (Risley 1894, ii)

Sixty years later, Geoffrey Gorer reconfirmed Risley's assessment, "the Lepchas became an 'inferior' subject race" in their own land (Gorer 1938, 36). The demographic data of Sikkim in the twenty-first century further substantiates the marginality of the Lepchas. The approximately 45,000 Lepchas make up 7.8% of Sikkim's total population. The 7,000 in Dzongu reserve area represent the main Lepcha population which continues to practice their indigenous shamanic religion (Little 2008, 230; 2010, 86). This miniscule number is often not mentioned in the religious demographics of Sikkim: 68% Hindus, 27% Buddhists and 3% Christians (Arora 2009, 97). The majority of the 45,000 contemporary Lepchas consider themselves Buddhist, Hindu, Christian or they hold modern secular beliefs. The notion of indigeneity cannot be defined by only ethnolinguistic factors. The absence of a people's ecological world and their indigenous ecological knowledge is a fact of uprooting. In this sense, the indigenous culture of the Lepchas is endangered though, at the same time, this endangerment activates the Lepchas' agentive responses.

In spite of their historical tensions, the Lepcha and the Bhutia people have made a political alliance in the twenty-first century that is predicated on indigenous environmental activism against the twenty-four hydraulic projects implemented and planned on the Teesta River (Arora 2007a, 3451). This is the second time in history that the two ethnic groups have allied themselves politically; the first time was when the Namgyal dynasty advocated the preservation of the Lepchas and the Bhutias as the indigenous peoples of Sikkim in the 1960s (Arora 2007b, 203). This move was intended to protect the political rights of the two groups as the indigenous subjects of an Indian protectorate. In comparison, the current alliance is not only aimed to ensure their political rights but also the environmental preservation of their indigenous homeland. In the process, Bhutia Buddhists are taking the lead with their full-fledged "sacred protest narratives" (Little 2012, 84) highlighting sacred Buddhist sites as well as the core geographical locations of Lepcha indigenous culture (Arora 2009, 95–96). The meanings of indigeneity have evolved from a pre-modern ontological state and corresponding modern identity politics to the current environmental activism with a discernible religious and spiritual reorientation to both a universal understanding of sacredness and a place-based indigeneity that intertwines ecology and human spirituality. Tibetan Buddhism is playing a significant role in articulating the sacredness of the synchronised indigeneity of the Lepchas and the Bhutias. This is an observable sign of the symbiotic indigeneity I wish to elaborate in the context of the New Himalayas.

The concepts of indigeneity in use since the mid-twentieth century have been centred upon what I call "political indigeneity," rightfully geared towards the protection of indigenous peoples' rights and justice. The widely cited definition of indigenous people put forward by the United Nations' Martinez Cobo Study is demonstrative of the inherent roots of such political indigeneity arising from global colonial history: "Indigenous communities, peoples and nations are those which, having a historical continuity with pre-invasion and pre-colonial societies that developed on their territories, consider themselves distinct from other sectors of the societies now prevailing on those territories, or parts of them" (UNHRC 2013). This UN-led definition has been instrumental in raising global awareness of the political conditions faced by indigenous peoples worldwide with such awareness manifested through governmental policymaking, international aid programs and political activism. At the same time, its side effect is that the political sense of indigeneity inadvertently suggests a global Other—an economically, politically and racially determined Other. The political approaches to indigeneity leave little room for ecological understandings of being indigenous.

While recognising the political challenges that Himalayan indigenous peoples face, I propose the Himalayan symbiotic indigeneity as a type of ecological indigeneity in contrast to the well-established political indigeneity worldwide. My emphasis on ecology is meant to highlight an ontological universal of humankind as an inhabitant of the earth, one who is implaced, environed and provided for in diverse ecological contexts. Given the intertwinement of the past and the present, tradition and modernity, coloniality and postcoloniality, empire and the modern state, in addition to ethnolinguistic territorial sovereignty and transboundary environmental flows in the New Himalayas, ecological indigeneity is synonymous with the symbiotic indigeneity. It gives a comparative emphasis to the notions and the documented experiences of indigeneity—not merely as an ethnically and geographically demarcated understanding but also as a transboundary, panhuman empathetic resonance or expression of what it means to be human, a relational species in the ecological sense. Its symbiotic nature entails four integrally connected meanings of indigeneity in the arenas of indigenous memory, indigenous experience, inter-indigenous collaborations and the indigenous inspiration and instrumentality in forging environmental humanities worldwide.

In the context of the New Himalayas, indigenous memory(ies) and indigenous experience(s) both suggest ecological rootedness and implacedness. The former refers to personally or collectively remembered ancestral indigeneity in specific geographical locations or relative to historical migrations away from one's ancestral homelands to elsewhere. The latter signifies continuously lived experiences in an ancestrally defined territory interwoven with exogenous forces of transformation. The geomorphology of the ancestral land may remain the same but its native ecosystem and ecological worldview have become hybridised and, thus, have undergone a co-becoming process combining the indigenous and the exogenous towards a third form of eco-ontological being. Modern ontology of a given indigenous experience is not a life story of its own but is relationally entwined and embodied with other kinds of indigenous experiences and ecological worldviews. This is what I mean by the "inter-indigenous collaborations" found in the New Himalayas, in which unlikely collaborators become collaborators for the sake of collective survival and continuity. The motivating force forging this collaboration of unlikely collaborators often comes from the radical alterations of dwelling places and ecological habitats by the exogenous forces of change—modernisation and globalisation. These changes generate precarious living conditions that, in turn, spawn what Anna Tsing calls a "contaminated diversity" that "implicates survivors in histories of greed, violence, and environmental destruction" and whose "units are encounter-based collaborations" (Tsing 2015, 33). In many places of the world, life without collaboration is destined to perish. The contemporary Lepcha–Bhutia alliance is a highly pronounced case of inter-indigenous encounters and collaborations as a form of the symbiotic indigeneity discussed thus far.

Zooming out from the Himalayan cases, I see a global landscape of indigenous environmental studies in which the inspiration provided by indigenous ecological practices and knowledge systems are instrumental in forging environmental humanities. Emerging as a vast interdisciplinary field, the formative phase of environmental humanities took place mostly in the Neo-European contexts of Australia and North America, where environmental humanists drew much inspiration from the wealth of indigenous ecological knowledge and environmental ethics. Australian environmental humanities has laid a significant foundation for the global development of the discipline in its diverse forms and contexts. Aboriginal ecological practices and wisdom are a "vital contribution" (Rigby 2019, 15) to this interdisciplinary innovation. They are visibly woven into the texts of the foundational scholars of Australian environmental humanities, such as Val Plumwood, Deborah Bird Rose and Thom van Dooren. Plumwood's "philosophical animism" (Rose 2013) and Rose's humanist articulation of the Aboriginal concept of Country (Rose 1996, 7–15), for instance, provided original impetus to an empathetic understanding of the living earth and to the collective search for an "earth language" (Rose 2013, 107) for the planetary environmental discourse. In this regard, indigeneity in its symbiotic forms is not merely ecological but is also transculturally intellectual.

Indigeneity in the symbiotic sense is a human ecological universal and its varieties are deeply entangled together, like a body of rhizomes, in the New Himalayas. Here, indigeneity and exogeneity are relative concepts. When both are in a symbiotic relationship, they are symbionts in a relational ontology simultaneously involving dependence, interdependence, hybridity and co-becoming for identified common purposes. The complexity of symbiotic indigeneity lies in its multiple edges, transitional zones, human power-dynamics and eco-political events. These factors are prone to contacts, tensions, invasions, counteractions, interactions and alliances. In these dynamic conditions, symbiotic indigeneity is not necessarily a harmoniously entwined evolution of different indigenous traditions and elements. It can also be incarnated in violent encounters or reluctant collaborations. A prolonged symbiotic relationship often induces a symbiogenesis of one or two new forms of indigeneity or inter-indigeneity to be more specific.

The anthropogenic Himalayas as a new-animist earth

In Arctic humanists' evocation of the "cryo-historical moment," ice tells a story of global climate change and symbolises the fugitive nature of human existence in facing natural or human induced environmental changes (Sörlin 2015, 128). Like those in the Arctic, glaciers in the Himalayas are an inherent part of the global cryo-history; however, as I write in the introduction, the Himalayas, as a "terrestrial ocean," hold more than ice. The region's high-altitude geomorphology encompasses mountains, valleys, rivers and plateaus that are nature's monuments and archives (Rudwick 2005, 539) signifying the earth's Deep Time with "no vestige of a beginning, no prospect of an end" (James Hutton's quotation in Rudwick 2008, 384). This beginningless time predates but overlaps and continues with human time. It is not the time of the past but the persistent presence of cosmic creative forces saturating everything and everyone. We are in and a part of Deep Time and it continues to exert its forces of change. Humans are thus geological beings besides being biological ones; we are a 200,000-year-old species—incredibly young in comparison with the geological age of the Himalayas traced back to 50 million years ago when the tectonic collision between the Indian Plate and Eurasian Plate began to take place. Our depth in Deep Time is shallow. Our understanding of Deep Time is even shallower as it only surfaced over 200 years ago when Hutton proposed it. The earth we know is largely a "human earth" in mythological, geological, ecological and economic terms; however, its current anthropogenic character is compelling us to sense and re-understand the earth as a planet with a multitude of more-than-human worlds and to re-articulate its maternal principle in a symbiotic language of human scientific endeavours and affective expressions. This is what the Himalayan environmental humanities practiced in this book is striving for.

In current interdisciplinary debates about environmental conservation in the Himalayas, there are two crisscrossing perceptual trends. One emphasises the Himalayas as a site of natural resources and another advocates the sacredness of Himalayan landscapes situated in different ethno-spiritual traditions. These two perceptions of the Himalayas can be respectively called "a utilitarian earth" and "a sacred earth." The former emphasises the utilitarian value of the Himalayas in material and aesthetic terms for human energy and resource extraction and tourist consumption (Ives 2004). The latter promotes the humanistic understanding of Himalayan physical environments in eco-spiritual terms as shown in the works of Drew (2017), Allison (2017), Spoon (2014) and Thinley Dema (in this book). This spiritually oriented perception of the Himalayas gravitates, then, towards the sacredness of the earth, a gravitation that has become a Tibeto-Himalayan character threading together a symbiotic Himalayas with diverse but interconnected indigenous communities and societies. This perceived sacredness bears both local and global political instrumentality in indigenous peoples' endeavours to preserve their human and natural heritages and in globally promoting environmental responsibility. At the same time, it is also effectively interfacing with the global trend of revering the earth as "our only home" (Alt 2020, 13) and as a sacred creation of the cosmos (Pope Francis 2015, 110). Tibetan Buddhism undoubtedly deserves a large portion of the credit for Himalayan and global environmental endeavours for a sustainable future.

While recognising the political and ecological efficacies of sacredness as perceived by the larger spiritual traditions like Hinduism, Buddhism and Islam, our Himalayan humanities would like to propose a third perspective of the modern Himalayas as "a new-animist earth" to emphasise the anthropogenically impacted sentient character of the earth besides its sacred, transcendental attributes from different eco-religious traditions. On the one hand, this conceptual perspective emphasises the sentience of the earth ontologically speaking through humans, animals, plants and formless beings under the stress of climate change and environmental degradation. On the other hand, it is intended to develop an understanding of the aliveness of the earth in the case of the New Himalayas through a combined effort of earth sciences and environmental humanism. I find that the idea of the Critical Zone in earth sciences is compatible with our animist understanding of the Himalayas (Smyer Yü 2020). To reiterate, in earth sciences, the Critical Zone refers to the continuous earth surface "between the top of the tree canopy and the base of weathered rock" (Dybas 2013, 1). It is understood as a "co-evolving entity driven by physical, chemical, and biological processes that sustain life" (Brantley et al. 2017, 842) signifying "the fusion of the biotic-abiotic complex" (Richter and Billings 2015, 2). Within this complex system, all organic and inorganic factors interconnect, interact and intertwine with each other. While the earth sciences regard the earth as mechanical and utilitarian, some earth scientists' personification of the earth is apparent in their consideration of the Critical Zone as "the living skin of the Earth" and as "an entity" (Brantley et al. 2017, 841). The entirety of the Critical Zone presents the most vibrant realm of life on earth. It is the sphere where animist cosmovisions and lifeways took root in diverse ancient human communities and continue to take new shape in communities in the twenty-first century.

My proposition of the Himalayas as a new-animist earth emphasises the animist union of the physical earth and biotic and abiotic. Biotic beings include both humans and nonhumans, with the latter referring not only to animals and plants but also to the indigenous worlds that contain supernatural beings, local deities and ancestral spirits, who are often embodied in geomorphic features of given regions and/or in the animating forces of environmental flows. The chapters by Erik de Maaker, Charisma Lepcha, Rongnyoo Lepcha and Mongfing Lepcha, Thinley Dema, and Kinley Choki demonstrate the animist intertwinement of humans, spirits and landscapes. What we learn from their case studies is that the animist nature of the physical earth is found in the shapes of rivers and mountains and in the felt presence of local deities and supernatural beings. This is where I find the conceptual efficacy of Critical Zone science in the new-animist approach to the studies of Himalayan environmental conditions and human and nonhuman responses.

Himalayan humanities and Critical Zone science are complementary in the sense that the current earth sciences do not address the intertwinement of the physical earth with the cultural dimension of humankind, while humanist approaches to environmental studies overall do not have enough engagement with the earth sciences. Thus, the idea of the new-animist earth is thus being conceived precisely for building bridges and filling gaps between the different sciences and disciplines in Himalayan studies to generate deeper understandings of the personhood of the earth in its diverse incarnations and expressions. It is the Himalayan version of a new-animist understanding of the earth, one in which the physical world is in fact an animated world of "listening people, speaking Earth"—"a community of related living beings" (Harvey 1997, 15). This symbiotic indigenous cosmovision allows for "the animating force of life, the fire of being that moves everything, even the constantly birthing universe, to life" (Hogan 2013, 25). The earth's interactive relations with humans and nonhumans are thus intersubjective and mutually embodying. The New Himalayas has its own living stories to tell. In the symbiotic indigenous cosmovisions, environmental changes in the Himalayas are triggering the anger of their placed-based deities (Thinley Dema and Kinley Choki in this book), strengthening the bonds of humans, the earth and the deities (Allison 2017; Smyer Yü 2014), and evoking the sacredness of indigenous landscapes (Arora 2007a; C. Lepcha in this book; Little 2018). Many of the chapters in this book are thus narrating how the animist, affective bonds of organic beings and inorganic matter together can generate a new planetary environmental ethics in the Anthropocene.

My conceptual coinage of the new-animist earth is expected to connect Himalayan environmental humanities with the global interscientific environmental studies of the earth as a "conative whole," which is "alive and ensouled" (Callicott 2010, 208) and, therefore, sentient. This modern recognition of the earth as a planetary being is no different from how the Lepcha see their homeland as a land animated with spiritual beings (C. Lepcha; R. Lepcha and M. Lepcha in this book), how Bhutanese and Tibetans perceptually "shapeshift" their sacred mountains from physical mountain bodies to revered and feared deities and supernatural beings, and how Australian aboriginals regard the earth they live with as "Country"—"with a consciousness and a will toward life" (Rose 1996, 7). Human mythical and historical memories as well as currently symbiotic ways of being indigenous, traditional, modern and global in the Himalayas all point to a common recognition of the earth not just a geological body but also as a planetary animated being. Humankind's journey with the earth is young but throughout our species' time here, we have experienced a profoundly felt sense of how the creative force of Deep Time works through the earth and all life forms it gives and receives. By proposing the Himalayas as a new-animist earth with a conative character, I would like to underscore the shared sense of the earth's psychological characteristics and moral considerability with peer environmental humanists beyond this book and to take an affective, animist approach to address the invaluable contributions indigenous ecological knowledge can make to the global discourse of human–earth relationality, environmental ethics and sustainability.

Transboundary water commoning

In the introductory chapter, I equate the Himalayas, inclusive of the Tibetan Plateau, as a terrestrial ocean or a water–land body that holds an immense amount of water in both solid and liquid forms feeding the habitats and the lives of at least 1.3 billion humans and countless nonhuman species in Asia. This is the planetary environmental significance of what I call the "Tibet-Himalayan water commons" seen from the perspective of the earth's hydrosphere. However, it is not an international commons or a resource domain governed by an international regime and legally accessible by all nations (Buck 1998, 6); instead, it is a water–land body that is cartographically sliced by the human nations of Afghanistan, Bangladesh, Bhutan, China (Tibet), India, Myanmar, Nepal and Pakistan. The central Himalayan region and surrounding foothills are divided up and "owned" by these nations and there are, therefore, no nationally un-owned common waters and lands in the Tibetan-Himalayas. When the Tibetan-Himalayan hydrosphere is factored into the human-conceived geography, the sphere of the environmental influence of the Himalayas is found in Central, East, Inner, Southeast and South Asia beyond the Himalayan Massif and the Tibetan Plateau. The meanings of water commons thus beg both human and the earth perspectives.

The case of the Tibetan-Himalayan commons that I attempt to build is a case of an inter-Asian water commons, with Tibet and the central Himalayas as the source region plus the river basins reaching the Indian and the Pacific Oceans. It is a supranational environmental commons that is situated in multiple human nations but transgresses and transcends human territorial boundaries and national identities. The human nations claim ownership of the partial commons within their territorial sovereignties; however, due to the fluid nature of water, this moving commons is *in* but not *of* the human nations. This is not to negate the biocultural fact that, as a terrestrial species, all human societies have place-based cultural interfaces with water within our ecologically niched habitats. For instances, it is common that a river has multiple names when it flows through different ethnolinguistic communities, such as the Yarlung Tsangpo/Brahmaputra in Tibet, India and Bangladesh, the Drichu/Yangtze in Tibet and Han China, the Dzachu/Lancang/Mekong in Tibet, Yunnan, and Southeast Asia, and the Nagchu/Nu/Salween in Tibet, Yunnan and Myanmar. One river can thus have two or more names of different human cultural origins. This human geography related to a river may afford different human communities' customary claims of owning different sections of the moving water body; however, in essence, the river has its own life and course of being, independent of the human cultural perceptions and political claims. What exactly then is this inter-Asian water commons? The answers can come from both human and nonhuman angles.

Seen from the human-centred utilitarian perspective, the Tibetan-Himalayan water commons can be considered “mobile commons” (Miller 2020), defined as “the governance of resources that cannot be physically contained within demarcated spaces and bounded regulatory regimes” and, thus, entails “the hybrid governance of transboundary resources,” “the governance of migratory taxa,” and “the governance of resources that move across property regimes in ecosystems that have geographically discrete and broadly recognizable natural boundaries” (Miller et al. 2020, 299–300). Therein, “governance” is the keyword, which Miller specifies as “environmental commoning” or “the active processes involving multiple state, private, and civil society actors that produce and maintain commons” (Miller et al. 2020, 298). In addition to its presence in Miller’s works, the notion of commoning, originating from the work of Linebaugh (2008), has been a processual understanding of commons evolving from a noun to a verb or from its substantive attributes to a set of collective actions meant to sustain the commons or exercise fair sharing of the resources that the commons holds. The meanings of commoning vary from context to context. In the work of Amin and Howell, commoning is articulated as a collective social force “against the forces of commodification,” which is pinned down as the neoliberal ruining of public resources and the common good (Amin and Howell 2016, 3, 6). In their study, J. K. Gibson-Graham and their co-authors emphasise the “tragedy of the unmanaged commons” in their capitalocentric discourse, as they write, “Commoning thus involves establishing rules or protocols for access and use, taking care of and responsibility for a resource, and distributing the benefits in ways that take into account the well-being of others” (Gibson-Graham et al. 2016, 192, 195). Commoning in Miller’s case study of the Mekong River is explored in a similar vein except that the Mekong River is a transboundary and multinational commons. It is thus quite clear that the ongoing studies of commons and commoning are largely human-centred endeavours towards the sharing of the commons as common pool resources in different national and political contexts. In those contexts, the environment is a passive object to be divided and shared among human/national stakeholders. Managing, governing and distributing with fairness are thus the key actions advocated to battle the monopolised commodification and consumption of the commons or to restore the order of unmanaged or mismanaged commons. Seen from the human inter-state angle, commoning is “inherently political” (Miller 2020, 488) but is politically disordered and ineffective in most parts of the Himalayas.

To elaborate further what I mean by “politically disordered and ineffective,” and given the unilaterally determined hydroelectric dam-building processes in the Himalayan nations over the last half century, the Tibetan-Himalayan water commons is an unacknowledged, cartographically sliced, heavily bordered, severely frustrated and highly fragmented transboundary commons. Except for the attempt by the Mekong River Commission (MRC), the Tibetan-Himalayan water commons is not an internationally recognised commons. Neither is it an unmanaged commons as it is one-sidedly managed and extracted for resources within the territories of the individual nations. As an inter-Asian water commons, its environmental flows have been fragmented, impounded, scarred, degraded and humiliated by the multitude of hydro-development projects within the boundaries of individual nations. China and India, the two largest Himalayan states, are known for their excessive dam-building projects. Two-thirds of China's 87,000 hydroelectric dams are built inside Tibet (Harrison 2018; Dechen 2019) and new dam-building projects are ongoing on the Dzachu (Lancang-Mekong River), Nagchu (Nu/Salween), Drichu (Jinsha/Yangtze River) and the Yarlung Tsangpo (the Brahmaputra) Rivers. To support the infrastructural and energy needs of its Belt and Road Initiative in Himalayan and Southeast Asian nations, China is exporting its hydraulic practices to Nepal, Myanmar, the Pakistan-controlled Kashmir region, Laos and Cambodia. Likewise, in its dam-race with China (Chellaney 2012; Gamble 2019), India plans to build 292 dams in the Indian Himalayas by 2030 (Grumbine and Pandit 2013, 36) on top of its several hundred existing mega-hydropower plants. By 2019, there were over "550 hydropower projects in existence or under construction" (Xu et al. 2019, 144) in the Himalayan region. Given the currently divergent hydropolitical and geopolitical positions of China and India, it is nearly impossible to have a multilateral, legal recognition of the Tibetan-Himalayan water commons in the foreseeable near future. As a matter of fact, Himalayan states have not yet established enough credibility and reliability with each other and with the global public for practicable measures of the transboundary governance of water.

Based on the case studies by Zainab Khalid, and Anwesha Dutta and Shailendra Yashwant in this book, environmental commoning is rather ironically possible in the absence of the needed support from the state when local traditional environmental knowledge and practices are revived, enacted and adaptively enhanced to face locally manifested global environmental problems and the anthropogenic impacts of inter-state hydraulic politics. In Khalid's work, *rajaki*, a locally based water management system, is clearly a type of effective commoning not invented out of the state's environmental management but evolving from a long history of the lived experiences and relationships the Hunzakutz have with the Himalayan glaciers. She writes, "*Rajaki* as a commoning exemplifies the ways in which humans implace themselves in a more-than-human world through their relational existence with the environment that is not based only on their human free will devoid of the physical earth in a pure philosophical and moral sense." Khalid's Hunzakutz case shows that commoning requires a preexisting community basis that treats the environment not merely as a utilitarian resource but that has embodied a locally unique human–nature entwinement as shown in the Hunzakutz's environmental knowledge and affective ties with their ancestral dwelling place. Likewise, the indigenous irrigation system *jamfwi* in Dutta and Yashwant's chapter demonstrates the indigenous Bodo people's long history in the region which has now been transformed into an India-Bhutan borderland. While the border of the two modern states separates the people on two sides and Bhutan's hydropower dam engenders floods downstream on the Indian side, the people living on two sides of the border innovatively enact their ancient irrigation system by combining modern digital communication methods prewarning each other the status of the dam's floodgates. The success of their commoning is not a result of a formal inter-state effort but rather of the people's own indigenous knowledge, initiatives and adaptive innovations.

Also shown in the case studies of the Bhutanese sacred landscapes, the trans-species relationships and the livingness of water bodies and mountains in the chapters by Thinley Dema, Jelle Wouters, Charisma Lepcha, and Rongnyoo Lepcha and Mongfing Lepcha, the basis for the understanding commons or shared habitats of humans and nonhumans is, in fact, not from a utilitarian origin but from the indigenous lived sense of the livingness of the earth. In my own work on the sentient quality of Tibetan mountains and water bodies, commons is not merely a physical property of the earth subjected to human utilitarian uses, but is an environmental actor who shapes the patterns of human dwelling, resource use and affective relations with the land. In other words, commons is an inherent part of the livingness of the earth, which responds to human use and human alteration of the environment in its own language, for example, replenishment and/or natural disasters. It thus reciprocates with humans as caretaker and provider but is not bound to humankind alone. It also exercises seismic and hydraulic violence to reset its own mode of being. At this point, the earth is doing its own work of commoning in the Tibetan-Himalayas, and not necessarily in favour of human needs and desires.

As the multinationally recognised and legally sanctioned Tibetan-Himalayan water commons is nonexistent, I find that the currently relevant actions for environmental restoration, conservation and fair sharing of water resources to be critically in need of recognition of the Tibetan-Himalayas as a world of many more-than-human worlds and as a unique part of the earth's livingness. Particularly for statesmen, policymakers and the concerned public, the imperative for establishing the human-regulated commons would reflect humans' critical understanding of the nonhuman environmental commoning of water from the Himalayas as the trans-boundary ethical basis of an intended multinational water governance. By "nonhuman environmental commoning," I mean that the earth has already done its own creative work, generative of habitability, before the inception of humankind, and that the rivers flowing from the Tibetan-Himalayas have never ceased their own version of commoning through rhizomically tying together different ecosystems, human societies and nonhuman habitats. Seen from the perspective of Deep Time, nature is not just a concept allowing humans make sense of the earth (Castree 2014, 5–6), but, more critically, has been independent from and yet inclusive of humankind. It began to interface with human worlds and experience human-induced alterations only within a small fraction of Deep Time. Nonhuman environmental commoning in this sense is a series of the integrated actions of nature through the geological, ecological, geophysiological and geomorphological forces of the earth. This nonhuman commoning is the ancient symbiotic and symbiogenetic journey of the earth (Margulis 1998), which has undergone a myriad genesis–extinction cycles of species and habitats. It requires a deep planetary vision to fathom the creative forces of the earth from more-than human perspectives and cosmological imaginations and to recognise the foundation of the flourishing of human diversity and biodiversity in the Himalayas.

Before the Tibetan-Himalayan sourced rivers were dammed for harnessing hydropower and before many native human communities were forcefully removed from their ancestral riparian homelands by the nation-states constructing the hydropower plants, everything and everyone—mountains, glaciers, rivers, wetlands, humans, animals and plants—had been co-commoning with one another in their rhizomically entangled worlds. The co-becoming of the rivers with diverse human societies has been a shared world-making journey in a relatively equilibrial manner. In co-becoming, the rivers have been the primary power of this multiregional, multispecies and multiworlding commoning. As presented by R. Lepcha and M. Lepcha, the reanimation of the origin story of the Teesta and Rangeet Rivers in their landscape paintings and ethographical narratives shows us not a fairy tale detached from the living world, but rather the creative forces of Deep Time present in the shapes of the two rivers in a set of affective and spiritual expressions. The awe-terror and the nourishing power of the earth saturate both the land and the human psyche. Commoning in this case is the commoning act of both nature and humans. The former gave birth to and shaped the two rivers as the life-giving source, while the latter internalises and expresses nature's creativity in the reverent language of folktales, indigenous spiritual practices and indigenous ecological knowledge.

Indigenous peoples' sense of the nature's commoning in the Himalayas has much in common with their counterparts in the Americas and Australia. For people in the highlands of Peru, as Malene Brandshaug writes, "Water is life, has life, and is considered to be both a commodity and a common good" and water commoning consists of "practices of a more-than-human commoning" (Brandshaug 2019, 538). Water-as-life stands as the foundation of human affective and existential relations with it as a gift of the earth. Water-as-life is the active "commoner" or commoning influence in a process that commenced the planetary hydrospheric flow along with the earth's ecological and geological creative forces long before humans and many nonhuman species emerged. It is "both a sentient being and a passive object as well as anything in between" (Brandshaug 2019, 539). For the Peruvian highlanders, water is simultaneously a commons, a commoning influence and a force of the earth's planetary commoning. Similarly, Australian Aboriginals regard their lifeworld as "Country," which "includes not just the territorial, land-based notion of a home *land*, but encompasses humans as well as waters, seas and all that is tangible and non-tangible and which *become together* in a mutually caring and multi-directional manner to create and nurture a *homeland*" (Bawaka et al. 2013, 186). Scholars of Country do not describe the caring acts of Country in terms of commons and commoning; however, in essence, as "a living entity with a yesterday, today and tomorrow, with a consciousness, and a will toward life" (Rose 1996, 7), Country shows itself as "an ontology of co-becoming" with both biotic and abiotic worlds, and humans and nonhuman beings, which is thus understood as "entangled becomings, creative and vital and always in the process of becoming through their connections" (Bawaka et al. 2013, 187). The earth in the forms of both land and water is the ultimate commoner and the generative force of all commonings in the Himalayas and everywhere.

The outcome of water commoning as nature's initiative taking the form of moving rivers and the subsequent generation of riparian habitats is a rhizomic world made of many worlds, with the creative principle of "the environmental freedom." Here, the rhizomic world is a metaphoric image of multiple habitats woven together by the different rivers originating from the same source region. The rivers in this world-making process are the veins of life threading together multispecies lifeworlds upstream and downstream. When the flow of the upstream part of a river is frustrated, the frustration reverberates through the entire river and riverine habitats up and down, near and far. The stories of the extinction and endangerment of species and human livelihoods due to hydraulic damming are common around world but the understanding of water commoning as an act of nature is scarce as we, scholars, scientists and policymakers, continue to be caught in the self-nominated role of the sole type of commoning influence, stakeholders and environmental ethicists. Thus, it is imperative to recognise the innate environmental freedom of rivers as a gift of nature that preconditions human commoning as it moves towards multienvironmental and multispecies equilibrium. As a creative principle, environmental freedom is not limited to the moving water bodies of rivers, but also encompasses the ecologically niched movements of humans and nonhuman species, for example, fish, trans-habitat megafauna, migratory birds and nomadic peoples, whose lifeworlds and migratory paths are inherently dependent upon moving water. Within the riverine worlds, the freedom of rivers is a critical prerequisite for the free movement of humans and nonhumans for such activities as trading, foraging, hunting, reproduction and exploration. Environmental freedom is thus an inherent creative force of the earth. It is an ethos of nature expressing its vitality embodied in both biotic and abiotic worlds. It facilitates the caring and nourishing character of the earth interconnected between and diversely embodied in different ecosystems and human and nonhuman habitats.

I evoke the imageries and notions of rhizomic riverine worlds and environmental freedom as a response to the emerging regional call for transboundary governance of water originating from the Tibetan-Himalayas. The current state of affairs is that the major transnational rivers have all been dammed multiple times and that more dam building projects are under way throughout the Tibetan-Himalayas. Centred upon the inter-state fair sharing of water resources, this utilitarian-based call for environmental commoning and transboundary governance of river resources (Miller et al. 2020; Grumbine and Pandit 2013; Samaranayake et al. 2018) is largely absent of the participation of indigenous constituencies. Distinctly absent are perspectives reflecting environmental ethics and affective human–nature relations grounded in local communities and societies that hold their native environmental knowledge of sustainable living with the affected rivers. It is ethically and scientifically unjustified to allege that nomads in headwater regions are "ruining the environment" by "overgrazing" (Zhao et al. 2014, 23; Zhao 2020) and to promote environmentally misinformed policies under the false dichotomy of "privileging indigenous livelihoods over endangered species" (Miller 2020, 476) as if an either-or choice had to be made to achieve environmental conservation goals. The danger of technocratic and statist approaches to transboundary commoning include potential and actual endangerment to the knowledge maintained by indigenous peoples and their companion species in their more-than-human worlds. Considering the near-tragic condition of the Tibetan-Himalayan water commons, scholars and scientists in Himalayan environmental studies are compelled to take a global comparative perspective to recognise the increasing worldwide presence of indigenous environmental knowledge in the currently advancing studies of commons, commoning and planetary environmental ethics. Eleanor Ostrom's Nobel-Prize-winning concept of commons owes much to indigenous lived-experiences and modes of governance of commons based on collective action (Ostrom 1990; Keohane and Ostrom 1995). The inception of environmental humanities as an interdisciplinary field is indebted to indigenous environmental histories (Rose et al. 2012; Gibson et al. 2015). The ongoing innovative endeavours for new planetary environmental ethics around the world show an overwhelming trend of revitalising indigenous ecological knowledge and environmental ethics (Grim 2001) and dismantling the artificially reinforced hyperseparation of indigenous and nonindigenous worlds (Plumwood 1993; Bawaka Country et al. 2016).

Restoration of environmental freedom as the basis of common ethics

Many of the authors in this book yield from their individual research projects ethical implications of the affective bonding of the Earth with humans and nonhumans. The ethical concern for the environment is "an intellectual as well as emotional process" (Leopold 1949, 225) as it pertains to our utilitarian and affective relations with land. By recognising the common connection between land, humans, nonhumans and intangible heritages embodied in the land, we are compelled to update our planetary environmental ethics in the Anthropocene, a human-induced geological age. In it, we are being given an opportunity to understand the life of the earth as a planet, the ultimate progenitor of all known life forms, a sacred creation, a deity, a sentient being and a kindred spirit. It compels us to see that sustainability does not merely come from our utilitarian calculations and management of the earth's finite resources but also, more critically, it is inherently interwoven with the moral, spiritual and eco-geological fabrics of human- and nonhuman kinds. Thus, environmental ethics concerning sustainability and forming human-initiated systems for commoning require a renewed, empathetic understanding of the creative power of the earth as a union of the biotic and the abiotic, which never stops seeking connectivity, participation, interaction, co-creativity, diversity and transformation. A set of clearly articulated and enacted ethics is the basis of establishing any meaningful commons.

It is quite discouraging to say that the Tibetan-Himalayan states and the affected downstream riparian nations in Southeast and South Asia are not yet ready to reach out to one another for a sustainable future of the transnational water commons. Two accounts support this international unpreparedness for this human–environmental commoning. One is the ongoing India-China border conflicts that continue to transform human sphere of the Himalaya into a zone of bilateral confrontation, and engender a global geopolitical perception of the Himalayas only as a Sino-Indian or Indo-Tibetan borderland (Fisher 1978, 2; Smyer Yü 2017, 25–27). This, therefore, overshadows the actual diversity of ethnolinguistic communities and their environmental knowledge of sustainable living in the Himalayas. Another is the shrinking civil society or the absence of civil liberty in countries like China, India, Myanmar and Pakistan where wider citizen participation would initiate and foster effective commoning agendas. Ideally, if India and China, the two most populous nations on earth, can take the initiative to call for trans-Himalayan collaborative water commoning—not merely as an inter-state endeavour but also a borderland initiative involving local riparian communities from the Hindu–Kush Himalayas to South Asia and Southeast Asia—it would be a transformative change for the environmental health of the rivers and generate multispecies flourishing. However, the reality is sadly far from this ideal vision of transboundary environmental commoning. Statesmen and state-commissioned scientists and policymakers have not yet recognised the personhood as well as the moral considerability of the earth in the Himalayas except by their geo-economic and geopolitical calculus.

What is immediately needed in the anthropogenically affected Tibetan-Himalayas is the basic awareness of the rivers' environmental freedom and how riparian peoples have co-lived with this freedom since ancient times. It is this environmental freedom that has given us a sense of the livingness of the earth with a life of its own and that has afforded us to see the earth as a biotic-abiotic union. It coincides with the late evolutionary theorist Lynn Margulis's understanding of the earth as a symbiotic planet, a system in which everything and everyone exist in physical contact. As she wrote, "We are symbionts on a symbiotic planet, and if we care to, we can find symbiosis everywhere. Physical contact is a non-negotiable requisite for many differing kinds of life" (Margulis 1998, 14–15). Thus, the environmental freedom of the Tibetan-Himalayan rivers is an inherent part of the earth's creative principle, which is nourishing as well as dispossessing, comforting as well as threatening. It has never gratified any species-centred possessiveness of the earth because it is not merely "a nursing mother promising nourishment" (Haraway 2016, 43). As a living entity, the earth has its own geophysiological and cosmic agendas that are not always compatible with the preservation of individual species. This is how the environmental freedom of the earth works on its own accord. It intertwines the physical earth, humans, animals, plants and even supernatural beings among many Himalayan cultures. It possesses a transcendental quality that horizontally transcends ecosystemic differences and human national boundaries. It is thus an environmental universal that affords us to understand the earth's moral considerability as a living entity before we negotiate for fair shares of what the Tibetan-Himalayan rivers can offer us. With sound environmental ethics based on a full understanding of the rivers' environmental freedom, transboundary water commoning would be prevented from becoming a new form of water-grabbing in the name of common good.

References

Allison, Elizabeth. 2017. "Spirits and Nature: The Intertwining of Sacred Cosmologies and Environmental Conservation in Bhutan." *JSRNC* 11(2): 197–226.

Alt, Franz 2020 "Introduction." In *Our Only Home: A Climate Appeal to the World*, by Dalai Lama and Franz Alt. London: Bloomsbury Sigma, 13–22.

Amin, Ash and Philip Howell, eds. 2016. *Releasing the Commons: Rethinking the futures of the commons*. London: Routledge.

Arora, Vibha. 2007a. "Unheard Voices of Protest in Sikkim." *Economic and Political Weekly*, 42(34): 3451–3454.

Arora, Vibha. 2007b. "Assertive Identities, Indigeneity, and the Politics of Recognition as a Tribe: The Bhutias, the Lepchas and the Limbus of Sikkim." *Sociological Bulletin*, 56(2): 195–220.

Arora, Vibha. 2009. "'They are All Set to Dam(n) Our Future': Contested Development through Hydel Power in Democratic Sikkim." *Sociological Bulletin*, 58(1), Special Issue on Development of Democratic Routes in the Himalayan "Borderlands" (January–April), 94–114.

Balikci, Anna. 2008. *Lamas, Shamans and Ancestors: Village Religion in Sikkim*. Leiden: Brill.

Bawaka, Country, Sandie Suchet-Pearson, Sarah Wright, Kate Lloyd, and Laklak Burarrwanga. 2013. "Caring as Country: Towards an Ontology of Co-becoming in Natural Resource Management." *Asia Pacific Viewpoint*, 54(2): 185–197.

Brandshaug, Malene K. 2019. Water as More Than Commons or Commodity: Understanding Water Management Practices in Yanque, Peru. *Water Alternatives* 12(2): 538–553

Brantley, Susan L., William H. McDowell, William E. Dietrich, Timothy S. White, Praveen Kumar, Suzanne P. Anderson, Jon Chorover, Kathleen Ann Lohse, Roger C. Bales, Daniel D. Richter, Gordon Grant, and Jérôme Gaillardet. 2017. "Designing a network of critical zone observatories to explore the living skin of the terrestrial Earth." *Earth Surface Dynamics*, 5: 841–860.

Buck, Susan 1998 *The Global Commons: An Introduction*. Washington, DC: Island Press.

Callicott, J. Baird. 2010. "The Conceptual Foundations of the Land Ethic." In *Environmental Ethics: The Big Questions*, edited by David R. Keller. West Sussex, UK: Blackwell Publishing, 201–210.

Castree, Noel. 2014. *Making Sense of Nature: Representation, Politics and Democracy*. London: Routledge.

Chellaney, Brahma. 2012. *From Arms Racing to "Dam Racing" in Asia: How to Contain the Geopolitical Risks of the Dam-Building Competition*. Washington, DC: Transatlantic Academy.

Country, Bawaka, Sarah Wright, Sandie Suchet-Pearson and Kate Lloyd, Laklak Burarrwanga, Ritjilili Ganambarr, Merrkiyawuy Ganambarr-Stubbs, Banbapuy Ganambarr, and Djawundil Maymuru, and Jill Sweeney. 2016. "Co-becoming Bawaka : Towards a Relational Understanding of Place/Space." *Progress in Human Geography*, 40(4): 455–475.

Crosby, Alfred W. 2004. *Ecological Imperialism: The Biological Expansion of Europe, 900–1900*. Cambridge: Cambridge University Press.

Dechen, Palmo. 2019. "Tibet's Rivers Will Determine Asia's Future." *The Diplomat* 1, 2019.

Drew, Georgina. 2017. *River Dialogues: Hindu Faith and the Political Ecology of Dams on the Sacred Ganga*. Tucson: The University of Arizona Press.

Dybas, Cheryl Lyn. 2013. *Discoveries in the Critical Zone: There Life Meets Rock*. National Science Foundation.

Fisher, James F. 1978. 'Introduction.' In *Himalayan Anthropology: The Indo-Tibetan Interface*, edited by James Fisher, 1–3. Paris: Mouton Publishers.

Gamble, Ruth. 2019. "How Dams Climb Mountains: China and India's State-Making Hydropower Contest in the Eastern-Himalaya Watershed." *Thesis Eleven*, Vol.150 (1), 42–67.

Gibson, Katherine, Deborah Bird Rose and Ruth Fincher, eds. 2015. *Manifesto for Living in the Anthropocene*. Brooklyn, NY: Punctum Books.

Gibson-Graham, J. K., Jenney Cameron and Stephen Healy. 2016. "Commoning as a Postcapitalist Politics." *Releasing the Commons: Rethinking the Futures of the Commons*, edited by Ash Amin and Philip Howell. London: Routledge, 192–212.

Gorer, Geoffrey. 1938. *Himalayan Village: An Account of the Lepchas of Sikkim*. Melbourne: Nelson.

Grim, John, ed. 2001. *Indigenous Traditions and Ecology: The Interbeing of Cosmology and Community*. Cambridge: Harvard University Press.

Grumbine, R. Edward and Maharaj K. Pandit. 2013. "Threats from India's Himalaya Dams." *Science*, 339: 36–37.

Haraway, Donna. 2016. *Staying with the Trouble: Making Kin in the Chthulucene*. Durham: Duke University Press.

Harrison, Clare. 2018. "The Great Walls Damming Tibet: The World's Water Tower." https://tibetnature.net/en/great-walls-damming-tibet/, accessed on October 24, 2020.

Harvey, Graham. 1997. *Contemporary Paganism: Listening People, Speaking Earth*. New York: New York University Press.

Hodgson, Brian 1874 *Essays on the Languages, Literature and Religion of Nepal and Tibet*. London: Trübner & Co.

Hogan, Linda. 2013. "We Call it Tradition," in *The Handbook of Contemporary Animism*, edited by Graham Harvey. London: Routledge, 17–26.

Ives, Jack D. 2004. *Himalayan Perceptions: Environmental Change and the Well-Being of Mountain Peoples*. London: Routledge.

Keohane, Robert O. and Elinor Ostrom, eds. 1995. *Local Commons and Global Interdependence: Heterogeneity and Cooperation in Two Domains*. London: SAGE Publications.

Leopold, Aldo. 1949. "The Land Ethic," in *A Sand County Almanac: And Sketches Here and There*, edited by Aldo Leopold. Oxford: Oxford University Press.

Linebaugh, Peter. 2008. *The Magna Carta Manifesto: Liberties and Commons for All*. Berkeley: University of California Press.

Little, Kerry. 2018. "Mythology as a Protest Narrative: A Lepcha Tale." In *Playing with Nature: History and Politics of Environment in North-East India*, edited by Sajal Nag. New York: Routledge, 341–352.

Little, Kerry. 2012. "Sanctity, Environment and Protest: A Lepcha Tale." In *Independent People's Tribunal on Dams, Environment and Displacement*, edited by Doma T. Bhutia. New Delhi: Human Rights Law Network, 83–94.

Little, Kerry. 2008. "Lepcha Narratives of Their Threatened Sacred Landscapes." *Transforming Cultures eJournal*, 3(1): 227–235.

Margulis, Lynn. 1998. *Symbiotic Planet: A New Look at Evolution*. New York: Basic Books.

Miller, Michelle Ann, Carl Middleton, Jonathan Rigg and David Taylor. 2020. "Hybrid Governance of Transboundary Commons: Insights from Southeast Asia." *Annals of the American Association of Geographers*, 110(1), 297–313.

Miller, Michelle Ann. 2020. "B/ordering the Environmental Commons." *Progress in Human Geography*, 44(3): 473–491.

Mullard, Saul. 2011. *Opening the Hidden Land: State Formation and the Construction of Sikkimese History*. Leiden: Brill.

Ostrom, Elinor. 1990. *Governing the Commons: The Evolution of Institutions for Collective Action*. Cambridge: Cambridge University Press.

Plumwood, Val. 1993. *Feminism and the Mastery of Nature*. London: Routledge.

Pope Francis. 2015. Laudato Si': On Care For Our Common Home. *Vatican*. http://w2.vatican.va/content/dam/francesco/pdf/encyclicals/documents/papa-francesco_20150524_enciclica-laudato-si_en.pdf, accessed October 25, 2020.

Richter, Daniel and Sharon A. Billings. 2015. "One Physical System: Tansley's Ecosystem as Earth's Critical Zone." *New Phytologist* 206: 900–912.

Rigby, Catherine. 2019."Weaving the environmental humanities: Australian strands, configurations, and provocations." *Green Letters*, 23(1): 5–18.

Risley, Herbert Hope. 1894. *The Gazetteer of Sikhim*. Calcutta: The Bengal Secretariat Press.

Rigzin, Ngodup Dokhampa. 1998. "The Hidden Valley of Sikkim." *The Bulletin of Tibetology*, No.1, 75–86.

Rose, Deborah Bird, Thom van Dooren, Matthew Chrulew, Stuart Cooke, Matthew Kearnes and Emily O'Gorman. 2012. "Thinking Through the Environment, Unsettling the Humanities." *Environmental Humanities* 1, 1–5.

Rose, Deborah Bird. 2013. "Val Plumwood's Philosophical Animism: Attentive Interactions in the Sentient World." *Environmental Humanities*. 3, 93–109.

Rose, Deborah Bird. 1996. *Nourishing Terrains: Australian Aboriginal Views of Landscape and Wilderness*. Canberra: Australian Heritage Commission.

Rudwick, Martin J. S. 2008. *Worlds Before Adam: The Reconstruction of Geohistory in the Age of Reform*. Chicago: The University of Chicago Press.

Rudwick, Martin J. S. 2005. *Bursting the Limits of Time: The Reconstruciton of Geohistory in the Age of Revolution*. Chicago: The University of Chicago Press.

Samaranayake, Nilanthi, Satu Limaye and Joel Wuthnow. 2018. *Waters: China, India, Bangladesh, and Brahmaputra River Politics*. Quantico, Virginia: Marine Corps University Press.

Samuel, Geoffrey. 1993. *Civilized Shamans: Buddhism in Tibetan Societies*. Washington, DC: Smithsonian Institution Press.

Smyer Yü, Dan. 2014. "Sentience of the Earth: Eco-Buddhist Mandalizing of Dwelling Place in Amdo, Tibet." *Journal for the Study of Religion, Nature and Culture* 8(4): 483–501.

Smyer Yü, Dan. 2015. *Mindscaping the Landscape of Tibet: Place, Memorability, Ecoaesthetics*. Boston: De Gruyter.

Smyer Yü, Dan. 2017. "Introduction: Trans-Himalayas as Multistate Margins." In D. Smyer Yü and Jean Michaud, eds. *Trans-Himalayan Borderlands: Livelihoods, Territorialities, Modernities*. Amsterdam: Amsterdam University Press, 11–44.

Smyer Yü, Dan and Wangmo Sonam. 2019. "A Historical Sketch of Buddhism in Tibet." In *Oxford Research Encyclopedia of Asian History*, edited by David Luden. Oxford: Oxford University Press.

Smyer Yü, Dan. 2020. "The Critical Zone as a Planetary Animist Sphere: Etho-graphing an Affective Consciousness of the Earth." *JSRNC* 14(2): 271–290.

Sörlin, Sverker. 2015. "Cryo-History: Narratives of Ice and the Emerging Arctic Humanities." In *The New Arctic*, edited by Birgitta Evengård, Joan Nymand Larsen, and Øyvind Paasche. New York: Springer, 327–339.

Spoon, Jeremy. 2014. "Everyday Buddhism and Environmental Decisions in the World's Highest Ecosystem." *JSRNC* 8(4): 429–459.

Tsing, L. Anna. 2015. *The Mushroom at the End of the World: On the Possibility of Life in Capitalist Ruins*. Princeton: Princeton University Press.

United Nations Human Rights Council (UNHRC). 2013. *The United Nations Declaration on the Rights of Indigenous Peoples: A Manual for National Human Rights Institutions*. Sydney, Australia: Asia Pacific Forum of National Human Rights Institutions.

Xu, Jianchu, Ram P. Chaudhary, and Pratikshya Kandel. 2019. "Sustaining Biodiversity and Ecosystem Services in the Hindu Kush Himalaya." In *The Hindu Kush Himalaya Assessment: Mountains, Climate Change, Sustainability and People*, edited by Philippus Wester, Arabinda Mishra, Aditi Mukherji, Arun Bhakta Shrestha. Cham: Switzerland: Springer, 127–165.

Zhao, Hai-di, Liu Shi-liang, Dong, Shi-kui, Liu Qi, and Li Xiao-wen 2014. "三江源人类干扰与湿地空间变化关系研究Relationship between Human Interference and Wetland Spatial Variation in Three-River Headwaters Region." *Wetland Science*, 12(1).

Zhao, Xinquan 2020. "Thirty Years of Guarding the Three Rivers Source: Governing the dark soil, protecting the grassland ecology and balancing nomadic development 三十载守望三江源——治理黑土滩、保护草地生态、平衡牧业发展." *Science Daily (科技日报)*.

Conclusion

Indigenous heritages and sacred earth

John Grim

Indigenous peoples are a significant global population of over 500 million peoples in Africa, South and East Asia, Southeast Asia, Central Asia, Australia, the Pacific region, Northern Eurasia and the Americas. The term "indigenous" in this context usually refers to small-scale societies around the planet who share and preserve ways of knowing the world embedded in particular languages, story cycles, kinship systems, worldview dispositions and integrated relationships with the land on which they live. This integrated approach is evident in the chapters collected here and is explored below using the term, *lifeway*. The United Nations Declaration on the Rights of Indigenous Peoples adopted in 2007 recognises the struggle of indigenous peoples worldwide for their land rights, cultural vitality and social justice. This political understanding of indigeneity on the global scale is particularly evident in indigenous cross-cultural conversations. The ethnolinguistic and cultural traits integrated in diverse lifeways have enabled many small-scale indigenous communities to meet the challenges of modernity. Indeed, this is a major feature of their resilience in the midst of tumultuous change.

Indigenous traditions are inextricably entangled in multiple global events—modernisation, climate change, human migration, the transboundary flows of commodities and ideas and the encounters of different worldviews and belief systems. Indigenous societies in the Himalayas are also undergoing rapid transformations as shown in all chapters in this book. How, then, shall we understand the meanings of particular indigeneities in this complex global context? Coming from different cultural backgrounds and interdisciplinary fields of environmental studies, the authors in this book intertextually relate to each other in presenting indigeneity in the Himalayas as deeply interrelational. Each author presents their own terms in this regard, namely: symbiotic (Smyer Yü), hybrid (Das), organic (C. Lepcha), entwined in the land (R. Lepcha and M. Lepcha), eco-spiritual (Thinley Dema), local (Deshar et al.), community-based (Khalid), animist (de Maaker), ontological (Wouters), a changing epistemology (Kinley Choki), under the statist transformation (Gamble), transboundary (Dutta and Yashwant) and international (Davis).

While recognising political, social and economic changes in the Himalayas, the authors also identify diverse and enduring land-based indigenous environmental ethics as central for global endeavours leading to a sustainable future. These also find expression in the UN Sustainable Goals as well as in environmental studies in different fields of social sciences and humanities. Even though the ethnolinguistic, ecological and national contexts of the case studies in this book differ, there is a shared recognition of indigeneity across the Himalayas. The embodiment of indigeneity is primarily in the earth which is culturally expressed in subsistence, affective, religious and spiritual dimensions. In this regard, indigeneity is tangibly synonymous with land understood as homelands, sacred sites, dwelling places and surrounding natural environments all undergoing changes. It pertains to the material manifestations of what Thomas Berry calls "the maternal principle" of the earth (Berry 1990, 75). In the intangible, intellectual sense, indigeneity is expressed in ecological knowledge which is discussed in several chapters. Indigeneity in this sense expresses human–earth relations in diverse ethical, religious and spiritual terms that are always embodied and intellectually grounded.

Sacred natural heritages

This conclusion reaffirms the critical importance of what can be called, religious ecology, as a suggested approach by many authors in this book. By religious ecology, then, a conceptual position is suggested that indigenous peoples, as well as those axial age communities that gave rise to the "world's religions," interacted with biodiversity and local bioregions with subsistence practices in which regard for the sacred was integrated into daily life. Religious ecology recognises the spiritual bonds of human communities with their local earth-scape. Considerations of religious ecology recognise the importance of places and practices in indigenous traditions that weave community memory into local places. These practices deepen the humanistic approaches this book promotes. Situated in our current human-dominated epoch, or Anthropocene, the authors present insights into culturally based indigenous knowledge and modern scientific understandings of diverse environments. Indigenous knowledge is increasingly integrated into scientific awareness in exploring sustainable futures. In the Himalayas, regional political realities foreground the entanglement of smaller-scale indigenous societies with the larger surrounding cultural and civilisational systems. Most importantly, the authors here describe ways in which indigenous understanding of sacred environments are increasingly coupled with modern politics of land rights, environmental protection and management of shared resources. These are salient points in the chapters by Charisma Lepcha, Bhagarbi Das, Zainab Khalid, Anwesha Dutta and Shailendra Yashwant. On the global scale, many chapters echo the assessment in a UNEP Faith for Earth Initiative: "Spiritual values for more than 80% of the people living on earth have been driving individual behaviors. In many countries, spiritual beliefs and religions are main drivers for cultural values, social inclusion, political engagement, and economic prosperity" (UNEP 2008).

Lived experiences of sacredness and sentience in the earth are clear in Himalayan indigenous traditions. Human-nature kinship relations among the Lepcha and the Bhutanese are particularly revealing about human reverence for the earth. In these lifeways, sacred mountains and waterbodies nourish human survival and flourishing. At the same time, they are diverse embodiments of divinity as spiritually recognised geological and ecological forces. Thus, sacredness is not an abstract idea but holds material vitality in the earth. Interestingly, Himalayan peoples' sense of religious ecology resonates with those of the Native Peoples of North America. In this chapter, I bring my studies to this comparative observation emphasising a lifeway approach. This comparison of Himalayan and North American Indigenous traditions emphasises conservation and sustainability practices as significant dimensions of these religious ecologies. Moreover, these practices are not isolated phenomenon but embodied experiences of a larger cultural system or lifeway.

The sacred sites of mountains and rivers in the Himalayas and North America can be called "sacred natural heritages." Here, nature is humanised with reverence, affection and historical importance. These sacred natural heritages are ontologically and conceptually inseparable from the human communities that are spiritually nourished by them. They are also exemplars of human stewardship of the earth. Significantly, indigenous peoples also recognise and emphasise the nourishment earth extends to humans, animals and plants. Similar to what I have observed in North America, indigenous peoples in the Himalayas also consider themselves as caretakers of the local earth. They understand their ancestral practices as knowledge of the environment directly related to sacred sites.

Sacred natural heritages can be differentiated as areas of land, water or biodiversity concentration that have special spiritual significance. The style of life of Himalayan human societies, or ethos, recognises earth as a living entity (Rose 1996, 7). Yet, as Kinley Choki writes in her chapter, the lived experience of indigeneity in the Himalayas is increasingly becoming "a changing epistemology." For example, indigenous lifeways and religions such as Buddhism and Christianity are no longer mutually exclusive. This is clear in the cultural histories of Bhutan, Assam, Sikkim and the Hunza Valley. These complex spiritual systems are being braided together positively. Another example arises when considering Pope Francis's Catholic encyclical, *Laudato Si': On Care for Our Common Home* (2015, 131). In that teaching document intended for more than Catholic readership, the Pope extensively advocates protection of indigenous peoples' traditions and environments.

Global environmental crises in the Anthropocene are bringing different human traditions to a recognition of the livingness of the earth. Ironically, this is a hopeful sign as we face the common threats of human-induced climate change, biodiversity loss and many environmental degradations. Among indigenous communities in the Himalayas and across the world, the resurgence of reverence and respect for sacred sites models for many human communities new understandings of the natural world. Broadly speaking, in indigenous communities when the sacred manifests in space-time, the primary focus of attention is not on the human but on spiritual and cosmological forces experienced in the land as sacred heritage of the earth and human ancestors.

These placed-based emphases on sacred natural heritage also have far-reaching dimensions. That is, they relate human communities out into the larger world. A once-known Penobscot Indian, a Native Peoples of the coastal woodlands in Canada, observed that his people's sense of land was integrated into the life of a place:

> When we speak of "Land," we mean Penobscot reverence for the land. "Land" means fiddleheads, deer, corn, muskrat, beaver, ermine, fox, bear, bobcat, eel, salmon, the pond, the ledges, the mountains. Sara's springs, birch, ash, sweet grass, poison ivy, Fort Dawson, Sandy Beach, lower village, upper villages. Oak Hill, the swamp, the burial grounds, laughter on a warm summer night, tears at an Indian wedding, coos of a little baby, dogs, cars, oil trucks, Indian drums, old people hiding all the secrets of wisdom, young people finding wisdom through acrid smoke and finding that is not the essence of all things, and that when you come to the land you can't burn it, you can't smoke it, you can't ignore it. Land is forever, and we can't sell it, we can use it, for the benefit of ourselves and our children, and our grandchildren, and even those generations that are not born.
>
> (Nabokov 2006, 19)

Sacred natural heritages are pan-indigenous practice across the globe. They manifest religious ecologies in diverse expressions of animism and shamanism, as well as syncretic practices that blend into the world religions. It is not simply the land as an objective reality that demands attention. Rather, a sacred heritage site is significant by virtue of the interactions of communities of human and more-than-human life within that place. For indigenous peoples, this interaction is mediated by intangible, felt spiritual forces of single or multiple deities as the prime mover of all organic and inorganic life on earth. Practices in relation to that place, whether ritual practices or subsistence practices, are attentive to the eco-spiritual dimension of lifeway, called here, religious ecology. The primary focus is not on the human exclusively, but in maintaining a set of spiritual, affective relationships that attend to the earth as reciprocal flows of giving and receiving. Ecology as interactive flows is a crucial lesson that twenty-first century environmental humanists raise for appreciative critical attention. Importantly, these researchers remind us that we cannot alone bring about the transformations we need by moving from one environmental issue to another with mind-bending insights from science and in fierce legal and policy-oriented fights. Rather, we need to re-embrace our own place-based, spiritually felt and affectively practiced ecology that attends to a broad range of environmental demands. Cultural sensitivity to place and practices grounded in the earth as sacred has endured among many indigenous peoples. But it has come at a high price of both internal fragmentation and external pressures of accommodation, extraction and destruction.

Consider a Hopi sacred boundary marker of their land at a butte south of Holbrook, Arizona in North America. The Hopi, as many of us know, are Puebloan peoples in the Southwest, who still reside in the oldest continuously inhabited site in North America, namely, Oraibi. Many Hopi also live on three massive sandstone ridges that stretch out like fingers from a high geological formation called Black Mesa. Theirs is a complex spiritual path based on maintaining proper relations with rain through sacred natural heritages such as *Tsimontukwi*, or as local non-Native peoples call it, Woodruff Butte. The place was also known by other indigenous medicine practitioners, such as the Dene/Navajo peoples, as *Tooji' Hwi' idzoh*, or "Line Extends to the River." Zuni people know this same place as

Dematsali Im'a, namely, a place for gathering a particular plant for ritual divination. These different indigenous place-names all refer to a mountainous butte that once rose 500 feet from the surrounding plain. In the late 1980s, it was bought by a gravel-mining operation and, despite requests from these native peoples, the granite rock core was blasted, crushed and mixed with boiled used tires for highway pavement. Hopi priests alerted to the imminent destruction of pictographs at the site arrived in time to see bulldozers demolish the place before their eyes (McLeod 2001). This occurred in the twenty-first century not in the 1800s. It reminds us that the forests of the Lepcha homeland were cleared by the British to make room for tea plantations, and how the riverine landscape of Assam underwent ecologically disruptive separations of water and land in the British agricultural development of the region.

Sacred natural heritages require protection along with the recognition of rights and responsibilities. These are often not found in legal and economic systems based on private property, individual rights, commodity markets and totalising national sovereignty. These Anthropocene rights have been asserted over local commitments to the living earth. In this book, four historical and current trends are noted in which Himalayan indigenous lifeways and lands have been transformed: namely, the physical transformation of land; the forced relocation of people by the state; the combined force of the state and exclusionary cultural customs; and, finally, outmigration due to economic globalisation. The British colonial deforestation of Lepcha land (C. Lepcha) demonstrates the loss of the Lepcha's indigenous botanical knowledge. The artificial separation of land from water in Assam (Das) shows the severe systemic oppression of the chadwellers in Assam. The Chinese state's relocation of nomadic Tibetans from their ancestral pastoral lands for environmental conservation project ignores their nomadic roles as the environmental stewards of their ancestral land (Gamble). The Sunar, a Dalit community in Nepal, are spatially marginalised to a flood prone habitat due to the socially accepted Hindu notions of purity and pollution (Deshar et al.). Many Hunzakutz men leave Hunza Valley for urban jobs in far-away places (Khalid). These systemic pressures on indigenous peoples require a profound, complex understanding of the interconnectedness of life in ethical terms. Such an environmental ethics are steadily taking shape among conservation agencies and indigenous communities, for example, in documents published by the World Commission on Protected Areas of the International Union for Conservation of Nature (IUCN Red List).

Founded in 1948 the World Conservation Union, or as it is more commonly called the International Union for the Conservation of Nature (IUCN), has gradually come to realise that "protected areas" must include human cultures and that conservation should do no harm to human societies. Thus, biodiversity conservation has become attuned to "Community Conservation Areas" in which indigenous lived experiences and environmental wisdoms about governance, equity, participation and benefit-sharing have been implemented in successfully sustaining both indigenous cultures and wildlife in some parts of North America and Australia. Meanwhile, it is also vital to recognise that indigenous communities have diverse motivations for protecting areas and resources, and native terms and concepts may identify agendas quite different from the objectives of an international organisation such as IUCN in describing protected areas. In this regard, local concerns need to become an integral part of international and national conservation efforts. Local agendas might express community needs as a wellbeing, or good life, that masks an indigenous ethical worldview. Thus, the sustainable use of natural materials manifests a religious ecology rather than overt environmental protection of land, biodiversity, or wildlife in itself. In many instances, indigenous peoples have been nature conservators and protectors of their homelands since ancient times. As ancient caretakers of the environment, indigenous leaders and communities accepted certain rights and responsibilities that can broadly be identified with governance, equity, participation and benefit-sharing throughout the community.

Often the painful historical acts of colonialism and imperial expansions deprived indigenous communities of these rights. In some settings, indigenous elders involuntarily abdicated their responsibilities through despair at their exclusion from the enclosure of their formerly shared indigenous commons whether mountain, forest, jungle, or prairie. Many indigenous peoples have experienced a deep alienation from their cultural heritage and relationships with local biodiversity brought on by the extractive, profit-driven ethos of the global market. Similarly, in the name of national security or development, governments have seized lands, biodiversity and resources. In this sense, it was not that long ago when well-meaning conservation agencies themselves bought land and moved indigenous peoples off that land to create new parks oriented towards protecting biodiversity and encouraging tourist income for the dominant nation-state (Dowie 2009).

Several organisations, like the IUCN, have manifested amazing transformations in recognising that indigenous peoples in Community Conservation Areas can be effective conservators and protectors of both land and biodiversity. This is a remarkable accomplishment in which lifeways and traditional religious ecological practices are encouraged and allowed to function. The IUCN actively promotes an understanding in which cultural diversity is acknowledged and supported. In this shared setting, biodiversity flourishes. Such Himalayan projects as "Coping with Uncertainties: Building Community Resilience and Ecosystem Based Adaptation to Climate Change in the Indian Himalayan Region" and Balkila Himalayan Watershed Project" (IUCN 2019, 2011), present examples of its integrated approach to local issues and global support.

This brings a final point, namely, the relationships of local and planetary sense of a sacred natural heritage. Relationships engendered by diverse indigenous traditions have both a historical character and lived experiential dimensions. The phrase, "historical character," is trying to describe cultural identity that comes from living with a place with a long human history as well as a longer eco-geological history in Deep Time. "Deep" here suggests a twofold meaning. One pertains to the geological and the ecological evolution of the earth as planetary, which obviously predates the birth of humankind over 100,000 years ago as Smyer Yü presents in his introduction. Another refers to the historical depth of humankind within our own evolutionary path. In this sense, "deep" is used to suggest that process whereby a people become conscious of themselves and their past in relationship to place and the biodiversity there. Thus, the conscious stories of community identity and meaning generated by human dwelling and practice in place stand in relation to more unconscious, or deep, experiences of that place in eco-geological terms. This relates to ecological imaginaries that suggest deep psychic attractors resident in place for indigenous wisdom keepers. Of late, for Native Peoples in the Americas and elsewhere, their story often centres on resistance as they struggle against absorption into neo-colonial, global economic and mainstream national systems of governance. Such struggles are not totally new since most indigenous peoples recall their ancestors' mythic efforts to survive the challenge of origin moments, or inappropriate relations with place on the part of their ancestors, or attacks by monsters that resided in those places. Native Peoples often speak of the land as manifesting the physical remains of those fights with monsters whose bodies became mountains, whose blood flows formed river courses.

Indigenous lifeways in their changing lifeworlds

The chapters in this book argue that humans are an integral part of the environment. The environment, in turn, is not physical only but is animate or animated with indigenous pantheons and ancestral spirits embodied in different landforms, waterbodies and forests. Humans and the environment entwine each other together. While the environment physically environs humans, humans co-become with the creative forces of the earth in both subsistent and spiritual terms. The intertwinement of lifeforms and the environment is a common feature of many indigenous societies across the world. The human diversity and the biodiversity of the Himalayas are a highly pronounced example of the human-nature union. Thus, indigenous lifeways can be seen as forms of traditional ecological knowledge and practices that are interwoven into the whole fabric of human existence that manifest the spiritual ethos of indigenous communities. The indigenous scholars in this book successfully recount their native ecological experiences in the environmental humanist approaches of their choice.

The authors also elaborately discuss the changing lifeworlds of the Himalayan indigenous peoples, challenged and transformed by a variety of externally introduced economic activities. For instance, the Bhutanese society is encountering the impacts of tourism and the caterpillar-fungi trade as discussed in the chapters by Kinley Choki and Thinley Dema. The regionalisation and globalisation of local livelihoods are engendering a growing dependence upon a cash flow originating from the global market system. The demand from the market clearly shows a pattern of the growing extraction of the caterpillar fungi, the growing wealth and the deteriorating attentiveness to one's land-based cultural heritage. Bhagarbi Das's case study of the Miya Poetry Movement among the chadwellers in contemporary Assam shows a longer ecological–political cycle of radical changes. Similar to the British colonial system, the current Indian state continues the agricultural practice of saturating the land with Himalayan-sourced water for the sake of economic productivity. Paradoxically, it condones an extreme

Assamese nationalism that denies the chadwellers' belonging to the land where they live, contributing to nation-building by an independent India. Yet, in both cases, Bhutanese and chadwellers in Assam assert their traditional indigenous environmental knowledge. Their voices of indigeneity resonate with worldwide revitalisations of indigenous environmental wisdom.

The lifeways of contemporary indigenous societies and the ecosystems in which they reside continue to be vital, interactive wholes. For example, the Bunun people of Taiwan have transmitted striking modes of singing based on their ancient exchanges with the birds of the island. We know from studies of other Proto-Malaysian peoples that this attention to birds is a central dimension of an ancient worldview shared among these diverse peoples from the Malay Peninsula, through the Indonesian and Philippine archipelagos, and on to Taiwan. This ecological knowledge was not an isolated way of knowing but embedded in religious ecologies. In indigenous settings, the sense of "community" is not exclusively human oriented, but extends out into the multispecies community of life grounded in the earth as the ultimate planetary living entity.

There is no doubt that the investigations of local indigenous peoples exemplified in this book also uncover insights into the problematic character of human values, namely, the potential for manipulation of societal desires, needs, insights and motivations by those seeking power over others. Religious ecology in this case also requires the insights coming from environmental humanist perspectives. We need multiple understandings of indigenous peoples' spiritual and affective relations with the earth and nonhuman species. Given the contexts of coloniality, neo-coloniality and modernisation in the Himalayan region, religious ecology in the Himalayas is also a political ecology and a public discourse of rights, namely the fundamental human rights to one's ancestral land, the rights to preserve one's land-based cultural and spiritual heritages, and the rights to revive traditional modes of livelihood compatible with the carrying capacity of the land. While the globe-wise advocacy for indigenous political rights has become loud and clear as shown in the UN Declaration on the Rights of Indigenous Peoples (UN 2007), rights also entail other sets of meanings immediately critical to indigenous societies.

In this view, the stress on rights language is less legal and more integral with what indigenous elders often call "cosmovisions" that integrate persons and places with the spiritual dynamics of the larger universe. That is, "rights" do not immediately spring from jurisprudential concerns of individuals and properties as much as being an outgrowth of sapient views of animals, plants and bioregions as vested with personhood and having needs for existence. That such rights can obligate individuals and communities to ritual actions, for example, carries forward an impulse to stand in harmonious relation with significant other-than human persons. This type of wisdom experience definitely has cosmological orientations. To trace this type of incipient cosmological thinking into the cultural expressions of the Himalayan region will be a meaningful study. Thus, investigation of the formative roles of Himalayan indigenous peoples allows us to have a more contextualised understanding of indigenous rights situated in traditional cosmovisions of human dwelling places and their surrounding natural environments.

Underlying this last emphasis on cosmology is a holistic and organic turn that has especially rich roots among Himalayan indigenous peoples. Again, lifeway is a viable term for understanding indigenous environmental knowledge based on their lived experiences and associated religious and spiritual practices. In this regard, religion should not be understood from a contemporary secular perspective as a separate act from subsistence practices, trade or social interactions. Rather, concerns for the Source-of-Giving-in-the-World in diverse indigenous societies are interwoven into the full range of life. Thus, religious activities in indigenous settings hold insight into that complex of relationships treated as cosmology. In many indigenous settings, this identification of the range of life, or ecology, stands in significant comparative relationships to cosmology. Thus, one can conceptually distinguish both religious ecologies and eco-spiritual cosmologies among the world's religions, including indigenous traditions. But religious ecology is especially more central when discussing indigenous ecosystems that are inherently entwined with cosmovisions, rituals, myths and communal experiences that celebrate the livingness of the earth.

The value of indigenous cosmovisions and land ethics for a sustainable future

The indigenous authors in this book clearly reveal their ancestral ecological worldviews and environmental ethics centred upon the oneness of humans, animals, plants and local spirits and deities. Land ethics are not just a set of utilitarian norms treating the physical environment as resources but, more importantly, entail a range of spiritual and affective knowledge of the earth. Their chapters resonate in comparative harmony with a North American native counterpart, the Lakota writer, Charlotte Black Elk. She describes her peoples' relationships with a sacred natural heritage, Bear Butte, as not only formed by an ancient religious ecology but also by their current relationships shaped by resistance to the practices and conceptions of dominant society. Charlotte Black Elk said:

> Our name for Bear Butte is *Paha Wakan*, sacred mountain. Bear Butte was not only a religious place, it was also a meeting place...Our full name for Bear Butte is *Sinte Ocunku Paha Wakan*, "the sacred mountain at the edge of the trail." And so it sits at the edge of the Black Hills [in the western part of South Dakota in North America]. When you look at the religious view that lands have—that all lands are sacred, but some areas are more sacred—you look at Bear Butte as a place for reflection. A place where no war party would attack someone who was there to pray. It was respected by all the Tribes. Other people came down—people who were our enemies came down to pray. We did not attack them. I think looking at that, the fact that it has remained a central part of the Lakota community, of people going back there continually, it shows that there has been a continuum in our religion—*the ideas of everything being a part of the whole* and at a place like Bear Butte, you would be at the center of the earth at the center of the Hoop [of creation].
>
> (Horse 1998, 103–102)

The added emphasis on the part-whole relationship is meant to point out again that among many Himalayan and indigenous peoples, the part and the whole are quantitatively different but are qualitative the same. That is, different humans emerge from the earth along with other forms of life, and continue to be nourished by the earth and live with an animist sense of shared life. Indigenous land ethics signify the kindred relationality of everything on the earth and under the heaven. Similarly, the Lepcha's origin story of the earth starts with *Rongnyoo* and *Rungnyit*, two personified rivers/river spirits (R. Lepcha and M. Lepcha). Like the Lepcha, the Lakota recall their cosmogonic story that relates how they originally came from life's larger sources. For the Lakota, they came out of a cave on Bear Butte; for the Lepcha, they recall their riverine ancestors in the flow of life. Similarly, ancient human–yak kinships related in the chapter by Jelle Wouters, narrate a vision in which the spiritual essence of life appears in ever changing physical forms (Descola 2013, 122). In other words, the same creative force of life on earth is omnipresent in its many physical manifestations. Thus, the spiritual meaning of life cannot be isolated or extricated from its material forms.

In the last 20 years, the fields of religious ecology and environmental humanities have hosted a diverse range of interests in re-exploring the spiritual and ethical values of indigenous ecological knowledge as a vital source of the global environmental endeavour among scholars, scientists, policymakers and the concerned public for building a sustainable future. In both fields, scholars are keen on building new environmental ethics to address the global and local environmental challenges. Scholars of religious ecology call for "a unified worldview or a single global ethic" (Tucker and Berthrong 1998, xxiv). Simultaneously, environmental humanists call for an "earth language" to articulate thicker meanings of humanity environed and nurtured by the earth" (Rose 2013, 107 and 2012, 2). Rather than the romanticisation and commercialisation of indigenous peoples' spiritual and ecological heritages; it is more important to be attentive to colonial and nation-states' modes of power that continue to disenfranchise indigenous environmental wisdoms. Our industrial fixations call for a new environmental ethics, subverting most of the arrogant anthropocentric fixations of the Anthropocene. Many of our environmental problems are caused by human-induced climate change and subsequent natural disasters such as floods, landslides and droughts in the Himalayas recounted in the chapters by Rashila Deshar, Zainab Khalid, and Anwesha Dutta and Shailendra Yashwant. In this process of re-identifying the environmental ethical values of indigenous knowledge, it is essential that we do not isolate them as local knowledge; instead, we accept the fact that, like larger human societies, indigenous societies are integral to the earth community. Therefore, their ecological knowledge and environmental ethics arise from an anthropocosmic perspective that reaches into the more-than-human world.

In the context of the New Himalayas, the authors in this book take a collective symbiotic turn to re-embrace the Himalayas as a union of nature in culture, humanity with animality, body attuned to spirit and material flow in the cosmos. This symbiotic turn joins the call to end these hyperseparations (Grim and Tucker 2020, 2; Plumwood 2012, 85) that have encouraged and reinforced the misconception of the earth as a lifeless physical entity and, therefore, have deprived the earth of its moral considerations. A faux-modern utilitarian-based worldview that flourishes humans over the more-than-human world has proved to be detrimental. It prevails at the expense of the environmental health of all life. It is clear that building new environmental ethics for our wounded earth requires the integration of traditional environmental knowledge and empirical science, as Susan Kutz and Matilde Tomaseli write of the Mi'kmaq of Turtle Island:

> Bridging multiple knowledge systems requires drawing on natural and social sciences' methodologies and constant consideration for the value systems of all knowledge holders, a process that is based on ongoing iteration and feedback. The Mi'kmaq principle of "*Etuaptmumk*"or "two-eyed seeing" captures the concept of bringing different knowledge systems together to increase our collective bread and depth of understanding: "learning to see from one eye with the strengths of indigenous knowledges…and from the other eye with the strengths of western knowledges…and learning to use both these eyes together, for the benefit of all".
>
> (Kutz and Tomaselli 2019, 1136)

The belated meeting of indigenous knowledge and modern science afford us opportunities to implement the multiple intelligences of the earth into a new set of global environmental ethics. The authors in this book are taking the lead in the Himalayan region to revive the ancient native knowledge for sustainable living and to critique an objectivising, environmentally insensitive ethos of global capital that rejects nature as alive and sees economic use as the ultimate form of human–earth relationships. Mindful of the returning of our traditional affective bonds with the earth, we may journey towards a new flourishing of the Earth community. The ancient Himalayan ecological knowledge systems in their current manifestations certainly have roles to play in the global arenas dedicated to the restoration of the environmental health and a flourishing future.

References

Berry, Thomas. 1990. "The Spirituality of the Earth." In *Liberating Life: Contemporary Approaches in Ecological Theology*, edited by Charles Birch, William Eaken and Jay B. McDaniel. Maryknoll, New York: Orbis Books, 151–158.

Descola, Philippe. 2013. *Beyond Nature and Culture*. Chicago: University of Chicago Press.

Dowie, Mark. 2009. *Conservation Refugees: The Hundred Year Conflict between Global Conservation and Native Peoples*. Cambridge, MA: Cambridge University Press.

Grim, John and Mary Evelyn Tucker. 2020. "Introduction: Ways of Knowing, Ways of Valuing Nature." In *Living Earth Community: Multiple Ways of Being and Knowing*, edited by Sam Mickey, Mary Evelyn Tucker and John Grim. Cambridge, UK: Open Book Publishers, 1–9.

Horse, Arvol Looking. 1998. "A Song from Sacred Mountain: Lakota-Dakota and Cheyenne Interviews." In *Readings in American Indian Law: Recalling the Rhythm of Survival*, edited by Jo Carrillo. Philadelphia: Temple University Press, 103–104.

International Union for the Conservation of Nature (IUCN) 2011. "IUCN Balkila Himalayan Watershed Project." https://www.iucn.org/content/iucn-balkila-himalayan-watershed-project-supported-nokia. Accessed October 10, 2020.

International Union for the Conservation of Nature (IUCN) "Red List of Threatened Species." https://www.iucnredlist.org/. Accessed May 23, 2020.

International Union for the Conservation of Nature (IUCN) 2019. "Building Community Resilience and Ecosystem Based Adaptation to Climate Change in the Indian Himalayan Region (CwU)." https://www.iucn.org/asia/countries/india/building-community-resilience-and-ecosystem-based-adaptation-climate-change-indian-himalayan-region-cwu. Accessed October10, 2020.

McLeod, Christopher, Director, In the Light of Reverence 2001. 72-minute film. Bullfrog Films. 2001. http://www.sacredland.org/in-the-light-of-reverence/. Accessed May 23, 2020.

Nabokov, Peter. 2006. *Where the Lightning Strikes: The Lives of American Indian Sacred Places*. New York: Viking.

Plumwood, Val. 2012. *The Eye of the Crocodile*. Canberra: Australian National University E Press.

Pope Francis. 2015. *Encyclical: Laudato Si' – On Care for Our Common Home*. Rome: Vatican Press.

Rose, Deborah Bird. 1996. *Nourishing Terrains: Australian Aboriginal Views of Landscape and Wilderness*. Canberra: Australian Heritage Commission.

Rose, Deborah Bird, Thom van Dooren, Matthew Chrulew, Stuart Cooke, Matthew Kearnes and Emily O'Gorman. 2012. Thinking Through the Environment, Unsettling the Humanities. *Environmental Humanities* 1: 1–5.

Rose, Deborah Bird. 2013 "Val Plumwood's Philosophical Animism: Attentive Interactions in the Sentient World." *Environmental Humanities* 3: 93–109.

Kutz, Susan and Matilde Tomaselli 2019 Two-Eyed Seeing Supports Wildlife Health. *Science* 364(6446): 1135–1137.

Tucker, Mary Evelyn and John Berthrong. 1998. "Introduction: Setting the Context." In *Confucianism and Ecology: The Interrelation of Heaven, Earth, and Human*, edited by Mary Evelyn Tucker and John Berthrong. Cambridge: Harvard University Press, xxxv–xlv.

United Nations 2007 "United Nations Declaration on the Rights of Indigenous Peoples." https://www.un.org/development/desa/indigenouspeoples/wp-content/uploads/sites/19/2018/11/UNDRIP_E_web.pdf. Accessed October 10, 2020.

United Nations Environmental Program (UNEP) 2008 "Faith for Earth Initiative." https://www.unenvironment.org/about-un-environment/faith-earth-initiative. Accessed October 10, 2020.

Index

For Product Safety Concerns and Information please contact our EU representative GPSR@taylorandfrancis.com
Taylor & Francis Verlag GmbH, Kaufingerstraße 24, 80331 München, Germany

www.ingramcontent.com/pod-product-compliance
Lightning Source LLC
LaVergne TN
LVHW021120300926
849160LV00010B/792
* 9 7 8 0 3 6 7 6 9 9 8 1 9 *